Property

The sixth edition of this extremely popular and classic textbook has been updated to reflect on-going changes in the field of property development. Attention is paid to the impact of the global financial crisis on the property development process and, in addition, to the increasing relevance of technology to the property profession.

While the successful style and format of the text has been retained, new chapters have been added and existing chapters updated and enhanced to guide lecturers and students in their teaching, reading and studying. Other new features in this edition include:

- Fully updated discussion points and reflective summaries
- Examples of contemporary best practice based on international case studies covering the UK, USA and Australia
- New chapters on 'Property Cycles' and 'Technology'
- Online materials for lecturers and students.

This fully revised edition of a standard text for all property development and real estate students will also be of interest to early career professionals and those pursuing similar professional degrees in the industry and in wider built environment courses.

Richard Reed is Professor of Property and Real Estate at Deakin University, Australia.

Sally Sims is Senior Lecturer in Real Estate at Oxford Brookes University, UK.

Property Development

Sixth edition

Richard Reed and Sally Sims

With a foreword by David Cadman

Routledge
Taylor & Francis Group

LONDON AND NEW YORK

First edition published 1978
by E & FN Spon, an imprint of Chapman & Hall

Second edition 1983
Third edition 1991
Fourth edition 1995

Fifth edition published 2008
by Routledge

Sixth edition published 2015
by Routledge
2 Park Square, Milton Park, Abingdon, Oxon OX14 4RN

and by Routledge
711 Third Avenue, New York, NY 10017

Routledge is an imprint of the Taylor & Francis Group, an informa business

British Library Cataloguing in Publication Data
A catalogue record for this book is available from the British Library

Library of Congress Cataloging in Publication Data
Reed, Richard, 1963–
Property development / Richard Reed and Sally Sims ; with a foreword
by David Cadman. – 6th Edition.
    pages cm
Includes bibliographical references and index.
1. Real estate development – Great Britain. 2. Real property –
Valuation – Great Britain. 3. Real estate investment – Great Britain.
4. Real property – Great Britain. I. Sims, Sally. II. Title.
HD596.R44 2014
333.73´150941–dc23   2014000793

ISBN: 978-0-415-82517-7 (hbk)
ISBN: 978-0-415-82518-4 (pbk)
ISBN: 978-1-315-76665-2 (ebk)

Typeset in Sabon
by HWA Text and Data Management, London

MIX
Paper from
responsible sources
FSC® C013604

Printed and bound by CPI Group (UK) Ltd, Croydon, CR0 4YY

Contents

List of illustrations   vi
List of tables and examples   viii
Foreword to the sixth edition   ix

1 Introduction   1

2 Land for development   42

3 Development appraisal and risk   77

4 Development finance   111

5 Property cycles   148

6 Planning   165

7 Construction   183

8 Market research   222

9 Computer technology   240

10 Marketing and sales   250

11 Sustainable development   275

12 Emerging markets   310

Appendix: CPI by country (2004–2012)   324
References   330
Index   336

Illustrations

Figures

1.1   Relationship between land uses   4
1.2   Office rent in the Chicago CBD   5
1.3   Decision pathway   9
3.1   Development timetable   89
4.1   Returns by land use (UK) 1982–2012   119
5.1   Characteristics of a typical cycle phase   150
5.2   Multiple cycles in a single market   155
5.3   Property market, financial and economic framework   156
5.4   Equilibrium in a real estate market   159
5.5   External shift in demand for real estate   160
5.6   Over-supply scenario due to lower demand   161
6.1   Adaption of the planning process in England   171
7.1   Overall development programme – Gantt chart   207
7.2   Cash flow: fees/construction   208
7.3   Financial report: building costs   209
7.4   Checklist to monitor primary activities   211
8.1   Market research approach from broad to specific   224
8.2   Relationship between market research areas   224
8.3   Cross-tabulation – land use by geographical location   225
11.1 Three-pillar model of sustainable development based on triple bottom
      line accounting   277
11.2 Four-pillar model of sustainability   278
11.3 Evolution of sustainability rating tools   295
12.1 Global integration and local responsiveness   320
12.2 Steps in international strategy formation   322

Plates

1.1   Location of the development   35
1.2   Fairglen Development – Phase I (completed) and proposal for Phase II   36
1.3   Southern elevation of a 3-bed house   38
1.4   Proposed layout for Phase II   38

2.1   Initial promotion of the Hoffman Brickworks development based on the
      historical significance of the site   73
2.2   An early stage of redevelopment   74
2.3   Converted kiln with the high profile chimney stack being part of the
      marketing process   74
2.4   Kiln redevelopment into residential accommodation   75
5.1   Wilbow Corporation logo   163
5.2   Completed dwelling in the Wilbow Corporation development   164
5.3   Well-planned Wilbow Corporation development   164
7.1   Example of modular home floor plan highlighting the individual
      modular units   215
7.2   Example of hybrid modular home floor plan highlighting the two
      modular units   216
8.1   Property development completed with 49 units ready for sale   237
8.2   Entire property development placed on the market in 2006   237
8.3   Due to the contamination issues the property was vandalised with
      graffiti   237
8.4   Due to the contamination issues the property was vandalised, with
      internal fittings and white goods removed   238
8.5   A deep excavation to remove the decontaminated soil   238
8.6   Eventual decontamination of the site finally took place in 2013   238
8.7   The promotion of a proposed property development after the
      excavation of the site   239
11.1 View of the library from Centenary Square   303
11.2 Aerial view of the new library development in Centenary Square,
      Birmingham   304
11.3 View of the library from the amphitheatre   306
11.4 Sustainable design features   307

Tables and examples

Tables

1.1	Fairglen Low Energy Housing	36
3.1	Investment yields and respective year's purchase (YP)	85
3.2	Development timetable	89
3.3	Normal S-curve irregular pattern of expenditure	97
3.4	Sensitivity analysis and effect on adjusted developer's profit	108
3.5	Level of developer's profit expressed as a percentage	109
4.1	Returns by land use (UK) 1982–2012	119
7.1	Examples of prioritised criteria by client type	184
7.2	Advantages and disadvantages of design-bid-build procurement	193
7.3	Advantages and disadvantages of design and build procurement	196
7.4	Advantages and disadvantages of management contracting procurement	199
11.1	Property development stages and key potential sustainability issues	294
11.2	Green Building Councils and rating tools	296
11.3	Library of Birmingham vital statistics	303

Examples

3.1	Residual valuation	80
3.2	Excel formulas	82
3.3	Residual valuation	93
3.4	Cashflow approach	96
3.5	Net terminal approach	100
3.6	Discounted cashflow approach	101
4.1	Developer's profit analysis	134
4.2	Developer's profit analysis	135

Foreword to the sixth edition

Reading my foreword to the last edition, which was published in 2007, I note that although I suggested I should no longer forecast future markets there was some merit in pointing towards 'another correction'. Some correction! Within the next two years major banks would implode and major economies would be all but broken.

Six or seven years later, it is not at all clear that the lessons of that period of financial collapse have been learnt. In the UK, banks have been bailed out, personal debt is once again being encouraged and the Bank of England is having to introduce special measures to prick another housing bubble. At the same time, the government is moving away from its commitment to what is somewhat disparagingly called 'the green agenda', and the realm of 'sustainability', referred to in the last edition, has shrunk back to mean little more than energy efficiency and carbon management.

In the meantime, in the real world, economic and environmental frailty remain, with expectations of lower than average expectations for economic growth and with climate change scientists now accepting that we have stepped over the plus 2 degree Celsius threshold and are now heading towards plus 4.

In these conditions, while the basic development skills referred to in this book remain the same, it is necessary to reappraise investment strategies. For example, the long-established rule of investment 'spread' is no longer sufficient. Now it needs to be matched by the much more difficult task of 'selection', which is to say that there is a much greater need to identify and invest in those particular places that provide resilient commercial and residential communities, robust local economies which draw people to them. Furthermore, at a time when the cost of travel is rising and the cost of technology is declining, patterns of mobility and location are being radically transformed, and keeping up with user demand will need much greater attention. As a consequence, the risks of development within an investment portfolio may now be less than the risks of holding on to an ageing and increasingly obsolete stock.

Amid this change, this sixth edition, prepared by Richard Reed and Sally Sims, has been completely updated, with two entirely new chapters, one on property cycles and another on computer technology. At the same time, the authors have sought to retain the straightforward language and presentation of the text that have ensured its place as a textbook for such a long time.

As one of the two authors of the very first edition of *Property Development*, which was published in 1978, I have seen many changes in the property industry, but more than ever before I now sense that significant change is both likely and necessary. I would, therefore, urge all who now come to study property development and investment, to question convention and to broaden the scope of their enquiry, so that there is a proper understanding of the interconnection of economy, environment and society. The problems we face cannot be solved merely by new technology. They require a reappraisal not only of value, but also of the values and principles by which we govern ourselves in each of these realms. Means are as important as ends. What we take to be true and the ways in which we then choose to proceed will shape the outcome of all that we do.

David Cadman, Aldeburgh, December 2013

Chapter 1

Introduction

1.1 Introduction

This book defines concepts and details processes involved when undertaking property development. It a practical book which describes the process of property development, enabling the reader to obtain a comprehensive understanding of the fundamental concepts and conceptual framework for property development. The content of this text is suitable for new property developers and students who are looking to identify and understand the important components of undertaking a property development, as well as for experienced property developers seeking to understand the theoretical concepts underpinning this discipline.

This is the sixth edition of the book – since the fifth edition of *Property Development* was published in 2007 major changes have affected the property development discipline. In a similar manner to many other areas of business activity, there has been a substantial move towards globalisation and now property development is undertaken on a global as well as national and local scale. The concepts and discussions in this text apply in many different regions and countries with emerging markets, as discussed in the final chapter. The trend towards adopting sustainability in development has continued and now directly and indirectly affects the type of property developments perceived as desirable in the marketplace. Sustainability is therefore firmly embedded within the text, in addition to the inclusions of a 'Sustainability' chapter. Property development processes have been significantly affected by advances in technology in terms of communication technology, advances in building materials and in construction processes. For example, the internet provides easy access to property search engines which has speeded up the globalisation of business and allowed best practice in sustainability and property development to be communicated rapidly around the world. Accordingly this edition includes a chapter on 'Technology' in the context of property development. Since the previous edition of this text was published the world has experienced an inevitable downturn in economic markets known as the 'Global Financial Crisis' (GFC). The status of the property cycles at any given point in time can adversely affect the success of any property development and the developer needs to be fully aware of its status. Accordingly a new chapter has been added to discuss the theory behind property cycles, as well as how to interpret and work around changes in supply and demand interaction.

The term 'property development' evokes many feelings depending on varying perspectives. The definition adopted in this text is that property development is 'a process that involves changing or intensifying the use of land to produce buildings'. It is not simply the buying or selling of land for financial gain, since the land is only one of the essential components in a property development. Other components include the building materials, labour, infrastructure, financial capital and professional services. It is a global activity as reflected in the coverage of this text with relevance to the UK and Europe, the USA, Asia-Pacific and the rest of the world.

Property development is an exciting and occasionally frustrating activity often involving the use of scarce resources and large sums of money to develop a product which is largely indivisible and illiquid. It is, at times, a high-risk activity involving a high level of planning and co-ordination to maximise the use of limited resources. As the development process is often lengthy and can take years from initial conception until completion, the performance of external factors, such as the broader economies at the local and national levels, are an important consideration in the successful completion of a development, especially as the assumptions made at the outset may have dramatically changed by completion. Success often depends on attention to the detail of the process, although success is not always judged in terms of profit and loss and for some it is measured in social, emotional or aesthetic terms. However property development is, for many, a worthwhile and very rewarding discipline.

The emphasis in the text is on the practical application of property development where the reader is taken carefully through all of the stages involved in the process. In each chapter a series of discussion points are provided to prompt the reader to reflect on the content of the previous section. Furthermore a number of illustrative industry-based case studies are included to demonstrate the application of certain development stages covered in the chapters. This text is intended for (a) those who already practise in this field and also (b) as an introductory guide to students and those new to the field. The aim of this book is to provide a framework for successful property development at a local and international level.

1.2 The process of development

Undertaking a property development is largely about the 'process' of developing a property. There are different perceptions about what the process actually involves, which is due in part to the actual country within which a property development is occurring and the actual location itself. A major property development in London or New York, for example, will be considerably different in many aspects from a major property development in a rural town in the southern hemisphere. While the underlying concept is similar, each parcel of land is different which ensures every property development has its own unique aspects and interesting development challenges.

In a very simplistic format, property development has similarities with other industrial production processes and involves the combination of various inputs at specific timings in order to achieve a desired output or final product. In the case of a successfully completed property development, the final product is the result of a change of land use and/or a new or altered building in a process which combines the factors of land, labour, materials and finance to produce a varying level of profit and risk. However, in practice the successful implementation of this framework ensures the process can be very complex if poorly understood, especially since a development often takes place over a considerable time period, usually years. When the property development is completed the end product is then unique, either in terms of its physical characteristics, its location or both. No other process operates under such constant public attention, nor in recent times has received so much interest in broader society. For example the international quest to construct the world's tallest building throughout history is clear evidence of the interest in new property developments, with this view further supported by a high international profile and demand by individuals to visit such properties.

The development process can be sequentially divided into these major stages:

1 initiation
2 investigation and analysis of viability
3 acquisition
4 design and costing
5 consent and permission
6 commitment
7 implementation
8 leasing/managing/disposal.

The individual stages may not always follow this exact sequence and at times may overlap or be repeated. For example, commitment (6) may occur at an earlier stage when the purchase contract is subject to planning consent and permission (5), normally preceding (6), which may then be formally confirmed at a later stage. Another consideration is whether the development is a speculative project or a design-and-build project. The sequence listed above is typical of a speculative development where an occupier had not been identified when the commitment (6) phase of the property development took place. Alternatively if the development was undertaken based on a design-and-build approach and was pre-sold to an occupier or pre-let to a long-term tenant, then stage (8) would precede stages (2)–(7).

1.2.1 Initiation

The first stage of development process typically occurs when either one of two events occurs:

- a parcel of land or site is considered suitable for a different or more intensive use than its existing use; or
- there is an increased level of demand for a particular land use, which in turn leads to a search for a suitable site.

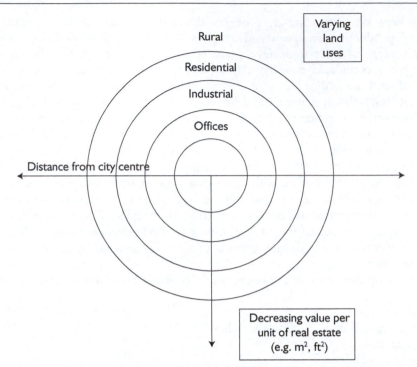

Figure 1.1 Relationship between land uses (Source: adapted from Von Thunen 1826)

There are underlying fundamentals for the property development process which are commonplace across different land use types. With reference to improving land, the amount of resources needed can vary from minimal structural improvements relative to the substantial land component in rural land, to major structural improvement in comparison to a minor land component in a city centre high density land use – for example, an office building. There are a range of land uses between these two extremes including residential, industrial, retail and office. The location for each type of development is usually driven by its proximity to cities, towns and transport networks. See Figure 1.1 for the underlying rationale for the location of each land use, where Von Thunen's original model published in 1826 provides a broad framework for understanding variations in highest and best land uses based on limited supply in the city centre and the hierarchy of higher yet diminishing returns for office, then retail, industrial and so forth. In a purely practical sense this model is very theoretical (i.e. not realistic with perfect concentric circles) and also ignores any limitations imposed by legal or planning regulations designed to protect against non-conforming uses.

The main focus in this book is towards the more intensive land uses in Figure 1.1 where there is increased pressure to develop the land to its optimal capacity. Emphasis is placed on land uses where property development is most likely to occur. For example, retail property has a relatively high level of obsolescence, partly due to constantly changing demand trends and consumer tastes, which in turn ensures the continual development or redevelopment of retail property including larger shopping centres. You will find examples included in this book of higher intensity land uses

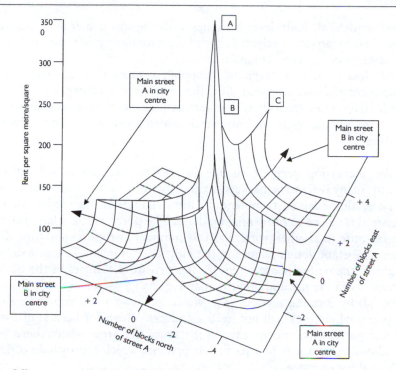

Figure 1.2 Office rent in the Chicago CBD (Source: adapted from Alonso 1964)

such as retail, office and industrial property although the development framework is applicable to other land uses including residential development. The diagram in Figure 1.2 has been adapted from a landmark rent-bid model for the high density Chicago office market (Alonso 1964) which closely examines actual property location and the rent per unit of measure. In Figure 1.2 the highest rent per spatial unit (e.g. square metre or square foot) is at location (A) at the intersection of Main Street A and Main Street B in the city centre. The rental level then decreases (B) in a location being one block east of this intersection, then decreases to (C) being two blocks east of the intersection and so forth. This type of rent-bid model can usually be easily adapted for most high density areas and can greatly assist property developers to analyse the levels of demand for various uses land in areas such as an inner-city centre. When this model is adapted for a residential development it would be affected by other variables such as access to leisure facilities and schools for families. In recent decades many city centres have re-emerged as mixed-use locations, combining office, retail and residential uses. In fact it is now commonplace for a property development to contain mixed land uses, such as an office building incorporating ground-level retail, or an industrial building with partial office space.

Increasingly such new mixed-use property developments have used creative planning and design strategies to combine multiple land uses. Examples include:

- Office/retail/residential: buildings which are predominantly office although with a large component of retail on the lower levels with residential on the upper levels.

- Hotel/residential: multi-level buildings with the lower half as a hotel and the upper half as private residential, both land uses having full use of all amenities including pool, gym and common areas.
- Retail/office/residential: regional shopping centres with a substantial office component and also a large medium density residential component.
- Residential/retail: especially with inner-city multi-level developments the bottom levels can include restaurants and retail outlets to cater for the residents of the building.

With the increasing pace of change in society and the added influence of sustainability, many real estate markets are now looking for buildings which are 'agile' and can be readily adapted to potential future changes in demand – for example from industrial land use to residential land use. The potential to 'adapt' the building from one land use to another, with relatively minimal disruption and time delay, has become an important design feature in the built environment. This process requires forethought in the design phase to ensure the structural features of the building do not constrain its future use and prevent adaptive changes – for example from 'office' to 'residential' by locating services (e.g. water, sewerage) and supporting columns in the building where they will not have an adverse effect for both land uses. Agile buildings are particularly sought after in the inner-city areas where there are often multiple adjoining land uses in a particular precinct which can include office, retail and residential for example.

The starting point for the property development itself can vary. Often it can be linked back to one of the stakeholders in the property development looking for an appropriate new site to meet their needs; for example a finance company may seek vacant land to construct a new office building so they can amalgamate their smaller expiring leases throughout town. Alternatively a stakeholder may be looking to redevelop or expand an existing site, such as a retail shopping centre, and need to acquire neighbouring residential houses for land required for a retail extension. These changes in demand may be due to any number of influencing factors resulting from the constant state of change that occurs within the market. Typically the factors which negatively affecting an existing land use or development use are usually referred to as a form of obsolescence such as economic, social, environmental, physical, legal, historical and so forth. Often it is impossible to disentangle which form of obsolescence affects a particular property and to what extent, as several factors frequently interact and negatively affect property which in turn will cause a decrease in value over time.

It is an essential concept in property and real estate markets for every parcel of land, in addition to any improvements which are affixed to and form part of the land, to be at its 'highest and best (legal) use'. (The reference to 'legal' use refers directly to statutory planning authorities who seek to separate non-compatible uses and maximise efficiencies in urban landscape through planning legislation.) Of course this scenario may vary substantially from the existing use. One example would be a residential area which has gradually become predominantly commercial over time although a small number of homes are still occupied. Due to this gradual change, the 'highest and best' land use for that particular area has also changed. This trend must be acknowledged by the property developer as soon as possible otherwise a new development may become obsolete relatively quickly, e.g. within five years.

1.2.2 Investigation and analysis of viability

In the preliminary stages of a property development it is essential for detailed market research to be undertaken. With developments in technology and the availability of decision support systems to assist the developer, including GIS and scenario modelling software, the risk of not developing to 'highest and best use' in the current and foreseeable market has been substantially reduced. The stage of fully investigating, evaluating and modelling the detailed process of a property development is crucial to the successful completion of the project. Errors or misjudgements here will often have an adverse effect on the financial viability either during each development phase or at the end. There have been many high profile property developments on an international scale which have made a substantial loss following completion. The process of undertaking a hypothetical development has been substantially assisted by computer software programs where detailed costings and potential scenarios via a sensitivity analysis can be closely examined. The critical importance of undertaking a rigorous investigation and evaluation cannot be overstated here if financial losses due to poor planning or undue haste are to be avoided.

The evaluation stage is arguably the most important part of the property development process. Simply explained, 'failing to plan is planning to fail'. This stage enables the developer to create the essential framework for the entire project management of the development which guides decision-making throughout the whole development process. A comprehensive investigation and evaluation will include detailed market research, both in general and specific terms, and examine in detail the financial viability of the proposal. There are proven methods for assessing the financial viability of a property development (see Chapter 3), although the data input will depend on the depth of the market research (see Chapter 8). The real estate market works largely on the premise that 'knowledge is power' so reliable and detailed property data is commonly not typically available for free. In many countries it is very expensive to gain access to this detailed information and it can be very tempting to bypass or reduce the costs associated with purchasing this data to save money. However, selecting the option of not purchasing the data would increase the unknown factors in the proposed development, thus increasing the underlying level of risk. Accordingly such risks are usually higher for the less professional developer who may be more inclined to forgo the additional expenditure of purchasing this data. One outcome from the process of undertaking the assessment of financial viability is the level of assurance about the success of a project if undertaken and completed. This evaluation should be approached using the minimum number of assumptions, although all forecasts will incorporate some assumptions based on known information and past experiences. There will also be a distinction here between public and private development projects and their respective drivers. For example proposed development in the public and/ or not-for-profit sectors will not have a core profit-seeking agenda, but rather will be aiming to minimise development expenses and be cost-effective and accountable. In contrast the private sector will be profit seeking and aim to maximise the financial return to shareholders, to whom they are accountable.

The evaluation process may also be used to assess the market value of a vacant site which has the potential to be developed. This approach is commonplace and an evaluation of a particular site may occur on a regular basis, often with the viability analysis concluding that the development is not profitable at this point in time. This may be due to external forces such as high borrowing costs of finance or lower final sale price of the completed project. In most towns and cities there will be sites which remain vacant because an analysis has confirmed a proposed development is not viable at that particular point in the property cycle. Over time the markets and economic climates change and most sites can then be developed at a later date.

Logically, the investigation and analysis of viability phase must be conducted prior to committing to the property development. Overlap here is not possible and delaying completion of this phase or underestimating its importance is the most common aspect of a failed property development. This stage also allows the developer to factor in any number of hypothetical scenarios which allows the developer to retain a degree of flexibility. This stage is usually approached using the combined resources of the developer's professional team including architects, quantity surveyors, accountants, planners and valuers/appraisers. The final decision to proceed with the associated risk is always with the developer, based on the findings from the detailed investigation and viability analysis. Even if the development is approved by the stakeholders and proceeds to the next stage, the viability analysis must be constantly re-evaluated and monitored throughout the process of undertaking the property development. This will ensure that all stakeholders can be kept informed about all aspects of the project as it progresses through each stage and adjustments can be made where necessary.

1.2.3 Acquisition

Following a positive outcome from the investigation and viability analysis stage, the decision to proceed is usually made. A considerable amount of preparation is required before the site can be acquired and the development started. Preparation prior to acquisition should include the following stages as shown in Figure 1.3.

(a) Legal investigation

In most cases the site will be purchased with the specific intention of undertaking the development project. One exception is where the site is already owned by the developer, although this is rarely the case. Prior to purchasing the site, the developer (i.e. the prospective purchaser) should fully examine all legal and planning aspects of the site. The starting point here is to investigate the type of ownership or tenure of the site, including the nature of the owners and their background, e.g. if they are located overseas or in the same locality. It is essential to identify any easements, outstanding encumbrances or liens on the title which could be potentially transferred to an unsuspecting new purchaser as they are often attached to the land itself. If the site is being acquired as a purchase in the open market between a willing buyer and a willing seller, these factors may adversely affect the time period required to transfer ownership of the land to the developer. Any time delays here could potentially delay the commencement of work on the scheme and extend the total development period by months or even years, which in turn can result in extended finance costs. Delays here can also result in the loss of a potential tenant (tenant risk) who will not remain committed to a property which is associated with a high level of future uncertainty.

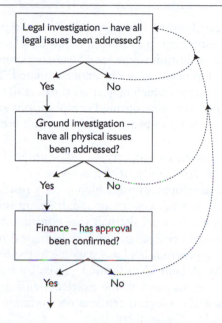

Figure 1.3 Decision pathway

If a site is being acquired by the public sector for the benefit of society, it may be acquired using compulsory purchase powers, so some of the details mentioned above (e.g. willingness of a seller) will not be relevant. However, when land is compulsory purchased it is still essential to examine the title deeds and identify any encumbrances and other characteristics which may impact the future use or value of the land. The assessed value of the property to be paid for compensation purposes will rely on this information. Although the use of such powers can be time consuming and costly, the vast majority of property developments which involve compulsory acquisition are completed with the co-operation of the original site owners, either by disposing of their interests through negotiation or by remaining a stakeholder in the development. The public sector may become involved in the initial acquisition stage – for example to assemble a large site with many occupiers and landowners as they can use their legal powers of compulsory purchase to ensure the tenure to the land is secured.

(b) Physical inspection and examination

A comprehensive on-site physical inspection of the site and all structural improvements (if any) is essential. This includes an examination and physical assessment of the capability of the site to accommodate the proposed redevelopment for an existing use or development for a new land use. Factors to consider include a detailed assessment of the site's load-bearing capacity (i.e. the potential foundations for a multi-level building), access (i.e. ingress and egress to and from the site), natural drainage and proximity to services (e.g. gas, electricity, water and sewerage) if not already in situ. All existing services (i.e. above-ground or below-ground electricity, water, gas, telephone and digital cabling) must be examined to assess their capacity to serve the proposed development. If there is a defined gap between (a) existing services and (b) the level of services required for the development, the cost and relevant steps required to overcome this shortfall should be estimated and built into the development plan. Additional indirect costs, such as a contribution to the local headworks for the supply of water or sewerage for a major development, cannot be underestimated. The physical inspection should also look for other factors which may affect the overall viability of the property development. This may include below-ground complications such as an easement for a water main or an underground railway, or above-ground aspects such as the height of the adjoining property.

Other attributes of neighbouring property should be noted if they have the potential to alter the perception or desirability of the proposed development – an example would be if the proposed development was a high density residential development but the adjoining land use was as a railway station, adult detention centre or a nightclub. The characteristics of each individual site will vary considerably but there should be an awareness of the potential presence of any archaeological remains, contamination or associated stigma from a previous land use, e.g. due to the removal of underground services and storage tanks. A land identification survey by a qualified land surveyor will normally be undertaken to confirm the exact dimensions and configuration of the site. Factors such as archaeological remains or contamination can hold up the development process or add additional costs.

(c) Finance

Unless the developer is in the rare position of using their own cash and equity to fund 100% of the project over the development period, there will be a need to borrow funds to finance the cost of the development. From the outset it should be remembered that costs associated with servicing the finance loan will have a major bearing on the overall viability of the property development, especially for projects conducted over an extended time period, e.g. a number of years. The terms and conditions associated with obtaining finance must be on the most favourable terms available in the marketplace for the developer before deciding to proceed. The subject of development finance is discussed further in Chapter 4. Typically the types of finance required by the developer are either short-term or long-term finance. Short-term finance is required to pay for costs expended before and during the development process itself. Long-term finance is needed to cover the costs associated with retaining ownership of the development after completion in situations where, for example, the developer retains the property for investment purposes, or possibly to seek a purchaser for the completed development if the build is speculative. The availability and conditions attached to finance provided by lenders varies, largely due to the risk associated with each developer's application and also the state of the financial market at that particular point in time. Part of the lender's decision will be based on other criteria including the viability and profit associated with the proposed property development which will be closely scrutinised, as well as the track record and credit rating of the developer.

1.2.4 Design and costing

In the preliminary stages of the property development process it is commonplace to consider some basic aspects of design and costing. This may be as simple as estimating the cost and feasibility of building different types of structures on a particular site. The design and costing aspects are interlinked and often an iterative process is used until an acceptable design within budget constraints (as per the brief) is provided. Clearly a building with a complex design is usually associated with a higher cost and vice versa. The design of each development is influenced by a number of factors, the client's brief, public perception and current architectural styles, but the aim of the developer is traditionally to maximise profit by maximising the development potential of the site. This is central to the success of the development and its importance should not be underestimated.

Design is an almost continuous process throughout the property development and runs parallel with other stages, becoming progressively more detailed as the development proposal increases in certainty. Following on from the preliminary investigation into design options, the developer may now have detailed knowledge of what design is required – this may be because the likely tenant or purchaser is known or an agreement (in writing) has been reached. In the case of a speculative property development proposal it is commonplace for the developer to initially consider a small number of conceptual proposals and to consult with real estate agents about the market's perception of each individual proposal, if completed. This process would lead to the professional team settling on a firm design brief for the property development project. The importance of groundwork is at times underestimated yet the groundwork undertaken for this brief will affect the ultimate final outcome and reduce uncertainty in the design phase. For particularly complex proposals it can save substantial time, resources and substantially assist the architect to produce the design which meets the needs of all stakeholders. It also has market appeal for existing and future prospective occupiers and tenants.

The preliminary design stages should be kept relatively brief in order to keep costs down before the developer is committed to the scheme. Due to the unique nature of property it is very rare to find two identical parcels of land, therefore the design itself only has relevance to that plot of land and for that developer. Nevertheless the preliminary designs should include enough detail to enable the quantity surveyor to prepare an initial and reliable cost estimate, which in turn will allow the developer to prepare the financial evaluation for a number of options. Although varying on a case-by-case basis, this initial brief normally includes the following:

- Architectural drawings with scaled floor plans showing the location (or multiple possible locations) of the proposed new building(s) and associated infrastructure on the site, together with relatively simple floor plans or sketches showing the internal arrangement of each level in the building on each floor.
- Plans of the main elevations or a cross-section of the proposed building(s), together with an outline specification of the required building materials and finishes, are often included at this point in time. These plans together with the initial cost estimate should enable the developer to prepare an initial evaluation of the development.

In due course a decision will possibly be made to submit a detailed planning application for the proposed scheme. This will require a full set of comprehensive plans showing the final layout, elevations, cross-section of the building and detailed design specifications. The developer now needs increasing certainty over costings to improve the accuracy of the financial hypothetical development valuation and appraisal which should include minimal unknown assumptions. Based on this information, the quantity surveyor will be able to make a detailed estimate of all building costs, which in turn will enable negotiations to commence with potential building contractors, either informally or formally by inviting tenders. Care during this phase will save time and expense at later stages of the development process, by reducing the element of risk and uncertainty.

These design and costing stages typically involve contributions from all members of the professional team with input from stakeholders such as the real estate agents and financiers. It continues throughout the entire construction phase of the property development. It is also the role of the developer to ensure there is fluent co-ordination between each stage of the development when producing the design and costings. As the major shareholder, the developer must ensure there are no delays to this process and project management software is frequently used to chart each stage of the development process, enabling steps to be taken to mitigate time delays and additional costs if there are unforeseen delays – for example a late delivery of building materials.

In most cases the final design will be very different to the initial design concept, which has often been through many design changes and alternations/modifications before the final drawings are complete. It is important that there is clear acknowledgement by all stakeholders of a point in time when modifications can no longer be made. This is commonly referred to as the 'commitment' phase, when the time for significant and potentially costly design changes has passed. The only exception here is where there are unforeseen errors in the original design brief or an external change, e.g. change in planning legislation.

1.2.5 Consent and permission

It is commonly accepted that a new development on a certain site may require a change of use from the previous land use, such as an ex-retail land use site which is to be developed into a medium density housing development. Since every proposed property development requires planning consent or permission from the local planning authority, a change of use is dealt with at this stage prior to commencing the property development. Details relating to the planning process are discussed in Chapter 6.

From an international perspective there are many different approaches to what occurs next. In many jurisdictions the developer may, where a building operation is involved, apply for what is termed an outline (or in principle) application before full approval is obtained. In this context an outline planning consent establishes the approved use of the site and the permitted size or density of the proposed scheme. The developer only needs to provide sufficient information to describe adequately the type, size and form of the scheme. However, outline planning consent, in isolation, is not sufficient to permit the developer to commence the actual development since a detailed planning consent is still required.

The detailed application which is usually submitted to the planning authority will explain all aspects of the proposal (with minimal assumptions, if any) and also include detailed drawings and information about the location of the structures on the lot, access to and from the site, design criteria, external appearance and landscaping.

Where the development requires a change of use, developers are required to submit the detailed application at the outset, effectively bypassing the outline planning stage. At times there may be multiple outline applications made if circumstances change before a developer purchases a particular site. If the design parameters change after detailed consent has been obtained, then further approval is again required from the local planning authority before proceeding.

The developer will often rely on previous experiences to make realistic initial estimates of the likely timeframe and costs associated with obtaining the appropriate permission during the evaluation stage. The task of seeking and being granted planning permission can become complex, requiring detailed knowledge of the relevant legislation and policies as well as essential local knowledge about exactly how a particular planning authority operates. The services of consultants to assist with this process may be necessary and cost-effective where planning problems are envisaged or encountered. Where permission is refused by the local planning authority the developer may appeal to the relevant government body, e.g. to the Secretary of State.

The developer may be required to enter into a contract with the local planning authority where a 'planning agreement' is negotiated as part of a planning approval. In some countries (e.g. the UK) these agreements used to be referred to as 'planning gains'; they deal with matters that cannot be covered as conditions to the planning approval. One example would be changes to local highways to improve access to and from the site when the development is finished. Planning agreements must be signed before approval is granted and often impose additional development costs (e.g. financial contribution to a park, off-street car parking) and therefore these agreements often affect the overall cost and viability of the scheme. It is not uncommon for a property development to be stalled or stopped at the planning stage.

A variety of other legal consents may be necessary before a development is allowed to commence. These include gaining listed building consent (e.g. the right to alter or demolish a 'protected' or 'heritage' building); the diversion or closure of a right of way; removing or re-routing existing infrastructure such as electricity lines; agreements to secure the provision of the necessary services and infrastructure; and in all cases where building operations are involved; building regulation approval. The prudent developer must clear all legal permission hurdles before full commitment to the development is possible.

1.2.6 Commitment

Prior to making a substantial commitment to the development, the developer must be completely satisfied that all the necessary preliminary work has been completed. Often this is referred to as 'due diligence' checks, such as ensuring the site has been decontaminated. The acquisition of statutory permissions must be satisfactorily negotiated before any agreements are signed, which in turn would make the developer liable for any financial commitment and outlay of money. After the work in the initial preliminary stage has been completed, the developer has a chance to reconsider the proposal before fully committing to the development. Now is the time to re-evaluate all the factors which may affect the success of the development. For example, changes to the economy may have affected demand for the completed product. One example could be where land is subdivided for a residential development and between the planning application stage and the completion of the first phase of the development, market preferences have changed with regard to actual house size. Even though the preliminary development phase was completed, it is still possible for the developer to reconsider the optimal size of the subdivided vacant housing allotments. This example highlights how important it is for the developer to access detailed reliable information and to take adequate time to reflect on the status of the project throughout the entire development process to ensure its viability.

The developer will endeavour to keep outlay or sunk costs to a minimum until all permissions are granted and the actual title (freehold or leasehold) has been transferred, giving full access to the site. Until this stage the expenditure will be primarily related to professional fees (e.g. planning consultant, architect) and staffing costs for those who are co-ordinating the project. Certain factors may affect initial costs, for example:

- professional teams working on a speculative basis in the hope of securing an appointment once the scheme commences; and
- developers acquiring land without planning permission (also referred to as 'land banking') with a view to applying for planning permission as some future date. Development contracts entered into are often referred to as Subject to Council Approval (STCA) and mean just that.

The next stage is for all parties to sign contracts for the property development. This includes the contract to purchase or lease (e.g. ninety-nine year ground lease) the land, secure all of the finance required, engage the building contractor and also confirm the professional team. Contracts with all parties will usually be formed within a short space of time, ideally within a matter of days, to avoid uncertainty and the risk of delays through prolonged negotiations.

1.2.7 Implementation

By the time a firm commitment has been made to a specific plot of land, the original design and construction aspects of buildings and the associated development costs can be spread out over the life of the project. Once all the raw materials deemed essential to undertake the development process are in place the implementation stage can begin. Note the additional flexibility associated with design and construction has been largely eliminated since the implementation phase has commenced. While it is important to maintain as much flexibility for as long as possible during the actual property development stages, there is level of risk associated – for example, too much flexibility may lead investors to consider the future of the project uncertain.

Throughout the implementation stage the primary aim is to ensure that the completion of the development and handover is within both (a) the allocated timeframe and (b) the financial budget as produced in the proposal, although an allowance for contingency permits a degree of flexibility for unknowns, e.g. delays in obtaining materials, variable weather conditions. No compromises are possible with regards to quality or taking 'shortcuts' to save time or money. Depending on the experience of the developer and the complexity of the proposed property development scheme, ensuring the implementation runs smoothly is often achieved by employing a project manager to co-ordinate the design and building processes. The project manager and/

or developer need to skilfully anticipate upcoming problems (if possible) and make prompt informed decisions to minimise delays and extra costs. Accordingly their qualifications, depth of experience and tacit knowledge come largely into play here.

It is commonplace for the developer to take substantial interest in both the running of the project and its overall promotion. The relevant property and real estate markets must be continually monitored throughout the property development phases to ensure the final product, when completed, remains aligned to meeting the market's needs. This might require slight adjustments to the specifications to reflect market changes. Where the development is for a non-profit organisation, the developer must aim to contain costs while at the same time maximising benefits to the final occupier. The construction and project management stage of the development process is dealt with in Chapter 7.

1.2.8 Leasing/managing/disposal

Although this is normally the final stage in the overall property development, these steps must be at the forefront of the developer's thoughts from the initiation of the scheme since the development will have no value unless it is sold or leased at the estimated price or rental value and within the period originally forecast in the evaluation. It is common for most tenants, either in the form of an owner-occupier or a lessee, to be locked in and confirmed as a tenant before the commencement of the property development or soon after construction has begun. This reduces, or eliminates, the need for the developer to be involved in any promotion to secure a tenant. Also a lessor who is contractually committed to leasing the completed development may possibly want to have some input at the design and construction stages to ensure a working environment which would best suit their staff. Increasingly, the need to secure a tenant for all or a substantial part of the development has been influenced by financial institutions which provide the finance for a high proportion of the completed development. It has become common for the lenders to insist on a set level of pre-commitment to the total scheme (which can range from 30% to 70% of the completed development either sold or let by tenants) before a lender will commit to forwarding the finance. This statement is very applicable after a major market downturn such as a GFC. The final disposal of the completed property development may be leased (often long term) or the transfer of the freehold interest. In the case of a major retail development there are many individual small leases, at times with single or multiple anchor tenants. In contrast, a small industrial development or office building may be leased to only a single tenant.

During the evaluation stage the letting and/or sales promotion strategy should be included in the development planning as early as possible and continually updated during each stage. A decision must be reached on the most appropriate time to sell or let the scheme. In some cases the development may be completed or virtually complete before seeking an occupier, although this decision is often influenced by other stakeholders, especially financiers or landowners if they have remained partners in the scheme. If a suitable tenant has not been identified by the completion or shortly thereafter, at times this will place the developer and stakeholders in a position where they are exposed to risk, e.g. financial risk due to holding costs with no income. Reputations and goodwill may also be negatively affected.

At the start of the process, the developer must decide what the final intention will be regarding the tenure or ownership of the completed project. The decision to take on the role of property investor and retain ownership of the completed development or alternatively to sell and realise a profit upon completion will depend on the motivation of the developer and what they perceive their role to be. This will also depend on other factors such as the condition of the prevailing property investment market at the time, as well as the ability of the developer to borrow funds over the long term as an investor without limiting their borrowing power for future investments. However, developers have to be flexible to accommodate changes in the investment market prior to completion of the scheme, especially when some larger developments can take many years from the initial conception to completion. Careful thought needs to be given to the final anticipated investment value at the initial evaluation and design stages, keeping in mind it is practically impossible to accurately determine the value of any property in the 'future'. If the decision is made to sell the completed development to an investor, importantly then the developer needs to fully research their requirements. The location, design specification and financial strength of the tenant will also be critical in achieving the best price for the investment.

To ensure the completion of the project runs as smoothly as possible, it is commonplace for the developer to engage the services of a real estate agent or a realtor to secure a sale of the property. The real estate agent should be employed as early as possible to advise on the design specification of the final product to ensure it meets the needs of the market.

The development process and the developer's responsibility does not cease with the completion of the construction process and final occupation of the building. There remains an on-going requirement for the developer to maintain a relationship with the occupier, even though no direct landlord/tenant relationship may actually exist. This enables the developer to keep up-to-date occupiers' requirements in general, and in particular to understand the post-occupancy shortcomings (if any) of the completed building from a management perspective. Management needs to be considered as part of the design process at an early stage if (a) the final product is to benefit the occupier/s and (b) the developer is to maintain and enhance their reputation. In reality, the financial success of the development cannot be accurately assessed until after the building is completed, let or sold. Often it may not be until the first rent review under the terms of the letting agreement (typically five years after commencement of the lease) that the true picture becomes clear.

Discussion points

- How do the individual stages in the property development process relate to each other?
- What is the sequence of events in a typical property development process?

1.3 Main stakeholders in the development process

So far the property development process has been divided into individual stages. In each stage there are a variety of important stakeholders who each contributes to the outcome of the process. Some stakeholders will be involved in more than one process or will remain involved for the entire development process. Each stakeholder will have their own perspectives and expectations, whereas an important role of the developer is to manage their diverse and often conflicting interests to ensure the project runs smoothly and reaches a satisfactory conclusion. With many diverse roles, the various stakeholders are discussed below in the approximate order they appear in the actual development process. However, their importance varies between each project and not all stakeholders will contribute to every development scheme.

1.3.1 Landowners

The landowner plays a critical role in the first stage of a development process. For example they may be engaged in initiating the development itself due to their desire to sell the land, or alternatively they may seek to improve the value of their land. At times it can be a combination of both of these drivers. On the other hand, a landowner may be unwilling to sell their land and can become an obstacle to a proposed development. Without the willingness of the landowner/s to sell their interest or participate in the development, no future development can take place unless it is possible to acquire the land through compulsory purchase powers. Often the landowner's motivation will affect their decision to release land for development and this is the same whether the landowner is an individual, a corporation, a public authority or a not-for-profit organisation. Each organisation may even take on the role of the developer, either in whole or in part. Broadly speaking, land ownership can be divided into three categories, namely traditional, corporate and financial, as summarised below.

a. Traditional landowners include the Crown Estate, the Church, aristocracy, landed gentry and occasionally the older universities – for example Oxford in the UK. They have a significant interest in land both in terms of area and capital value. They are not motivated entirely by the economic return on investment (ROI) and their reason for land ownership is often linked to social, political and ideological constraints rather than purely a return on capital.

b. Corporate landowners are related to the term 'corporate real estate' or CRE. This group own land which is complementary to their main purpose, which is usually the provision of some form of production or service. In this category there are a wide variety of landowners including rural farmers, manufacturers, industrial companies, extractive industries, retailers and a variety of service industries. This can also include public authorities, such as central, local and nationalised industries which own land that is complementary to providing a particular service or product. Understanding the motivation for land ownership within the group can be relatively complex since their attitude to land ownership is directly linked to the core reason for their existence, namely their product. In addition they may be constrained by their legal status and will not always be seeking to maximise the return (ROI) on their land holding or real estate since this is a secondary to their core business function and there may be little apparent economic advantage in releasing land for development. If an organisation in this group is forced to sell their land due to a compulsory purchase or resumption scenario, at times they are compensated for expenses related to relocating and temporary disturbance which affects the operation of their business activities.

Note that little or no financial allowance is usually made for the fact that they are unwilling sellers. In addition there are often intangible losses to commercial businesses which are difficult to evaluate. This scenario would be different if the landowner/s were residential occupiers.

c. Financial landowners view their landownership simply as an investment and will treat land in the same way as equities/stocks and shares. Therefore, the landowner/s is often more willing to co-operate with the proposed development if the return on their land is financially adequate and commensurate with their level of investment risk. A financial landowner will have a clear motive for financial gain and consequentially they are likely to be the most informed type of owner regarding financial variables such as land values and the profit/risk level in the overall development process. Some of the major organisations in this category are financial institutions, typically pension funds and insurance companies which own a substantial proportion of real estate when measured by capital value. Financial landowners may directly adopt the role of the developer, or alternatively enter into a partnership arrangement with a developer. Major property companies generally hold large property portfolios and also undertake property development and therefore may assume the dual role of both landowner and developer.

d. Landowners have historically played an important role in shaping the way land has been developed. They have influenced spatial layout, the type of buildings, infrastructure and the design of the landscape. An example would be Grosvenor Estates in London. More recently, the introduction of legislation and the planning system has reduced the level of influence a landowner can have on the type of development itself, but they can still influence the location and planning of the development.

e. An additional challenge for development is where there are several landowners involved in a single property development. In this scenario every landowner must agree to the development proposal and the greater the number of owners and the smaller their holdings, the more difficult it is to assemble a site for development. In these situations it can take years for a property development to come to fruition, requiring great patience on the part of the developer.

1.3.2 Developers

Private sector development companies come in a variety of forms and sizes ranging from single-person entities up to global multi-nationals. In a similar manner to most privately owned organisations their main objective is to return a profit and maximise financial returns for its shareholders. This is the same way as any other private sector company operates, generally regardless of the actual product or service they offer.

Developers primarily operate as either traders or investors. Due to limited resources such as equity and access to borrowed funds, most small development companies act as traders and have to on-sell the properties they develop. Very few developers have sufficient resources to retain ownership of a finished development project. The traditional pathway for a developer is to grow in size and build goodwill in the marketplace. This process gives a smaller trader type of developer the chance to evolve into an investor-developer, thus enabling them to retain profits for investment purposes. By contrast, some of the larger property development organisations, as measured by capital assets, hardly undertake any new development, preferring to specialise in managing their property portfolio and carrying out only refurbishment and redevelopment work. Most residential developers operate solely as traders, historically developing and on-selling for owner occupation or to private landlords. Many residential developers also become substantial landowners after undertaking successful developments over time. The reasons behind this trend are linked largely to the smaller exposure to risk with this type of development due to the wider cross-section of possible purchasers, compared with retail or office developments, and the smaller financial outlay required when undertaking residential developments.

As the type of developer varies, so do their developments. For example some developers operate only in specific geographical regions, others specialise in developing offices or retail schemes. Others prefer to spread their risk by producing different types of developments in different locations and countries. Nevertheless property companies formulate their individual development strategies and mission statements in accordance with the interest and expertise of their directors, their perception of the prevailing and future market conditions, as well as following the strategic direction they desire their organisation to pursue when looking forward.

1.3.3 Public sector and government agencies

Government organisations and the public sector are rarely directly involved in the development process. Usually government organisations are primarily concerned and involved with developments for the sole purpose of their own occupation, community use and/or the provision of infrastructure. However, governments are public entities which have to balance competing resources with political influences and so are typically constrained by their financial resources and limited by their legal powers. In addition, in most developed countries, government bodies are required to be transparent and are usually publicly accountable for their actions. Their core aim is to meet the overall needs of the community they serve.

The degree and type of government participation and involvement in the development process will depend largely on whether the government seeks to (a) encourage development or (b) control development in order to maintain standards. Also many government authorities undertake their own economic development activities which are designed to encourage and attract property development and investment to regenerate their immediate area and support economic growth. Some of the more proactive government bodies act as a catalyst to the development process by supplying land, and in some instances buildings and infrastructure, to increase the economic development of their area.

Recently there has been a broader acceptance of PPP (public-private partnerships) as a means of undertaking major property development, without the financial outlay that previously was required by governments. This has also been accompanied by relatively long leaseholds issued to private organisations, such as a ninety-nine year ground lease. It is a common arrangement for a government authority to retain freehold ownership of a development site but to grant a long leasehold interest to the developer. Often this is accompanied by a share of rental growth through the payment of ground rent by the private investor to the government body.

Most government policy restricts public body intervention in the development process unless it can be clearly demonstrated that private market forces have failed to deliver an adequate level of property development proposals, particularly in locations already targeted for economic development. For example a government may have an urban regeneration initiative which is administered through a dedicated governmental department. The role of such a department is largely to facilitate development for the benefit of society while also attracting investment in partnership with the private sector. A government body will at times be able to assist developers with particular development roles including site identification, site reclamation, the provision of infrastructure and perhaps financial grants if government needs are sufficiently high and the financial resources are available.

1.3.4 Planners

The underlying objective of most planning systems is to monitor and control the use of land in alignment with the public interest by encouraging development which is harmonious and to prevent 'undesirable development' such as incompatible land uses adjoining each other. Generally speaking, planners can be divided into two broad categories: politicians and professionals. The politicians, usually on the advice of their professional employees, are responsible for approving the development plans drawn up by professionals in accordance with the current legislation, policy and influenced to a varying degree by factors such as the size of the development, the number of objectors (if any) and the current status of the government. Planners are responsible for determining whether applications relating to permission for development proposals should be approved or refused. On the other hand, professional planners are responsible for advising the politicians and administering the system and its day-to-day operation.

The underlying basis for determining planning applications is laid down by statute secondary legislation and supported by guidance notes and publicly available local and regional plans, such as strategic development plans. Individual planning applications are determined in the light of these development plans, government policy and advice, as well as previous and the unique circumstances surrounding each individual application. Nevertheless in reality there are often gaps and conflicts in the planning guidance, leading to uncertainty, which in turn causes developers to employ planning consultants to assist them in direct negotiations with the planners. In turn this process saves time and therefore reduces overall exposure to risk. Property developers need to be fluent with the planning process and to know what type of land use is acceptable, what density of development is permitted or possible, as well as what design standards are required in order to obtain planning permission. A successful development application is usually achieved by undertaking prior consultation and negotiation with the relevant government authorities before lodging the planning application. In some instances this may involve agreement by the developer to provide infrastructure or community facilities in the case of a large development, sometimes referred to as 'planning gain'. Examples include the provision of off-street vehicle parking, recreational park facilities or perhaps a financial contribution to the government for town water or sewerage headworks. In the context of tight public spending, a planning agreement can be seen as a useful means of securing benefits for the community without making any financial outlay. However, the issue of planning gain has been controversial at times and there is a limit to how much profit a developer can afford to lose to acquire planning permission and still ensure the development remains financially viable.

Planning authorities, both national and international, differ widely in their policies towards property development. For example, planning authorities in locations with low economic activity may seek to encourage development activity by placing only minimal restrictions on development proposals, particularly those that will increase employment opportunities. However, planning authorities in areas of high economic activity mainly see their role as imposing higher development standards, ensuring sustainable development practices are adopted, and even slowing down the pace of development in order to achieve a better balance of uses and improve the design of buildings. In some localities this results in an increased level of conflict between property developers and government planners, leading to an increasing use of the appeals system. In some instances this conflict is caused by the politicians ignoring the advice of the professional planners, which may also result in a high profile planning application being the subject of media attention.

1.3.5 Financial institutions and lenders

It is unusual for a development to be entirely financed with a developer's own capital, and therefore financial institutions have a critical role in the development process. The term 'financial institution' has traditionally been used to describe pension funds and insurance companies. Now there are many different types of financial intermediaries who supply a large amount of money in return for a secure mortgage or lien over a property. Financiers include banks and companies with the primary aim of lending money. Other financiers include pension funds, insurance companies, clearing houses and building societies who can all provide finance for property development purposes. Various other hybrid financiers have emerged in the 21st century and have developed some innovative funding vehicles including mezzanine funding. Another type of financier is a private syndicate, usually a small number of individual investors who collectively provide funding for a development without the use of an intermediary such as a bank. Generally there are two main types of money needed to complete the development:

a. short-term finance or 'development finance' to cover the costs during the development and construction process; and
b. long-term money or 'funding' to cover the cost of retaining ownership of the completed development as an investment.

A developer may not always want to retain ownership of a completed project and may seek to dispose of the development in the long term which will allow the repayment of the short-term loan and hopefully realise a profit.

Financial institutions are generally motivated by financial gain. However, unlike most developers, financiers usually take a long-term view and understand that investment in property is a relatively long-term investment. This is in direct comparison to other investments such as the stock market. Financiers need to achieve capital growth in order to meet their financial obligations to their shareholders, so financiers seek to minimise risk and maximise future yields. The yield on any investment is the annual income received from the asset expressed as a percentage of its total capital cost or value. Property and real estate is only one of a number of investments that institutions invest in, although this proportion is surprisingly low. For example in many countries the total proportion of property value in the entire portfolio may represent only 5–15% of their entire portfolio of investments. It is commonly accepted that in the case of property investment, the financier will receive a lower initial income when compared with a fixed-interest investment but this will be compensated by long-term growth.

A financier may provide both short-term and long-term finance to a developer by what is called 'forward-funding'. In other words they agree to purchase the development on completion while providing all of the finance in the interim. In this scenario almost all of the risk passes to the developer who will, in the majority of cases, provide a financial guarantee. Alternatively they may act as the developer themselves to create an investment, and therefore all of the risk is theirs but they do not have to provide a profit to the developer. Some financiers only purchase completed and fully let developments as they perceive carrying out the actual build phase of a development as being too risky. In order to be persuaded to take on the risks associated with development, rather than purchasing a completed and let scheme, they need a higher return (yield).

Whether acting as a developer, financier or investor they all adopt relatively conservative and risk-averse policies where possible, although each stakeholder differs in their individual criteria and strategic directives. Nevertheless each stakeholder is fully aware of the benefits of diversification with regard to reducing exposure to risk, and will try to achieve a balanced portfolio of property types rather than focusing on one particular land use or development only. In addition most developers, financiers or investors also aim to spread their investments across different geographical markets. They will seek properties or developments which meet their specific organisational criteria in terms of location, quality of building and tenant characteristics (i.e. financial strength). Therefore, it is essential for developers to develop proposals to meet the strategic objectives of the financiers rather than only for the occupiers. Financial institutions seek to purchase a building which will have the widest possible tenant appeal. Often this policy ensures their advisers will adopt a low-risk conservative view and recommend the highest specification for the development project. This can potentially lead to over-specification and less sustainable buildings.

The availability and level of funding to the developer depends on factors including the risk profile of the developer, their accepted industry track record as well as the availability of funding in the marketplace at the time of the development. For example if the proposed development is not considered to be 'institutionally acceptable' or the developer is not prepared or unable to provide the necessary financial or personal guarantees, it is possible for the developer to approach the banking sector for funding. On the other hand, if the property development is being undertaken in a period of rising rents and capital values the developer may prefer to use debt finance to maximise the potential profit on completion. In some countries the financial lending market has been deregulated and therefore the lending market is somewhat competitive and able to provide a variety of lending products. This includes providing finance for both short-term and medium-term loans. Actual lending rates will of course depend on variables such as the availability of finance at the time of the development, the risk profile of the developer and the base lending rate of the central lending bank, e.g. government bonds.

Banks also aim to make a profit from the business of lending money. In order to reduce the exposure of their shareholders to risk, financiers will usually register their interest on the legal title. Property is attractive as security as it is a large identifiable asset with a resale value. In addition most external parties would conduct a title search which could confirm the existence of an outstanding loan for the property. As financiers have an interest in the successful completion of the development they will take a close interest in the attributes of the property and will wish to ensure that the property is well located, the developer has the ability to compete the project and that the scheme is viable. With reference to corporate lending, the financer is primarily concerned with the risk profile and strength of the development company, its assets, profits and cash flow characteristics. In certain instances and depending on variables such as the size of the loan and the financier's level of exposure to risk, a financier may be in a position to take an equity stake in the property development scheme.

Residential developers who focus on building owner-occupied homes usually only require short-term development finance, often over months rather than years. The amount of money they need to borrow is comparatively small in contrast to commercial developments. Loans of this nature are often provided by a bank where the process is relatively straightforward. As with all developers, the ability to raise finance is based on variables including their 'track record' (i.e. credit rating) and the level of risk to the lender.

For public sector developments, similar sources of funding are much more difficult to obtain due to a higher level of accountability and transparency by public bodies. Another option for local government bodies is to apply for funding through broader grants. For example it has been increasingly commonplace to allocate substantial funding for urban regeneration projects in inner-city areas or locations where major regeneration projects require significant investment to enable a change of use, such as from ex-light industrial to medium density residential. This type of funding is usually acquired via a competitive tender process which is often targeted at creating private, public partnerships with development undertaken in partnership with the private sector and the community. The use of private-public-partnerships (PPP) has become popular with some governments because it can potentially bypass some of the common problems associated with obtaining finance while still retaining ownership of the land and continuing as a major stakeholder in the development. The private party which leases the land from the government also remains a major stakeholder in the development but their motives are typically profit-seeking; therefore, they will be interested in the profit and risk potential of the development. If planned and co-ordinated correctly then both parties can usually benefit from this arrangement.

Developers may also obtain financial assistance from the various government agencies in the form of grants and rental guarantees. However, the developer has to prove that the project would not go ahead without such assistance and that it will provide either new infrastructure or create employment opportunities for the local population.

1.3.6 Building contractors

Building contractors are employed by developers to undertake the task of physically constructing the development scheme and their prime objective is direct financial gain and profit. There are many different forms of building contractors and construction companies and there is also a considerable variety of contractual systems in order for a building contractor to satisfactorily ensure the timely completion of a new building.

Some development companies employ their own building contractors. For example residential developers tend to employ all of the necessary expertise in-house rather than tendering and outsourcing to a third party contractor. Development companies may keep their contracting division at 'arm's length' as an entirely separate profit-making centre. However, there has been an observed general trend towards an integrated approach especially with regard to residential developers.

A builder may also take on the role of developer, e.g. as a house builder. However, they will also be exposed to additional risks associated with the overall development process. Where a builder is employed only as a third party contractor, the financial profit is related to the building cost and length of contract. If the agreement is based on a design-and-build contract then a contractor will take on a design role which will involve a greater element of risk in relation to the responsibility for cost increases. Larger contractors, including international organisations with the relevant expertise, may take on the role of a management contractor with responsibility for managing all of the various subcontracts for the developer in return for a fee. In a scenario where the builder is also the developer (i.e. 'developer-builder') then a larger return is required due to the risk involved. However, when combining the building and development profit it is often acceptable for an overall lower level of profit. For builders or construction companies which employ a relatively large labour force, an additional objective may be to ensure continuing full employment for the workforce. Sometimes the only way to achieve this is by reducing tender prices and therefore profits; however, this is only a short-term survival strategy and not a realistic long-term business practice.

Building contractors have a specialised role in the development process which commences at the time of both maximum commitment and maximum risk for the developer. Accordingly it is essential for a prudent developer to ensure the capability and capacity of the contractor(s) to undertake the proposed work, seeking the right balance between accepting the lowest tender and ensuring a high quality of performance. It is not in the contractor's or developer's best interest to have a situation where the contractor is unable make an acceptable profit from the scheme. Clearly it is not in the developer's interest for the contractor to become bankrupt or to compromise on quality.

1.3.7 Real estate agents

Many terms are used to describe agents who sell real estate or real property and act as intermediaries between buyers and sellers. The more common of these terms include 'commercial agents', 'estate agents', 'real estate agents' or 'realtors'. Quite often this agent can be instrumental in initiating the development process and/or bringing together some of the main stakeholders in the process. Typically agents are skilled at networking and often rely heavily on technology to assist in this task. They also bridge the gap between the developer and the occupier, unless the developer uses 'in-house' staff to perform the agent's role and there is no need for the occupier to be represented by an agent. Nevertheless agents play an important role in the development process and they are often involved in every stage of the process from acquisition of the site through to the final sale of the completed product. Agents are able to perform this role due to their detailed knowledge of both the property market in terms of demand and current rents/prices and their 'personal' contacts with developers, occupiers, financial institutions and landowners. It has been said that the property industry is all about 'people'.

The agent's motivation for providing a professional service is to make a financial profit from the fees charged to their client, be it a developer or an occupier. In the case of introducing one party to another (for example a landlord to a tenant) they will receive a small fee but only if the transaction is completed and this fee is usually a percentage of the value of the transaction. An agent may also be a catalyst for the initiation of the development process by either finding a suitable site for a developer or advising a landowner to sell a particular site due to its development potential. Unless an agent is retained by a developer to specifically identify a suitable site for a particular land use (e.g. a specific location and land area/shape), an agent will often take the initiative by identifying suitable sites and 'introducing' them to developers. This process has been assisted greatly by internet technology which enables an agent to send email alerts to all interested parties as soon as a property is placed on the market for sale.

Agents will do their homework and only introduce sites to particular developers who they consider have the appropriate expertise and resources to both acquire the site and successfully complete the development on time. In a competitive real estate environment an agent will not waste time with a developer which does not have the potential to proceed. Quite often the agent will negotiate with the landowner on the developer's behalf and advise the developer on all matters relating to the evaluation stage. If the acquisition proceeds then the agent may be paid a fee for introducing the site, which is usually a percentage of the purchase price. The real estate agent may also secure appointment as the letting and/or funding agent for the development scheme. This is often a lengthy and time-consuming process for the agent; however, the financial rewards can be high. If an agent acts for a landowner then they would advise on both the potential achievable land value and the likely market for the site.

Agents may also be used by a developer to assist them in securing the necessary finance for a development scheme due to their knowledge of the requirements of the financial institutions or banks. Many financial institutions also retain an agent to advise them generally on their property investments and development funding; such agents may also specifically find development opportunities for their client to fund. The institution's agent will normally advise their client throughout the development process and act as one of the letting agents on the scheme. Some of the larger or more specialist firms of property professionals may undertake a consultancy and act as financial intermediaries, arranging funding packages with banks and other institutions in return for a fee related to the size of the loan.

Agents are widely employed by developers as letting or selling agents providing an essential link between the developer and the final occupier (either the tenant or new owner). In performing this role the agent should be involved from the start of the development to enable them to advise the developer on the occupier's viewpoint. Unless the agent's organisation is sufficiently large to have a specialist marketing department, it might not be possible to provide both comprehensive advice on the state of the market during the development and a forecast upon completion. A developer will need to commission market research to obtain more detailed knowledge of the specific market for the completed development. Some developers may employ an in-house research team, although the advantage of an agent is knowledge of the market in general and their contacts with potential occupiers or their agents.

Developers, landowners, occupiers, financiers and property investors may at some stage use the services of a valuer, chartered surveyor or appraiser to assist their decision-making process. Chartered surveyors, valuers and appraisers are employed by many commercial and residential agents to enable them to undertake professional work alongside their agency work. Developers will usually require an independent assessment of the hypothetical market value of the property to confirm their own opinion of value, where they have insufficient knowledge themselves or when it is required by a financier of the development. Independent and in-house valuers are also used by financial institutions and banks to evaluate proposed development schemes for which they are considering making loans or granting mortgages, including the asset value of any security being provided by developer. Financiers will often employ building surveyors to check on the construction phase of a development to ensure it is being built to the right specification, as well as to certify drawdown of the development loan. In the public sector the local authorities, central government and taxation departments (e.g. Inland Revenue) traditionally use valuers to advise and check on any development-related work.

1.3.8 Professional team

The development process is complex and most developers do not have the comprehensive range of skills or expertise to carry out a major development by themselves and need additional skills. Therefore, developers employ a range of professionals to advise them at various stages of the development process. These professionals include those listed here.

(a) Planning consultants

Planning consultants are employed to negotiate with local planning authorities to obtain the most valuable permission for a development, particularly with large or sensitive development proposals. If a planning application is refused they may be employed to act as expert witnesses in presenting the case for the developer. Planning consultants can also advise a landowner to ensure that the sites within their ownership are developed to their highest and best use in accordance with the legal requirements of the planning scheme. At times this may involve negotiation with the local planning authority at plan preparation stage or subsequent representations at an enquiry into the development plan. In performing this role, planning consultants can be important initiators of the development process and help to assist with the progress of the planning scheme.

(b) Market research analysts/economic consultants/valuation surveyors

An increasing emphasis has been placed on the role of market research analysts and their ability to gather and analyse relevant data, which in turn will reduce the overall chance of error and minimise exposure to risk during the development phase and after completion. It is therefore most important they are employed at the evaluation stage to provide a detailed market analysis in terms of the demand and supply levels of the type of development being proposed. Following the global financial crisis most lenders insist on a comprehensive market analysis by a professional third party when evaluating their lending risk for development finance. A lender or financier will often have their own in-house researchers who constantly assess the relevance of their funding criteria and lending policies in light of the changing economic environment.

(c) Architects

Architects and designers are employed by developers to design new buildings or assist with the refurbishment of existing buildings. Their knowledge of design, building material and construction methods also places them in an excellent position to administer the building contract on behalf of the developer, if required, and to certify completion of the building work. In the case of refurbishment work, building surveyors are usually employed to survey the existing building and advise on alterations and provide contract administrations services. Architects are normally responsible for obtaining planning permission if a planning consultant is not employed. With reference to a refurbishment, the building surveyor will usually perform or assist with this task. Architects are paid on a 'fee' basis, usually a percentage of the total building contract sum. It is important that the architect is employed at the earliest possible stage to ensure that all design work has been approved and completed prior to the commencement of construction. It is also important to employ architects with the appropriate experience, reputation, resources and track record in the industry. Due diligence needs to be undertaken by the developer in this area as an architect can make or break a development on many levels. A developer should ensure the architect has the right balance of skills to produce both (a) good architecture and valued design and (b) a cost-effective and workable design that is attractive to occupiers. This balance is often hard to achieve and therefore it is important for the developer to produce a clear architectural brief from the very beginning. Problems and complications start when there is poor communication between the architect and the developer.

Larger architectural firms usually offer a comprehensive service including project management, engineering, interior design work and landscape architecture. This may be an effective 'one stop shop' approach for some developers and improve communication channels. However, most developers actually prefer to assemble their own professional teams to achieve the appropriate balance of skills by bringing together the most effective and experienced team. Some larger development companies will use their own in-house architects and design professionals with the added benefit of existing communication links and familiarity with the developer's needs and objectives.

(d) Quantity surveyors

Quantity surveyors (QS) are in simple terms 'building accountants' who advise the developer on the likely costs of the total building contract and its associated costs. Their role can include costing the designs produced by the architect, administering the building contract tender, advising on the most appropriate form of building contract (procurement), monitoring the construction phase and approving individual stage payments to the contractor. A quantity surveyor needs to know the location of the proposed development since the cost of building materials and labour differs between each location. In addition the actual timing of the purchase is critical as this will affect the costs of purchasing goods and services since prices typically rise with time. More recently quantity surveyors have become increasingly more involved in the administration and management of design and build contracts. Like architects their fee is based on a percentage of the final contract sum of the completed development. The choice of quantity surveyor should be based on appropriate experience and reputation. Based on their own experience and according to their due diligence checks, the developer should be able to identify a quantity surveyor who works well in partnership with architects and other members of the professional team to produce cost-effective designs for the final development. Also a good quantity surveyor will be able to provide the developer with cost-effective ideas as alternatives to those proposed by the architect.

(e) Engineers

Structural engineers are employed by the developer to work with the architect and quantity surveyor to advise on the design of the structural elements of the building. They will also participate in the supervision of the construction of the structure. Civil engineers will be employed where major infrastructure works and/or groundwork is required. On large and complex schemes there is often a requirement for mechanical and electrical engineers to design all the services within the building, whether new or refurbished. Engineers are usually paid a percentage fee based on the value of their proportion of the total building contract.

(f) Project managers

Project managers are employed to manage the professional team and the building contract on behalf of the developer. Project managers are usually only employed on the larger and more complicated schemes. They are often from other professions associated with the property industry including architects, chartered surveyors, quantity surveyors and civil engineers. For many projects the actual developer will assume the role of the project manager and co-ordinator or alternatively rely on in-house staff or another member of the professional team. Project managers should be appointed before any of the other professional team or the contractor so that they are in a position to advise the developer on the best professional team to be assembled for the development project. Their fees can be either an agreed salary or based on a percentage of the total building contract sum. In addition, there is often an added financial incentive for managing the scheme within the agreed budget with delivery on time. For 'design and build' schemes it is often possible for developers to perform the role of project managers themselves when occupiers seek to employ the expertise of a developer in constructing their own premises.

(g) Solicitors

Solicitors are needed at various stages throughout the development process, starting with the initial acquisition of the development site through to the completion of leases and contract/s of sale to third parties. In addition, solicitors are often involved with the legal agreements covering the funding arrangements entered into by the developer. Furthermore if the developer has to appeal against a planning application via the court system, then both solicitors and barristers may be involved in presenting the developer's case at an inquiry. With some schemes, collateral warranties (guarantees) are required by purchasers and the solicitor will be involved in this process. Collateral warranties are defined as supporting documents to a primary contract where an agreement is needed with a third party outside the main contract. This could be the architect, contractor or subcontractor who will need to provide a guarantee to a funder, tenant or purchaser that it has fulfilled its duties under a building contract and accepts liability for their performance.

(h) Accountants

Specialist accountants may be employed to advise on the complexity of taxation-related issues that can have a major cost impact on a development. A knowledgeable accountant will often reduce the added taxation expense and can identify the best financial structure and holding vehicle for the developer.

The above list of the various professionals and specialists employed during the development process is not meant to be finite; however, such professionals are normally part of a successful development team. There are a considerable variety of other specialists who may be necessary, depending on the characteristics of the proposed development, including its size and complexity. Other professionals may include highway engineers, land surveyors, soil specialists, archaeologists, public relations consultants and marketing consultants. This list of specialists highlights the variety of skills that are required within the development process which are unique, as much as each development is also unique.

1.3.9 Objectors

Objectors have the right to provide an input into the viability of the proposed development based on social and community considerations, so it is important to be aware of the role of objections to the development process. There are two categories of objectors who can potentially cause delay and possible abandonment of development projects.

The first group of objectors may be purely 'amateurs' and self-interested neighbours of the proposed development who usually live near to the proposed development. They are often referred to as NIMBYs ('not in my back yard') and, where organised, they can achieve considerable obstruction to the progress of development. The rise of social media has greatly enhanced their ability to raise the profile of the proposed development to a broader audience, frequently through the use of dedicated websites. The second category is the well-organised professional, permanent bodies at local, regional or national levels. At a local level they may be referred to as 'amenity' societies who take an interest in every proposal affecting their local environment and heritage. This may also occur at the regional or national levels and the degree of involvement depends on the size of the project. Often these bodies have considerable influence with the local planning authorities and tend to be always consulted on major development applications. These organisations are well informed and have a good knowledge of the planning and development processes.

The developer must be aware of the interest (or likely interest) of these objectors which are at times legitimate and well intended. Accordingly the developer must be prepared to either accommodate (which may result in some form of compromise on the development scheme) or alternatively refute their opposition. Ideally any negotiations resulting from objections to the proposal should be carried out before a planning application is made to avoid lengthy delays. Opposition to a proposed development can be costly to a developer, by imposing higher standards and costly alterations or lengthy delays resulting in additional holding costs. At worst, opposition can lead to the complete abandonment of a development proposal that may have otherwise been sound. At times a large-scale or sensitive development may become part of a political discussion and receive substantial media attention, often resulting in intervention at government level in the development proposal. Where possible, a prudent developer needs to account for objectors when evaluating the likelihood of their development proposal receiving planning permission.

1.3.10 Occupiers

Unless the final occupier of a building is the actual developer or is known early in the development process, the final occupier is not considered to be a major stakeholder in the development process because they are unknown until the development is completed and let/sold. The demand for accommodation triggers the development process and influences both land prices and rents, to which developers respond. The occupier is a major stakeholder in the process and their future requirements must be researched at the beginning of the process. In the past some developers have tended to produce buildings in accordance with the requirements of the financial institutions, where the needs of the occupier have been overlooked. Following the global financial crisis it is important to ensure that property development is viable and meets the needs of the future occupants, which in turn will lessen the exposure of the lender to risk.

When the final occupant of the property development is known early on in the development process, then the occupier becomes a major stakeholder in the overall development. Accordingly the building will be constructed in accordance with the occupier's future needs and requirements, which at times can be unique to a specific occupier. This is particularly applicable to the occupants of industrial property developments. In some instances the developer may need to persuade the occupier to compromise on their requirements in order to provide a more standard and flexible type of building. This will broaden the future appeal of the building in the wider market if it has to be offered for sale or lease. At the same time the developer will also be concerned about maintaining the financial value of the building as security for loan purposes.

At times an occupier, either tenant or owner, may regard the buildings they occupy as an overhead and incidental to their core activities as providers of a service or product. Although some companies employ an in-house property team, including a facilities manager, many occupiers tend to fail to adequately plan for their future property requirements far enough in advance. They simply react to changes in their business as they happen. The property requirements of occupiers are influenced by both short-term business cycles and long-term structural changes underlying the general economy, including events such as the global financial crisis. Such factors can influence occupiers at a specific level or across the business sector in which they operate. Their demand for accommodation in the real estate market is also influenced by advances in technology where such changes affect both operational working practices and their physical property requirements.

Occupiers have been frequently criticised by agents and developers for not knowing exactly what they want. However, many companies are becoming far more knowledgeable about the role of property within their business and frequently employ their own facilities manager to advise on their accommodation requirements and building specifications. Different occupiers have different real estate requirements and priorities, particularly in the case of office space, making the developer's task of producing a building suitable for as many tenants as possible very challenging. The response of the financial institutions is to seek the highest quality specification with a layout to suit the widest possible range of tenants. As a consequence an occupier may be forced to occupy a building which compromises their requirements in terms of location or specification.

Developers and the financiers are taking more account of the needs of the final occupants. For example many tenants seek a degree of flexibility in their leasing agreement terms, especially in uncertain economic times. The occupants prefer to have the option to extend the lease at their discretion, which is opposed to the preference of the developer and lender who prefer a longer lease with upwards only rent reviews and no options to end the lease. A balance must be struck here and this is often driven by conformity with the broader real estate market. Occupiers have been increasingly seeking 'sustainable buildings' in the 21st century which follows on from the initial trend for 'green buildings' in the latter part of the last century. Developers need to be aware of this trend and ensure their buildings are future proofed and have incorporated sustainability aspects into the property's design and marketing. Occupiers are seeking sustainable buildings for a variety of reasons including corporate social responsibility (CSR) which appeals to customers, a better workplace for staff and potentially lower operating costs, e.g. providing natural ventilation instead of air-conditioning.

Discussion points

- Who are the main actors in the property development process?
- For a large-scale out-of-town retail development, which parties do you think would need to be part of the developer's team and why?

1.4 Economic context

Property development does not occur in an isolated vacuum. Every property development is located within inter-related different markets based on geographical location and also economic considerations in local, regional, national and global markets. Property development stakeholders have no control over the market, as shown quite dramatically by the global financial crisis which highlighted the ability of one country to directly affect volatility levels in global markets. The demand by occupiers, either tenants or freehold ownership, is a factor of supply and demand interaction in the broader market. The level of occupier demand is a reflection of short- and long-term changes in the economy where the availability and cost of development finance is also linked to conditions in the wider economy. The economic climate is important to developers as the local economy helps to determine the market for an individual scheme and the wider economy affects general property market conditions and the confidence of occupiers, investors and developers. Of course, the longer a development takes, the more complex the task of trying to forecast future demand and investment returns.

1.4.1 The local economy

Research suggests that most demand for an individual office or industrial development is drawn from a small geographical area around the scheme. The ability of a retail scheme to attract national retailers depends largely on the spending capacity of the local population. Since local economic conditions will help determine how much development and what type of development is appropriate in a particular location, it is in the interest of any developer to look beyond the individual scheme to the wider economy. The local economy can be a useful indicator of the likely viability of any development project and can be used alongside the development evaluation. However, it must be remembered that the focus is on the completion of the project in the future, so a downturn in the market today may result in a market upturn close to the completion of the project.

1.4.2 The national economy

At any given time the state of the national economy has a direct effect on the real estate market from both a supply and demand perspective. The developer must be aware of the implications of changes in the national economy and must factor in any changes which will affect their business plan. The strategic approach is to anticipate these changes as much as possible and adjust accordingly. Fortunately there are many organisations which analyse data and forecast changes in the marketplace and much of this information is now either free or available to purchase.

Supply considerations

Competing developers will enter the marketplace when there is the potential for a positive return on their investment. One way in which the state of the national economy affects the viability of a property development is the effect it has on the cost of borrowing. For example the level of interest rates is set by the national government and the amount of interest on loans to finance a development can be the largest single cost, especially where the project has a lengthy construction phase. High interest rates can make or break a development. Alternatively a government may restrict the level of investment from overseas investors for a particular development or place limits on the proportion of ownership of residents in a foreign country. This can result in lack of equity for a proposed development which in turn may not meet the lending requirements of the financier regarding the loan-to-value ratio (LVR).

Demand considerations

The national economy can directly affect confidence in the broader market and the need for property accommodation. For example, if national unemployment levels are high there will be reduced interest in purchasing homeownership, with households unable to borrow funds due to a lack of stable employment. However, this will then increase demand within the rental market. High unemployment also reduces disposable income, so retail spending may be down. Government measures such as reducing national taxation rates or the introduction of a stimulus package can help mitigate the impact on property.

At the national level, factors might be outside the control of the developer but a skilled developer can interpret changes in the national economy and produce a development which will be of value in the future, when completed.

1.4.3 The global market

The increased use of technology and the instant availability of information changed the nature of the property market and it is now interlinked on a global basis. The global financial crisis was one example where the behaviour of the economy in one market (e.g. Greece) adversely affected the entire global market very quickly from an economic perspective. The global financial markets are now very closely related so uncertainty in one market can adversely affect other markets. A developer must now consider broader changes in the marketplace, regardless of the size of their development. Some organisations are moving part of their business off-shore (e.g. to India) to reduce operating costs and reduce risk. This can have a direct effect on employment levels and demand for a certain real estate product. Property developers could be considerably affected by changes in the global economy and need to be aware of prevailing trends and forecasting changes.

Note: Loan to Cost (LTC): a new and important trend in financing development has recently emerged as an alternative to the traditional LTV financing. Instead of basing the finance loan for a development on the predicted value of the development once completed, the lenders base the amount they are prepared to lend on the development cost (LTC). This reduces both lender and borrower's risk since the cost to build can be fairly accurately calculated, unlike the value of the development upon completion. Although it is likely that this recent trend has emerged in response to the global financial crisis, the associated reduction in risk means that LTC is, arguably, a safer option than LTV for calculating development finance. While the main benefit of

LTC is to reduce the financial risk associated with a new development, it may not be particularly good news for the developer as the cost to build may be significantly lower than the predicted value upon completion and therefore the loan amount will be much less than if offered on a LTV basis, thus leaving them in a situation where they need to find more equity to fund their development (CREFC 2014).

Discussion points

- How has the economic context that property developers operate within changed over the past decade?
- Which of these drivers were within the control of a developer and which drivers were not?

1.5 Reflective summary

This chapter outlined the strategic framework for property development and introduced the stakeholders, each of which has different objectives. Collectively they operate within the overall context of the local market, the national and global economies. Development is affected by the status of property cycles at any given time, which can adversely affect the viability of a new development. The development process may be initiated by any of the main stakeholders identified but it can only take place with the consent of the landowner. An exception is when compulsory purchase powers are used. The lender/financier is a critical stakeholder in the property development process and assumes a large proportion of risk, therefore practically becoming a partner in the development. As a development proceeds through the various stages both the developer and the lender become increasingly committed and therefore flexibility is reduced, exposing both parties to greater risk prior to completion. Before a developer makes a commitment to acquiring land and signing a building contract they need to undertake due diligence. This includes obtaining all of the necessary consents, carrying out the necessary investigations and securing the finance at an acceptable lending rate. In addition a thorough financial and market evaluation should be carried out with the best information possible to establish the project's viability in light of the status of the occupier (either to rent or own) market. The ultimate success of the completed development will depend on many factors including the skill and experience of the developer, the state of the local, national and global markets, as well as a myriad of other variables which are often outside the control of the developer. An underlying key to a successful development is closely linked to the ability of the developer to be a 'visionary'. This includes knowing when to proceed with a development, or alternatively when not to proceed.

Plate 1.1 Location of the development © 2013 Google; Image: © 2013 Digital Globe; © Infoterra Ltd and Bluesky

1.6 Case study – low energy housing development

Fairglen Low Energy Housing is located at Loggans, a suburb of Hayle near St Ives in Cornwall, UK (see Plate 1.1). Phase I of this scheme consists of a development of affordable housing being 12 energy efficient houses. The format is 2 × 4-bed houses, 10 × 3-bed houses.

Further details are given in Table 1.1.

Location

The development is located on a 1.2 hectare brownfield site at Loggans which was formally used as an agricultural nursery. The site was originally part of the estuary. Loggans is within 0.5 miles (0.8 km) of two supermarkets, a retail park and an industrial estate. A primary and secondary school are within 1.0 mile (1.6 km). There is good access to public transport (165 feet (50 metres) to the nearest bus stop, 1.0 mile (1.6 km) to the train station), main roads and a sandy beach (approximately 2 miles (3.2 km) away. The site is surrounded by other residential development and was an ideal location for additional housing.

Background

This low energy housing scheme was initially discussed with planners in 2003 as a pioneering residential scheme designed to test the viability of building low energy homes for open market sale. A proposal was submitted in 2005 but due to its

Table 1.1 Fairglen Low Energy Housing

Client:	Percy Williams and Sons Ltd, Falmouth Road, Redruth, Cornwall, UK.
Architect:	Phase I: John Stengelhofen. Phase II: Chris Richards of Lilly Lewarne Practice Chartered Architects, Truro, Cornwall.
Development:	Phase I – Twelve units (10 × 3-bed semi-detached, 2 × 4-bed detached units).
	Phase II – Sixteen units planned (11 × 4-bed detached homes, 5 × semi-detached units). A change to this plan has recently been proposed to reflect changes in the market and is likely to consist of smaller homes comprising a mix of semi-detached and terraced 3-bed homes.
Cost:	Phase I – £2.5 million (cost included part of the groundwork and infrastructure required for Phase II).
Build time:	Phase I – undertaken and completed in 2008 (planning permission took two years).
	Phase II – currently under consideration (see Plate 1.2).
	Sustainability rating: Phase I achieved Code for Sustainable Homes – level 4 rating[1]

1 The CSH assesses the drawings and specifications for a house at the design stage and awards points out of 100 in nine design categories: these include factors such as 'energy and CO2 emissions', 'water usage', 'waste', 'ecology'. Level 4 = 68 points; Level 5 = 84; and Level 6 = 90. Retrieved from http://www.homebuilding.co.uk/advice/key-choices/green/code-sustainable-homes#sthash.kf48VJvY.dpuf (accessed 15/12/13))

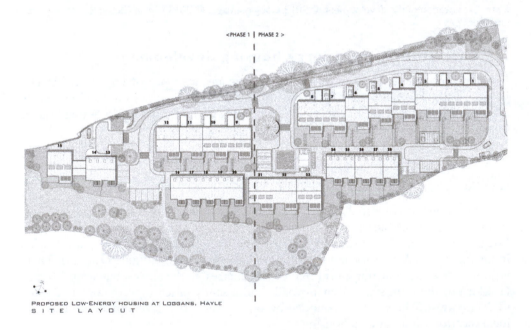

PROPOSED LOW-ENERGY HOUSING AT LOGGANS, HAYLE
SITE   LAYOUT

Plate 1.2 Fairglen Development – Phase I (completed) and proposal for Phase II (Printed with permission, © Lilly Lewarne Practice Architects, John Stengelhofen)

contemporary design which was driven by low energy and low CO_2 emissions, the planning committee was reluctant to grant permission. Permission was eventually granted some 126 weeks later in May 2007. Part of the 'Section 106' agreement was for the provision of eight affordable houses. This was negotiated down to five affordable houses. The first phase of the development was completed in 2008, just as the property market collapsed (see Plate 1.2). Five of the homes were released as affordable housing to Devon and Cornwall Housing Association. All have been occupied since 2009. Phase I was generally considered successful both in terms of performance and knowledge gained by the design team.

Following the return to a more buoyant market, a design and access statement was prepared for the second phase of the development in December 2012 (see Plate 1.2).

Site issues

- Ground conditions: The land originally formed part of the estuary valley and therefore contained alluvial mud and sand which changed to soil underlain by broken rock. The southern edge of the site is a flood plain and unsuitable for building. This part of the site formed a small wildlife area maintained by the residents and was also an integral part of the sustainable drainage strategy (SUDS) scheme. Surveys and enabling work was undertaken for both Phase I and Phase II in 2005.
- Contamination: A full land survey did not reveal any contamination.
- Mining survey: Cornwall is mineral rich and had a significant number of mines. A survey found no evidence of mining activity which would impact the development of the site.
- Trees, ecology, bats and protected species: No protected species were identified with the exception of bluebells and hedgerows.
- Highways and vehicular access: The main site access road was constructed in Phase I and was adopted by Cornwall Council under a Section 38 Agreement. In preparation for Phase II the main site entrance was widened. All other roads and parking areas serving Phase II will be maintained by the residents' company.
- Drainage: A site-wide SUDS was prepared for the site in 2005 and the infrastructure, including attenuation ponds, was installed as part of Phase I.

Design of the scheme

Phase I: The design was driven by low carbon rather than minimal energy use. The building fabric was of solid construction, including concrete floors to provide a high thermal mass, timber framing to the north wall, high levels of insulation, aluminium roofing, masonry and timber external finishing. The site is on a south-facing slope. Two east-to-west facing terraces were built, enabling each home to benefit from solar gain (see Plates 1.3 and 1.4). Large southeast-facing windows were incorporated into the rear of each unit to maximise the collection of solar energy. In addition to improving the thermal efficiency of the building fabric through insulation and reduced air flow from windows and doors, a combination of renewable technologies were adopted to reduce reliance on traditional fossil fuels. Reducing heat loss by making the building virtually airtight increased the risk of poor quality stale or damp air within the home. To overcome this problem a heat exchange ventilation system was incorporated.

Plate 1.3 Southern elevation of a 3-bed house (printed with permission courtesy of Percy Williams and Sons Ltd, © Graham Gaunt)

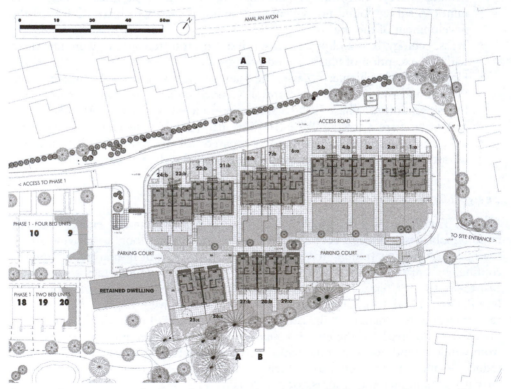

Plate 1.4 Proposed layout for Phase II (printed with permission, © Lilly Lewarne Practice Architects, Chris Richards)

Under-floor heating removed the need for wall-mounted radiators, improving the interior design. CO_2 emissions were predicted to be 63% lower than those homes built to standard building regulations in 2008.

Sustainability measures (Phase I)

- Insulation: The typical home loses over 40% of its total heat through the roof (Percy Williams 2008). Two layers of Rockwool insulation was fitted to take the insulating thickness to 350 mm.
- Passive heating: By combining good building design with effective insulation, you can collect and store the sun's energy to heat your home both day and night, throughout the year. Passive heating reduces heating costs and condensation (Smarter Homes 2013). The homes were all designed to benefit from solar gain and were positioned facing an open southeast aspect which allowed active solar energy collection in sun spaces located on their front facades. The heat was captured then distributed around the house by a mechanical ventilation unit which also incorporated heat recovery to maximise energy retention. The sun spaces (glazed rooms) were designed to capture solar energy and provide free space heating. However, they also can become a potential net energy loss if the space is not properly managed (i.e. shut up at night and on overcast days to keep the warmth in the main part of the house). The additional benefit was plenty of natural light, reducing the need for artificial lighting.
- Rainwater harvesting: A rainwater harvesting system was incorporated to supply filtered rainwater for toilet flushing, to washing machines and via outside taps for garden purposes. This helped to limit the impact on the environment and increase resilience to drought and flooding. This resulted in a dramatic reduction in the amount of water required by each property which also reduced water bills. The rainwater harvesting system filters rainwater into a 600 litre underground tank located close to each house; the water is passed through a second filter before being sent to an 'active switch' pump within the building. The pump raises the water pressure to mains level, and will also allow the mains water to supply the system should there not be enough in the rainwater tank.
- Mechanical heat recovery: The heat exchange ventilation system provides fresh air to the house while extracting stale or damp air from areas such as the bathroom or kitchen. The incoming fresh air is slightly preheated by the heat exchanger, thus avoiding any cold draughts. Air is constantly flowing through the house, maintaining the heat and providing a constant flow of fresh air. The benefit is that it significantly reduces heat loss through ventilation and provides excellent internal air quality. The only disadvantages observed by the developers is that the system loses efficiency if the windows are left open during times of heating and air filters must be changes approximately every six months.
- Wood burning stoves: Wood burning stoves were installed as an additional source of heat. These provide a carbon neutral fuel source and can be relatively inexpensive to run if wood is sourced carefully. The only disadvantage is the lack of controllability so they can either be inefficient as a heat supply for the whole house or over-powerful, resulting in the need to open windows to allow cooler air in.

- Ground source heat pumps (GSHP): A GSHP pumps heat out of the ground using a closed loop pipe concealed within an 80–100 metre-deep borehole under each house. As the water passes through the pipe it absorbs the heat from the ground which is then increased to a temperature of approximately 50 degrees Celsius and used for under-floor heating. This helps to keep the interior temperature very stable. The developers found the benefits to be the long life expectancy (approx. twenty-five years) and reduced maintenance costs, which they estimate saves about £400 per annum (as at 2012) since no annual service is required. As energy bills are likely to rise in the future, occupiers will see additional savings from using GSHP over conventional types of heating. The system is computerised and completely automatic, unlike most conventional heating systems. The only disadvantages observed from using GSHP was the reliance on electrical energy input which may not be efficient or low in carbon emissions and the large (and occasionally noisy) piece of machinery which must be located inside the house (usually inside a cupboard).
- Solar hot water panels: Solar thermal is an efficient and low maintenance system and used to produce hot water for use in the home. No disadvantages were found with incorporating this into the build (see Plate 1.4).
- Solar photovoltaic: Solar PV panels were installed and it was estimated that they would provide around 1.2–1.3 kW (kilowatts) per hour and contribute about one third of the electricity requirements for each property. Their cost should be repaid to the owners within eighteen months to two years due to the energy savings they should provide. The savings have been closely monitored by the Low Carbon Buildings Trust (2009) and recent information from the architects' brief for Phase II of the development suggests a slightly lower figure of energy production (approx. 20–30%) of total house requirement (see Plate 1.4). The main disadvantage was the high cost of the solar panels at the time of the development.
- Additional factors:
  - Government grants: At the time of construction the development benefited from a range of government grants which were designed to promote renewable 'bolt on' technologies which provided the developer with an opportunity to test a range of renewable options. The developer found all technologies used in Phase I to be successful to varying degrees.
  - Occupiers' behaviour: To maximise energy efficiency within the house, the occupier was required to actively adapt the house, depending on the season and ambient temperature. The largest impact on temperature was caused by air flow from the sun collection room into the main house. Occupiers could control heat loss or maximise heat gain by opening or closing the door to the sun collection room. The developer found that even with minimal occupier intervention, the houses still achieved significantly above what was considered normal for standard homes at that time.

Results

- Energy saving: The technology included proved to be a success with total energy costs (heat, lighting, water, electricity use from appliances) reported to be as low as £700 per annum. (Many homes in Cornwall are heated by oil-fired boilers and a typical annual bill would be around £1,400.)

- Materials: In terms of embedded carbon, the aluminium roof was not ideal. Local slate would have improved the rating but unfortunately local supplies were either prohibitively expensive or unavailable at the time of construction. Slate could have been imported from China but would have been transported to the UK via ships burning crude oil and as a consequence increasing the embodied carbon footprint of the development.
- The use of GSHPs is a success due to their simplistic design and minimal maintenance. By contrast, rainwater harvesting has proved to be more difficult than first thought and any future installation will be modified.
- Costings: The cost of the units proved to be excessive (no figures are provided). Due to the pioneering nature of this development, the complexity of the design and the use of new technology, the developers encountered many problems that had to be resolved as the development progressed which increased the build cost. These units were placed on the market at £250,000 for a 3-bed home and £350,000 for a 4-bed. Due to the property market crash, they sold for substantially less than the market price.

Phase II

Despite the problems encountered in building this pioneering development and the poor market conditions at the time the development was completed and released for sale, Phase I was generally considered a success. Building on this success, plans were drawn up for Phase II in 2012 (see Plates 1.2 and 1.4). Some modifications were made. First it was found that although new build energy efficient housing had become more acceptable to buyers within the housing market, much of the demand came from young couples and families who generally found it more difficult to obtain mortgages to purchase large 4-bed homes. The architects found that those who could afford large 4-bed homes were often put off by the terraced design typically associated with energy efficient housing, preferring a more traditionally designed detached house, despite higher running costs through poor thermal efficiency. Second, the demographic who were concerned about environmental issues and therefore likely to be the target market for Phase II expressed a desire for larger gardens than those available in Phase I.

The architects have responded by making changes to the original plan. They now propose that Phase II consists primarily of 3-bed units (2-bed units were considered but due to the additional cost associated with incorporating sustainable measures into the build, the market price would be too high for the general market and therefore they were not considered to be financially viable). Removing the 4-bed detached homes from the revised plans meant that the density would increase from 23.3 units per hectare to 25 units but would allow for two additional units bringing the total up to 30 units.

Chapter 2

Land for development

2.1 Introduction

The essential element in every property development is a geographical location. The acquisition of land, or space above land in the case of a strata title development, is usually the developer's first major decision when seeking to undertake a property development. The land may be previously undeveloped or include buildings or other structures which are attached to the land. When undertaking a property development, the purchase of the land should not take place until after a full analysis and evaluation of the proposed development project has been undertaken, as discussed in Chapter 3. This chapter discusses the initiation and purchase stages of the property development process. The selection of a site fundamentally affects the nature and success of a development. Poor site selection cannot be rectified after the decision has been made and no amount of careful design or promotion can totally overcome the disadvantage created by a poor location or a lack of demand for accommodation at that location even when the value is pitched at below market price for the final product. Land is unique and every site has its own characteristics. Site identification and acquisition can be a very long, frustrating and unpredictable process as there are many factors, some outside the developer's control, which affect the successful acquisition of a viable site.

The objective of this chapter is to discuss approaches to be used by a developer to identify and acquire a site suitable for development. There is also a discussion about the role of landowners and other stakeholders in the process. As much as each site is unique, often a unique approach must be undertaken to search for an appropriate site. While technology has greatly enhanced the availability of information, instantly accessible to the developer, industry networking and an understanding of the basic skills related to site identification are still closely linked to a successful development.

2.2 Identification of development site

The first step before identifying a development site is to confirm the search parameters by defining the aims, nature and geographical area of search. The overall strategy and aims of the development company will form the basis for the identification of sites and potential development opportunities. Some companies restrict themselves to a specific type of development in a particular location. The obvious advantage is that they can benefit from previous knowledge about certain types of development or location. The availability of resources such as the amount of finance they can obtain, a skilled workforce or materials will also be a major consideration.

Within this overarching strategy a developer needs to accurately define the search criteria which will identify the specific geographical areas and their exact real estate requirements for their search. This criterion relates to the size, nature and location of sites. In addition, the geographical area of search for sites depends on a number of factors which include variables such as:

- the risk profile of the developer
- their knowledge about a particular location
- the forecast status of the market both now and in the future
- the potential to spread risk across a number of locations in a portfolio
- the availability of development finance, and
- the results of detailed market research into supply and demand considerations.

The location of the development company's office will also be considered because the further the development site is located away from the office then it is more likely that the management of the project will be less effective. However, this can be overcome by establishing a temporary office nearby if the size of the development is sufficiently large. If the developer is operating within his local area, the developer will usually have already established good contacts with real estate agents, occupiers and the local planning authorities and therefore will be able to consult with them on the proposed development. In a scenario where a development site is a considerable distance from the developer's office and in a different region, a good working relationship will need to be established with local agents. Often it may be prudent to carry out the development with a local development partner as there is no substitute for local tacit knowledge. Larger national and international development companies benefit from economies of scale and are often able to spread their exposure to risk by spreading their development activities over different locations. For example if an oversupply of accommodation occurs in one location it is restricted to only some projects and does not affect the entire development portfolio.

The process undertaken by a developer for the identification of a suitable site may be largely influenced by the way in which the developer sources finance for the development project. If a developer intends to seek finance from a particular lender or financier, then the developer needs to be aware of their preferred locations when lending for property investment. In previous development booms the location of a speculative development sites was less important in the process of obtaining finance, mainly due to the widespread availability of funding for loans. However, since the global financial crisis the lenders have tightened their lending criteria and therefore place increased importance on the attributes of the property developer and the proposed completed development. Nevertheless, within any economic climate obtaining funding is likely to be more successful on well-located sites – such sites will have the widest appeal to potential occupiers with associated reduced risk. Regardless of the state of the real estate market at any given time, a prudent developer should always seek the best location appropriate to the proposed highest and best use. An alternative, or second choice location could affect the long-term success of the project and the developer should consider this additional risk carefully at the evaluation stage.

When identifying the optimal location for the geographical search a developer will focus on the completed development and therefore the developer's perception of occupier demand, supported by reliable market research, is a critical factor. In this role the developer's skill, knowledge and experience are important in identifying areas of potential growth where market forces will provide increased demand for accommodation and, providing there is no unexpected downturn, demand should exceed supply by the time a development project is completed. Proactive forward-thinking developers may commission research at a strategic level to identify trends in the market and areas of potential opportunity, which also aligns with the concept of a successful developer being a 'visionary'. Market research should seek to identify current and projected levels of supply and demand of various types of accommodation in a particular area in addition to short- and long-term trends in rent and capital values. From a value perspective, access to services and transport are usually sought after by residential and commercial occupiers. Accordingly good direct access to road networks and public transport schemes are some of the more obvious factors which often influence levels of demand for accommodation in a particular location. Developers will seek to purchase a site in a location which is likely be affected by future population growth and an expansion of urban areas.

As part of any market research the developer should identify and examine factors which influence occupiers in their choice of location. Different factors affecting decisions about the choice of location for each type of land use are now explained further.

(a) Residential development

All humans have a basic primal need for shelter which will provide protection against the extremes of the external environment. The form of shelter varies substantially between different locations and regions depending on variations in climate, weather, crime, government policy as well as the availability of land and construction material. Most importantly, residential real estate (as with all real estate) usually has an associated bundle of property rights which provide options to the owner/s. Increasingly the family home, in addition to providing shelter, is recognised as having an additional benefit, being a wealth asset or an investment in a family's asset portfolio. Wider acceptance of homeownership also provides many direct and indirect benefits for individual households and broader society including lower crime rates, a sense of 'place' as well as improving social cohesion and sustainability (Forster-Kraus et al. 2009).

To achieve homeownership status in western civilisations the procedure for a large proportion of the population is to purchase their home with the assistance of a loan or mortgage from a financier, e.g. bank, credit society. In order to meet loan repayment obligations it is an essential requirement of the lender that the borrower is in employment. Following on, in order to reduce transit time between home and work it is accepted that the optimal location for most residential property, especially in urban cities, is in relative proximity to the homeowner's place of employment. This scenario forces prospective homeowners to typically search for residential real estate with nearby transport links and, with increasing global commitments to sustainability, a requirement to be close to reliable public transport networks. From a planning perspective there have been increasing debates about options for urban cities to expand 'up' (i.e. high-rise condominiums) or 'out' (i.e. urban sprawl into outlying suburbs). Nevertheless residential densities continue to increase due to an expanding global population base and in many regions this has placed pressure to live in smaller living spaces, such as high-rise accommodation in Hong Kong or New York. However, this type of residential accommodation is not suitable for all urban cities and occupants, accordingly careful market research needs to be conducted to assess the demand for different types of dwelling from a forward-looking perspective. Consideration must always be given to the life cycle of a residential property and ensuring today's property development will not become obsolete due to rapidly changing demand in a particular location.

(b) Office development

Due to the evolution of many cities from a marketplace to a town to a city, the centre of these urban areas has traditionally remained the most sought-after location where there is limited land supply aligned with the highest level of demand (see Figure 1.1). Accordingly the traditional location for office space has been in the centre of the cities where transport hubs are easily accessible to and from residential areas located on all sides of the city. Some cities have become so congested there has been a trend to develop office 'hubs' which are located on the city fringes in closer proximity to workers and often on cheaper land. However, planning approval and the risks associated with a location away from the core office precinct (i.e. city centre or downtown) require careful assessment.

The proximity of reliable roads, rail and air is vital in the consideration of locations for office development. The choice of location for an office occupier is determined by a wide range of such diverse factors including traditional locations, proximity to other related professionals (e.g. other financiers, lawyers), transport links, staff availability, quality of accommodation, availability of parking and individual preferences by decision-makers, e.g. traditional location.

With the advent of internet technology, including smart phones and tablets with associated high speed connectivity, many companies have partly reduced their requirements for office accommodation in contrast to previous needs. Many organisations encourage staff to work from home or in transit (i.e. 'tele-commuting'); therefore, each office worker does not necessarily require an individual dedicated office desk which is vacant most of the time. The concept of 'hot-desking' where several employees share a desk or workstation is designed into some office accommodation and this is particularly common with active and mobile staff who spend most of their time out of the office. There continues to be much debate about how such changes in working patterns and advances in information technology will affect the location of offices in the future. Earlier concerns that tele-commuting would severely reduce demand for office space have been dispelled although office buildings must be agile to adapt to the changing future needs of occupiers due to the extended life of an office building (in comparison to other land uses).

(c) Retail development

Retail is an integral and essential component of 21st century society and development takes place within a hierarchy of shopping locations and a diverse range of retail accommodation types. This hierarchy of shopping locations consists of regional centres, district centres, local centres and superstores/retail warehouses and is often a combined social outing, regardless of the increasing competition from online retailers. A particular shopping area will be classified within the hierarchy by reference to its general demographic characteristics and the size of its catchment population. The catchment population is typically calculated by reference to the size of the population living in close proximity to a specific location for a proposed shopping centre.

Prior to making a decision to develop a site, a retail developer will carefully analyse the catchment area surrounding the proposed scheme and undertake scenario modelling. In relation to regional and district retail centres the catchment will be modelled in terms of (a) travel time to/from the retail centre and (b) the potential target population and its demographic characteristics, e.g. age, household income, disposable income, spending patterns. An analysis will also be made of competing shopping centres (existing and planned) to evaluate the impact of the proposed scheme on the proposed retail centres and vice versa. In carrying out this analysis the developer is assessing the trading potential of the proposed scheme within the broader retail market with the emphasis placed on the return on investment. The analysis is conducted from the point of view of both the individual retailers and the individual shopper as the overall viability of the scheme will depend on this.

An example of an evolved retail category is retail warehousing which is sometimes referred to as 'big box' or 'out-of-town' retailing. A developer of this type of land use has a detailed knowledge of the locational requirements of each retailer through their working relationship with them and associated in-depth research into the industry. Many of the national and global retailers who operate in out-of-town stores carry out their own developments. With this approach the development division of the retail company will identify potential sites and the retail division will consider the potential trading position.

A critical factor in identifying the optimal location of any proposed town centre retail scheme, whether it is a single shop unit or a major shopping centre, is pedestrian flow characteristics. Studies can be carried out to model the pedestrian flow which is heavily influenced by car parks, bus and railway stations, pedestrian crossings and the location of major stores (referred to as 'anchor' tenants or stores). The precise location of a shop or store is directly linked to the level of rental value and it is crucial the developer makes the correct decision here. Retail shops and shopping centres are compared in the market on a ranking system which is somewhat similar to the ranking undertaken for office buildings. For example different retail locations are classified using terms such as 'prime' and 'secondary' with reference to characteristics including pedestrian flow and proximity to competing retail stores.

(d) Industrial development

Industrial development is closely aligned to real estate zoning restrictions designed to separate non-compatible uses from each other, e.g. industrial and residential developments are rarely planned in the same location but retail and residential uses are encouraged. Therefore, industrial development occurs in designated and separated locations due to the side effects of transport noise, potential pollution (e.g. noise, air) and the large parcels of real estate typically required. There are various types of industrial property, each of which have different locational characteristics. Industrial property can be categorised into many different industrial land use categories including general industrial, light industrial, service industry and warehousing/distribution.

General industrial land use commonly refers to larger land parcels which have good ingress/egress for large trucks as well as dedicated parking on hardstand for trucks, often equating to up to 50% of the total area of the parcel. The developments are often large-scale industrial buildings with a small office component. The emphasis is placed on the ability to provide accommodation for a wide range of industrial activities (e.g. heavy industry) where each industrial land use is too narrow to have separate industrial categories. The category of light industrial typically refers to real estate used for processes which involve the manufacture of goods with minimal or no environmental impact. There are various types of occupiers who require light industrial property. At one end of the scale there is the traditional manufacturer as opposed to companies in 'high technology' industries who require office and research and development facilities alongside production facilities. Often this type of development is often located in a dedicated area (e.g. Silicon Valley in California) as opposed to a business park which focuses on administrative processes or an industrial park which focuses on manufacturing. Light industrial buildings are developed to a specification and also in a location in demand by potential occupiers.

Industrial warehouses are large industrial units occupied by retailers, manufacturers and distribution companies. An occupier's needs have increasingly become more diverse and often highly specialised due to technological advances. A large proportion of the warehouse development is carried out on a 'design and build' basis rather than on a speculative basis to suit a potential new tenant. Sites suitable for a warehouse or distribution centre must be in a location with good access to major transport infrastructure. This also suits the 'just in time' (JIT) approaches adopted by many companies where stock is not stored for extended time periods but moved quickly in and out of the warehouse. In most areas service industry accommodation is perceived as semi-retail and provides direct services and sales to the community.

There are certain variables which influence the location of industrial premises. Industrial occupiers need to locate in areas close to their markets and supplies of raw material and with good access to major transport routes including road, rail and sea. Companies which employ a high proportion of office and research and development staff will often have similar locational requirements to occupants of office property, i.e. attractive landscaped built environment and the availability of quality housing in the nearby proximity.

Once a developer has established the parameters of the geographical search, a strategy and brief for the site finding process can be produced. It is important to define the size of target sites by calculating the total land area of the potential development scheme and the preferred land-to-building ratio. The next stage is to identify preferred locations in a town, city or region.

2.3 Brownfield and greenfield sites

In order to increase the efficient use of land in urban cities an increasing emphasis has been placed on urban regeneration and the use of existing vacant sites. The same cities typically are experiencing higher levels of peak traffic congestion where commuters are transiting past, often in slow moving traffic congestion, vacant and unused sites in order to reach their outlying homes or destinations. The need for increased urban regeneration is generally accepted by all stakeholders and is also linked to increasing the density of future developments and encouraging the higher use of public transport use by residents. The aim of urban regeneration is often supported by major changes in government policy to promote the re-use of existing land, often with a different previous land use, known as 'brownfield' land. The development of brownfield land is actively promoted while greenfield development (on previously undeveloped land) is encouraged if 'land banking' or storage is occurring. Urban regeneration, or 'urban renaissance' as it is sometimes referred to, is increasingly undertaken in all developed countries. Brownfield land located in or close to the inner-city appeals to property developers as it is considered to be in a good location. In general these sites have become available due to a decline in manufacturing; however, attention must be paid to associated challenges such as contamination. An example of conversion of brownfield sites often occurs when a major redevelopment is undertaken, such as developing an Olympic village in a previously disused industrial area.

Brownfield land redevelopment can however be complex and involve substantial risks. For example many of the best brownfield sites may have already been redeveloped and the remaining sites are often poorly located or the costs of cleaning up contamination are prohibitively high. In many global cities it is now probable that there are locations where there is an insufficient supply of brownfield land for residential and commercial property development, thus leading to increased densities or pressure to rezone outlying land for development. There has been a broad trend to apply the same principles of increasing densities to new greenfield developments, although allowing densities not as high as the inner-city development densities. In turn this will reduce the amount of land required to accommodate the development.

2.4 Initiation

After undertaking research and defining a strategy for site acquisition, the next step for the developer is to actively seek and identify potential development sites. This can be achieved in a number of ways; however, theory and practice often differ. A developer may have a well thought out and thoroughly researched land acquisition strategy but actually achieving that strategy will depend on numerous factors, many beyond the control of the developer. This is where the property development process is unique and a lot depends on the opportunities available. Above all, a developer's ability to acquire land (and existing buildings which form part of the land) is dependent on the availability of land at any particular time. But the availability of land is dependent on many factors including the state of the real estate market, planning policies and physical factors. In addition every proposed property development will also depend to some degree on the motives of the particular landowner. The developer, landowner, real estate agent and government bodies sector are the main stakeholders involved in the development initiation process. The landowner may take an active or passive role in the process.

The developer needs to understand the dynamics in the real estate market. For example, in accordance with standard economic theory there is likely to be more land supply available during a time when land values are rising rapidly. The availability of new land will be influenced by the allocation of land within a local planning authority's 'development plan' and the perceived chances of obtaining planning permission in respect of unallocated areas of land or land allocated for other uses. Although land may be available on the market and is allocated within the development plan for the proposed use, it still might not be suitable for development due to certain physical factors, e.g. potential to flood, which is becoming a more frequent occurrence in some countries. The lack of necessary infrastructure, such as road access and services, might make a development scheme unviable. Also the physical ground may be contaminated or unstable, therefore ensuring it is cost prohibitive when seeking to undertake a profitable development. It is important to closely examine the various ways of initiating the site acquisition process.

2.4.1 Initiation by the developer

In most instances the development will be initiated by a developer who will identify a potential development opportunity and make the first approach. The search for land will often be undertaken via an internet search, being on a subscriber list for new 'for sale' properties or in regular email contact with a real estate agent/s. For a small developer there can be no substitute for directly approaching a potential seller. A larger developer may employ an in-house team, an agent or a planning consultant to actively identify a development site/s based on the criteria set out in the site acquisition strategy.

- In-house land buyers: Many developers, particularly those who specialise in a certain type of development (e.g. retail shopping centres) and land development companies, employ their own staff who specialise in buying land (also referred to as 'in-house purchasers'). Their job is to identify and acquire sites in accordance with the development company's strategy. They will require a good knowledge of the target area and relevant planning policies. These searches will sometimes be made via the internet but also in person, by car or on foot to identify potential sites. If the site identified is not for sale the next step is to find out who owns the land. There are a number of ways of achieving this including asking local agents or literally knocking on the door. In some countries privacy legislation has been introduced which has restricted the release of personal information by government bodies to third parties. Another option is to investigate whether a planning registry is in use, or alternatively a body which records historical planning applications and approvals.

- Employing a planning consultant: A developer may employ a planning consultant to carry out a strategic study of a particular geographical area to identify suitable land within the planning context. A strategic study of an area will involve examining planning documents, i.e. development or strategic plans and (where they exist) local plans covering that area, as well as discussions with local authority planning officers. The study will usually identify sites which have been allocated in the strategic development plan but not yet developed, commenting on their suitability and availability for the proposed use. A report will be made on each site describing its characteristics, planning history and details of the landowner if known. The study will also identify sites that have not been allocated but where there is a good chance of obtaining planning consent for a property development by negotiation or on appeal. Often the best time to carry out this study is when the development plan is in its draft or review stage as there is then a chance to influence the allocation of land by presenting evidence at the public inquiry. Accordingly it is of vital importance that developers know the timetable of every review and draft publication of the development plans relevant to their area of search. The study should advise the developer on which individual sites should be pursued.
- Employing a real estate agent or realtor: A different approach is possible when a developer employs a real estate agent or approaches a number of real estate agents to identify prospective sites in a particular area. The developer will need to brief the agent/s as to their requirements in terms of the nature and size of sites. A good agent will have an in-depth knowledge of the local area and its relevant planning policies, so very often a local agent/s is usually employed or approached. A developer is usually in contact with a number of agents and it is important to develop good relationships to build trust and ensure the agent stays loyal. If the real estate agent is retained directly by the developer, a fee will be payable if the latter is successful in acquiring a suitable development site which was identified by the agent. For example, this may equate to 1% of the land price; however, it is often a matter of direct negotiation and depends also on the real estate agent's involvement in the latter stages of development, letting and funding of the scheme. Another advantage of using a real estate agent is that they become the developer's eyes and ears. Through their knowledge of the area, they know who owns a particular site and its history. They can anticipate whether a particular site may be coming onto the market for example. With occupied buildings they may know when leases will expire and therefore when possible redevelopment opportunities might arise. They will keep in contact with local landowners via established networks and therefore usually know exactly when a site might be coming onto the market.

It is advantageous if sites can be identified as early as possible since it gives a developer a chance to negotiate directly with the landowner and secure the site before it is released onto the open market. A developer's ability to acquire a development site 'off market' will depend both on the developer's negotiating abilities and on the current state of the market. When the market is booming and land values are rising rapidly, the landowner will be strongly advised by agents to put the site on the open market. A negotiated deal may not be possible if the landowner is a government authority, as they are publicly accountable and need to demonstrate that the highest price in the market has been achieved.

Developers may also identify sites in some less obvious ways. A developer may acquire an entire organisation as part of their acquisition strategy and secure a site or an entire portfolio of properties. For example a development company may purchase a particular retail chain as a means of securing 'prime' sites in a built-up area. The developer may retain ownership of the property assets and either (a) on-sell the operating part of the business to a third party, (b) move the business to other leased premises, or (c) close the operation of the business. Developers also may acquire individual properties or entire portfolios through direct approaches to other developers or property investment companies regarding their corporate real estate assets.

2.4.2 Approach via real estate agent

Although real estate agents (simply referred to from this point forward as 'agents') may be retained exclusively by a developer to find sites, they will often take the initiative and introduce opportunities to developers directly. The opportunity may be a site already on the market or a site that is likely to come on to the market shortly. If the introduction to the developer ends in a successful acquisition of the site, then the agent will expect a fee from the developer, unless they are retained and instructed by the landowner. The fee is typically 1% of the land price but may be negotiated. It will also depend on whether the agent continues to be involved with the scheme through marketing, letting and/or funding instructions.

An agent will introduce a site to a particular developer who is most likely to be successful in acquiring that particular site and undertaking the development. Some agents remain loyal to a certain developer because that particular developer is an established client and they have built up a good working relationship over time. The agent will look at the experience of development companies, their track record and financial status, when making their decision about who to introduce a particular site to. The most likely candidates are those developers who are active in the particular market at the time, whether it be, for example, retail shopping centres, commercial offices or industrial units, and who generally have a history of successfully bidding for sites.

A development company, depending on its size and financial status, may receive introductions regarding prospective development sites on a daily basis when market conditions are favourable. It is particularly important to set up a register of sites that have been introduced to the company because it is likely that different agents will introduce the same scheme to different people within the same company. It is important to avoid duplication of agents, otherwise two acquisition fees might be payable for the same property.

When introducing a site to a developer, the agent should provide enough detail to enable an initial decision to be made by the developer as to whether or not to pursue the potential development opportunity. Ideally this information should include a site plan, location plan, planning details, and details of the asking price and terms. It is the introducing agent's responsibility to assist the developer throughout the acquisition process. The agent should be able to provide advice on the local property market, rental values and information on existing and proposed schemes of a similar nature to assist the developer in the evaluation process. The agent is also frequently relied on to negotiate the land price on behalf of the developer.

This approach to finding a site is a two-way process between the developer and the agent. The developer must establish and maintain a good relationship and regular contact with local, regional and national agents. It is important to provide those agents with details of site requirements to avoid a situation where site opportunities are continually rejected and the agent gives up and moves to working with a rival developer. At the same time, agents should provide a good service to their developer clients to ensure the business relationship continues and increase their chances of receiving letting and funding instructions associated with the property development. Other property professionals such as solicitors, planning consultants, valuers, architects and quantity surveyors may introduce opportunities to developers. It can be said that property development is more about who you know than what you know and the skill of networking cannot be understated, e.g. via LinkedIn. Professional networking websites assist with maintaining networks and developing additional links with other professionals.

2.4.3 Landowner initiatives

A landowner may take an active role in initiating the development process via a decision to sell their land or enter into partnership with a developer. This may be because the landowner is 'asset rich–cash poor' and lacks the access to finance, skills and knowledge to develop their own land or property to its highest and best use. Understanding the drivers behind the landowner's decision to sell can aid the negotiation process and speed up the sale.

An obvious source for identifying development sites for sale is advertisements, whether on a site board, via direct mail, in the media or in property publications such as *Property Weekly* and *Estates Gazette* in the UK, *Real Estate Times* in Asia, or *Real Estate Magazine* in Canada. More often the starting place is now the internet with most agents listing available property 'for sale' on their own website. These sites are designed to be very easy to navigate with most including a search engine with filters, which allows the user to define the characteristics of the sought-after property and reduce search times. In addition it is often possible to subscribe to a particular agent's website who will send you notification of a new property 'for sale' based on your requirements. The benefits of using this method to search for land are that (a) potential purchasers are instantly advised about a property which has just been placed on the open market, and (b) it reduces the need to constantly revisit real estate internet websites in case they have been updated.

Other advertising mediums where 'for sale' property can be found are local and regional newspapers, both of which usually have real estate sections. A developer may also receive particulars of a site for sale direct from a landowner or their agent. Sites which are advertised in the open market will automatically involve the developer competing in the open market to purchase the site. However, the level of competition from other developers will depend largely on how the site is offered to the market and the associated conditions of sale. Various methods are available to bid for the land including informal tender, formal tender, a competition process, auctions and open 'for sale' listings. The method of disposal is at the discretion of the landowner after considering advice from the agent – this decision will depend on market conditions and the motives of the landowner. The developer may be in competition with any number of other potential purchasers or there may be a selective list of bidders. The different approaches to sale are discussed further below.

(a) Informal tenders and invitations to offer

Informal tenders or invitations to offer involve inviting interested parties to submit their highest and best bids within a certain timeframe. This usually involves all parties who have expressed an interest in the site and the invitation to bid may include an indication of the minimum price acceptable. For example, it might state that offers to purchase over a certain amount are invited. In addition it will state any conditions that are attached to the bid, e.g. payment in cash only or no opportunity to inspect the building prior to the auction.

The important point from the developer's perspective is that the bid made is subject to any necessary conditions. After a bid has been accepted by the landowner, the developer then has the ability to renegotiate the price if there is some justification to do so before the contract is agreed by both parties in writing. There is always a risk that the landowner may not accept a revised price and may offer the lot to one of the other prospective purchasers who also made a bid. Generally speaking, developers prefer 'informal' to 'formal' tenders as they allow bids to be made on the developer's own terms. However, the more conditions a developer attaches to a bid then the less likely it is that the bid will be acceptable even if it is the highest received. The landowner as a general rule will accept the highest bid unless the conditions attached to it are unacceptable or the developer's financial standing is questionable. After receiving the bids, the landowner may negotiate with several of the prospective purchasers before making a decision in an attempt to vary conditions or the level of the bids.

(b) Formal tenders

A formal tender binds both parties to the terms and conditions set out in the tender documentation, subject only to contract. It involves an invitation to interested purchasers and or the entire market to submit their highest and best bids by a certain deadline. The invitation will set out the conditions applicable. The document will usually state that the landowner is not bound to accept the highest bid. Developers, as a general rule, do not favour formal tenders as it reduces their flexibility and hence increases their risk. The exception to this would be a situation where all the possible unknowns had been eliminated, e.g. where a detailed acceptable planning consent was in place, a full ground and site survey had taken place and the site was being sold with full vacant possession.

(c) Open 'for sale' listings

A landowner may decide to list their property for sale on the open market at a certain price. The listing may be with one particular real estate agent, although multiple agents are often used. Selling on the open market is also used to dispose of property which did not sell at auction or via the tender process. The advantage to the landowner is that there are usually no upfront costs for marketing the property, unlike the costs involved in a tender or auction process. The downside is that achieving a sale can take many months or even years. Accordingly, open 'for sale' listings are often priced above the market's true value in anticipation of the asking price being driven down during the negotiation process.

(d) A competition process

Competitions are used by the landowner when financial considerations are not the only criteria for disposal of the site. Therefore, competitions are mainly used by local authorities and other public bodies seeking to choose a developer to implement a major scheme. They are also used in a more informal way by other landowners seeking development partners. For example a landowner may want to obtain planning permission before disposing of the land – therefore, the developer may be selected on the basis of planning expertise. Alternatively, the landowner may not wish to dispose of the land and will seek a property developer to project manage the scheme in return for a profit share. This has become increasingly common and is commonly referred to as a private-public partnership (PPP).

As the majority of competitions involve government bodies and other public bodies, the emphasis is placed on these public authority competitions. Government authorities and other public bodies will invite competitive bids on a tender basis, whether formal or informal, and the bids will normally be judged on a financial and/or a design basis.

As a first step, the authority will usually advertise their intention to set up a competition and invite expressions of interest. Alternatively, the government authority may choose a selection of developers to enter the competition. If the former method is adopted, then developers are usually invited initially to express their interest in becoming involved. They will usually be asked to provide details of their relevant experience and track record, financial status (usually a copy of their company report and accounts), the professional team if appointed and any other information which is relevant. For example, a developer may own adjoining land to the competition site or may have been involved with the subject site for some considerable time.

The government authority will then assess expressions of interest and compile a shortlist of suitable developers to enter the competition. This may or may not be the final selection process, and bids may be invited from those shortlisted in order to compile a final shortlist. The number of selection processes will depend largely on the number of interested parties and the complexity of the competition. If the authority is requiring each interested property developer to submit both financial and design bids for a relatively detailed design, then the number of developers shortlisted for the final process should be no more than about three. Many competitions involve property developers in spending large sums of money to submit bids – in these circumstances extended shortlists are not favoured.

It is important that a development brief is prepared to provide guidelines for the competition. The development brief should state the basis of the competition and the criteria adopted for choosing the eventual developer. The development brief will set out the requirements of the government authority with regard to such matters as total floor space, pedestrian and vehicular access, car parking provision, landscaping and any facilities which the authority considers desirable in planning terms. The authority may include a sketch layout or outline sketch drawings illustrating the development required, but in the majority of cases it is the developer's responsibility to suggest design solutions. The brief should state how flexible the authority will be in assessing whether the bid meets its requirements. It is very important that a developer finds out whether they will be penalised for not strictly adhering to the brief. As a general rule, developers who follow the guidelines set out in the brief will be looked upon favourably, unless a developer proposes a better solution to that outlined in the brief. It may be that through their ability and expertise a developer may produce a higher financial bid by proposing an additional area to lease than that envisaged in the brief while still producing a good and sensitive design. Every competition is different and it pays the developer to study the development brief in depth and look at all possible angles that can be used to advantage.

Overall, developers find competitions the least attractive method of acquiring development sites, mainly due to the lack of control by the developer (e.g. negotiating skill is irrelevant) and the resources and financial outlay needed. Competitive bids involving designs necessitate time and expense on the preparation of drawings and financial bids which will be irrelevant for any other sites if the developer does not win approval for that site.

(e) Auctions

In some regions and countries the preferred method of sale is via the auction process. Some development sites are sold at auction in this manner. Alternatively, other regions use auctions as a sale method of last resort and often for unusual sites. For example a government authority may use an auction to dispose of disused railway embankments and land with no, or limited access. This way the market actually determines the final sale price. An auction may be used to sell real estate investment opportunities where the leases are due to expire in the next five years and there is obvious redevelopment potential. A developer conducting a search for potential development site/s should regularly look through auction catalogues for development opportunities.

At auction the highest bid secures the site, providing that the reserve price has been exceeded. With this type of sale the control rests with the landowner who dictates the conditions of sale. The landowner will instruct the auctioneer of the reserve price which is effectively the lowest price acceptable. If the reserve price is not reached then that particular lot is withdrawn and typically placed back on the open market for sale. The auction will set out both the standard and special conditions of sale relating to each particular lot. Once a bid has been accepted, the successful bidder exchanges contracts at that point by handing over the deposit, together with details of their solicitor (where applicable). Therefore, if a developer plans to acquire a site at auction, he must ensure that a thorough evaluation and all other preparatory work is carried out before the bidding starts. Occasionally, the lot may be acquired prior to auction by direct negotiation with the landowner.

While there are many different approaches to selling a site, tenders and auctions are common methods of disposal chosen by landowners when market conditions are good. However, a developer will generally prefer to obtain a site off the market, avoiding competition at all costs. If a developer enters a number of competitions and tender situations, they could all be lost or all or some could be won. However, there is no certainty and the developer may become very frustrated, wasting a lot of time and money in the process. Securing the appropriate site for an acceptable price is based on the developer's ability to judge the right opportunities to pursue and the right level at which to submit a financial bid. However, in many instances it may be simply a case of luck or being in the right place at the right time. The site acquisition process can be very competitive, especially since a developer is naturally looking for sites in areas where there is demand for new development. It must be appreciated that even the best thought-out acquisition strategy may not be achieved in the way or in the timescale first envisaged.

Discussion points

- What options does the developer have to seek and identify potential sites for development?
- What are the main differences between brownfield and greenfield sites?

2.4.4 Local authority initiatives

In many regions the public sector is now less directly involved in the development process due to recent changes in government policy. Often the emphasis of government policy is to enable development and facilitate private sector involvement in development with minimum interference from government bodies. However, government authorities still have an important role to play in initiating the development process, commencing with the responsibilities of the planner through the planning system. They may also facilitate development by promoting or participating in development opportunities themselves.

Local governments are restricted by the scope of their legal powers, the availability of finance and the need for public accountability. Some local authorities are more active than others, depending on the political party in overall control of the authority and whether they wish to actively encourage development within their area. It is important to examine the various methods by which local authorities influence the availability of land for development.

(a) Planning allocation

The allocation of land within a government planning authority's development plan establishes the framework for the permitted use of land and therefore directly establishes its potential value for the purposes of development. In formulating planning policies in the development plan a local authority has to balance the demands of developers against the wider long-term interest of the local community. A developer will examine the development plan relevant to the areas identified in their search for sites. At the same time the local government authority in their role as planner can influence the availability of a particular site by allocating a specific use to it in the development plan.

However, it must be stressed that allocating a site in a development plan does not make it automatically available for development. The developer and landowner must be able to agree terms and the site must be suitable in physical terms for the proposed use. Even if it is available it may not be developed because the location of the allocated land does not meet the requirements of occupiers. If the developer and/or landowner disagree with a particular allocation within a development plan, in preference to their own site, then they can present their case to a planning inspector at the public enquiry into the development plan.

(b) Land assembly and economic development

Local authorities may make land available for development by assembling development sites for disposal, which may involve using their statutory compulsory purchase powers to acquire land. Their ability to take on this enabling role obviously depends on the amount of land under their control and their attitude towards encouraging development. Some authorities in areas of economic decline and high unemployment are active in encouraging private sector investment. For example the planning department may work with developers to bring forward sites for development using their land acquisition and development powers to deal with physical constraints on development. In economically prosperous authorities an activity may be restricted to involvement in prestigious sites such as enclosed retail centre shopping schemes. However, positive participation by local authorities in making land available is not just limited to land assembly, whether by agreement or compulsion, but may include site reclamation, the provision of buildings, the provision of infrastructure/services, the relocation of tenants and general promotion of their area as a business location. Any or all of these activities tend to be described by the general term of 'economic development'.

The need for a local government authority to become involved in 'economic development' depends on the initiatives taken by the private sector and whether market forces alone meet the expectations of the local authority for the development of their area through the creation of employment opportunities. Another consideration is the extent to which local government authorities can undertake 'economic development' which is commonly restricted by government policy. One approach for a local government authority is to use a proportion of their capital receipts (i.e. proceeds from the sale of land or buildings) for new capital investment. The remaining balance of funds is used to redeem debts or as a substitute for future borrowing or set aside to meet future capital commitments. A local authority's ability to raise money through capital receipts is taken account of when the government sets their credit approval limit, i.e. the extent to which they can borrow money. The reference to capital receipt usually extends to the receipt of rent (e.g. occupational rents and ground rents) and the receipt of reduced rent in lieu of some benefit (which must be fully valued in monetary terms). In addition, temporary financing by local government authorities such as the acquisition of land pending disposal to a developer will count against their credit approval limit if the period between acquisition and disposal runs over a year. This extended timeframe is not unusual in property development processes.

Some local government authorities, particularly those in the inner cities or in areas of high unemployment, may receive additional funding from the government. However, over time and in particular since the global economic crisis, the access to government assistance is becoming increasingly competitive and local authorities are being forced to explore more innovative ways to achieve economic development aims.

When a local authority does become directly involved in land assembly it can also benefit a private sector developer, but it will almost certainly extend the timeframe of the whole development process. A developer may require the assistance of the local government authority when a particular site identified by the developer is owned by a number of different landowners. This situation is quite common with medium or high density urban and town centre locations. The local government authority may have allocated the site for comprehensive redevelopment in the relevant development plan and therefore is willing to work with the developer to achieve their planning aims. In such a situation the developer may experience difficulties in negotiating reasonable land values with the various landowners as they effectively hold the developer 'to ransom' by demanding unrealistic prices, due to the fact that a particular landholding is vital to the proposed development due to the 'special value' with adjoining lots.

The landowner may be unwilling to sell their landholding as their motivation for occupation of the land is their business. On the other hand a particular development site might be land-locked with access under the control of a landowner who is seeking a price well above the market value because of their advantageous position. A government authority who assists with land assembly in this way will try to reach a compromise agreement by negotiation. If this is unsuccessful then a government authority may have the option of making a compulsory acquisition or purchase order, subject to prevailing legislation. There are strict rules and regulations surrounding compulsory acquisition and compensation, as well as the obvious implications from a political perspective for elected government officials. Overall, the whole process of compulsory purchase is often a very lengthy and drawn-out process, involving the local government authority agreeing to compensation values with all of the individual interests affected by the compulsory acquisition. The number of interests involved may be in the hundreds in the case of an inner-city redevelopment and even in the thousands in the case of a new road or railway line through an urban area. Notice must be served on all interests involved and details of the scheme publicised. In most cases the compensation payable by the acquiring authority is the value of the land which would be realised if it was sold on the open market at its highest and best use for full market value. In addition there may be other heads of claim forwarded by the affected party including disturbance, relocation and removal expenses, cost of adapting new premises, loss of profit and so forth. In many cases the disturbance compensation may be more than the value of the land. In the event of disagreement between the parties the matter is referred to the court system for adjudication.

In return for assisting a developer with site assembly, it has become commonplace for the local government authority to require the provision of social facilities, a financial contribution towards other government assets (e.g. park contribution) or even to participate in the financial rewards of the eventual development scheme. In the past some local government authorities in their role as planners have used the threat of their compulsory purchase powers in a negative manner to achieve some material benefit in the form of 'planning gain' or amendments to planning applications.

When a local government authority disposes of land to a developer they usually produce a development brief which outlines their forward vision of how they would like to see the site developed. It is important that the brief is flexible and not too detailed in order to allow the developer freedom to react to prevailing market conditions. Where compulsory purchase powers are used the land assembly process may have taken several years, by which time market conditions could have completely changed. It is important from a developer's point of view that sites are disposed of on as clean a basis as possible. In other words any problems or constraints which exist with regard to the legal title, services, planning and access should be tackled from the start.

(c) Infrastructure

The provision of supporting infrastructure is critical to the site acquisition process and local authorities play an important role in ensuring its provision. Infrastructure is a term used to describe all the services which are necessary to support a new development, i.e. good road ingress and egress to the site and the general locality, water and sewerage provision, open space and parkland, schools and retail shops.

2.4.5 Site access and additional infrastructure

Access to a site via existing or the proposed provision of roads is important in assessing viable locations for property development. While proposals for a new road will generate pressure for development along its route, a new development will also create new traffic pressures on the existing road network. The existence of infrastructure is absolutely critical to the viability of a particular development scheme – so much so that it directly influences land values and the highest and best use of the site. If the necessary infrastructure does not exist to support a development then a developer will take account of the cost of its provision in the evaluation of the land value.

Local government authorities are largely responsible for deciding the level of infrastructure required and securing its provision. In performing this role they have to determine who is ultimately responsible for the cost of its provision. Due to government control on spending, local government authorities often negotiate agreements with developers to secure the provision of new infrastructure if it is required to support the development, e.g. the provision of a roundabout to link the development scheme with the existing road network or the provision of public open spaces and parks for the public to access. The assessment of future infrastructure requirements at a strategic level is the responsibility of different government departments or associated bodies. Their assessment of future requirements will vary depending on the infrastructure they are responsible for, for example the road network (either/or local and major roads), electricity, sewerage, water supply, parkland or waterways.

Many government authorities adopt a more pro-active approach to the provision of infrastructure because they recognise that new roads open up land for development. Land is often assembled at the same time as the new road so that the government authority can benefit from enhanced land values by packaging sites for disposal to the private sector. There is often a debate about the pressure for development caused by new roads, particularly in prosperous or environmentally sensitive locations and in light of sustainable development policies adopted by many countries. In some cases the developer may be required to make a financial contribution to pay for improvements to existing roads in order to accommodate the additional traffic caused by a new development. This potential increase in road traffic also presents an opportunity for the government to reduce the need to travel by car by influencing the location of development schemes relative to public transport networks or existing roads. From a sustainable perspective they may encourage development which is easily accessible using low carbon (CO_2) forms of transport such as cycling, walking and public services.

Many local government authorities are actively promoting the use of public transport as a viable solution to traffic congestion and a shortage of car parking spaces. This usually includes a standard bus service; however, in some locations this has been expanded to include a light railway or tram system in order to ease traffic congestion (see for example the tram services in Manchester and Nottingham in the UK or in most cities in the Netherlands). In order to secure the necessary public funding for such transportation systems, government regulations usually stipulate that private sector contributions have to be secured in advance. Often developers and landowners with sites that will directly benefit from a new nearby proposed transport system will be approached to provide a proportion of the required capital. However, there is a realistic limit to how much developers can pay as any additional financial contribution will be automatically deducted from funds the developer had allocated to purchase the property. The same scenario applies to financial contributions made by developers for road and other infrastructure improvements.

2.4.6 Public-private partnerships with developers

Many government authorities prefer to retain a legal interest in the development scheme by only granting a long leasehold interest (e.g. ninety-nine years) to the developer instead of selling the freehold of the site. In return they receive regular payment for a ground rent at a percentage rate usually linked to the long-term success of the scheme and at the end of the lease the land and any buildings or improvements revert back to the authority's control. For example in the UK the income from this type of lease is often called a 'peppercorn rent'. This type of arrangement is usually referred to as a 'public-private partnership' (PPP) and overcomes the problem of the government making a financial outlay to develop the site. Alternatively the government may only retain an interest until the property development has been completed, which is undertaken by granting a building licence for a nominal premium to a developer to enter onto the land and complete the development. Under this arrangement the government authority sells the freehold to the developer on completion and, therefore, also receives a financial benefit from any increase in property value. Although the arrangements above are presented to the public as a partnership arrangement, this is not a true partnership as the private sector bears the majority of the risk while sharing in only some of the rewards.

Property developments involving a local government authority can often result in a lengthy and costly competition to choose a development partner. The size of such schemes involve a significant risk for the private sector, with substantial sums of money being expended before funding for the development is secured. The legal agreement between the government authority and a developer often takes a long time to negotiate and is usually subject to the securing of finance by the developer. Many developers are forced to withdraw from such schemes due to lack of funding or because the initial evaluation of the scheme has changed significantly due to the amount of time that has lapsed. A further complication is that there is often conflicting interests between a local government authority's social objective and a private developer's profit objective. Also the local authority has a number of overlapping roles in the development process which may directly conflict with each other, such as being both the planner and the landowner. Nevertheless a local government authority is run by democratic (and arguably transparent) processes which are often lengthy and inflexible compared with the relatively quick decision-making approach of the private sector. Sometimes it may take a while for a property developer to gain confidence among the elected members of the local authority, only for changes to occur in the personnel and the political party in overall control following government elections.

Some government authorities, as an alternative to the above arrangements, have entered into joint ventures with developers via companies limited by share or guarantee. Also some have formed wholly owned subsidiaries, for example, enterprise boards to carry out economic development initiatives. This can also apply to those companies which are essentially extensions of a government authority. As a general rule the government wants to ensure that local government authorities remain accountable to the public and will have policy or legislation which restrict government bodies from using joint companies as a means to avoid capital expenditure restrictions. Many developers prefer joint venture arrangements as they are familiar with them and they allow quicker decision-making. In addition they allow more flexibility in securing funding for the scheme as the property developer passes some of the risk to the local authority, who then shares in any decrease in the value of the completed scheme.

Discussion points

- In what ways do government authorities assist the development process in respect of land for development?
- Why have PPPs become a viable option for property development stakeholders?

2.5 Site investigation

Before acquiring a site the developer should undertake a number of very important investigations. These investigations will influence the terms of the contract to acquire the site and the price the developer is willing to pay. Although landowners, particularly local authorities, will provide as much information as possible, it is up to a property developer to satisfy themselves that there will be no unexpected surprises once legally binding contracts for the acquisition of the site have been drawn up and exchanged. The investigations listed below are of vital importance when a developer is acquiring a particular site. The investigations may reveal that the proposed scheme is no longer viable due to the physical state of the ground and the cost of remedying the problems.

2.5.1 Site survey

A site survey needs to be undertaken by qualified land surveyors to establish and/or confirm the extent of the site and whether the boundaries agree with those shown in the legal title deed. The location of buildings, fencing, hedgerows and even a 'boundary' wall on a site does not necessarily confirm the location of the actual site boundary as stated in a legal title deed. For example there is not always an obvious boundary when the site adjoins community or state owned parkland. To avoid future disputes over ownership, physical inspection of the site and comparison with the legal title deed is essential. The need for a site survey is also of vital importance where a development site is being assembled by bringing together various parcels of land in various ownerships. In this instance the survey needs to establish that all the boundaries of the various parcels are correct in both length and alignment and that the entire site (consisting of multiple individual properties) is actually being acquired. It would be a disaster if the developer discovered, half-way through undertaking a property development, that a small but vital part of the site had not been acquired. In this scenario the developer would then have to negotiate with that landowner from a very weak position, effectively being held to 'ransom'. The site survey also establishes the contours and levels of the site. If any existing buildings on the site are to be retained, a structural survey will need to be carried out.

The legal search of the title deeds need to establish who is responsible for the maintenance of the boundaries. It will also identify any encumbrances on the site, such as a sewerage or electricity easement under the ground which may prevent the construction of any improvements on the land above (see also section 2.5.5). Another example would occur when there is an access or 'right of way' easement over the site to an adjoining lot. In addition, the access arrangements to the site need to be checked to ensure that the site boundary abuts the road. If a public highway exists, the solicitor needs to check whether it has been adopted by the local authority and is maintainable at their expense. If access to the site is via a private road then the ownership and rights over that road need to be established.

2.5.2 Ground investigation

Unless reliable information already exists as to the state of the ground then a ground investigation needs to be carried out by appropriate specialists. The purpose of the ground investigation is to assess the suitability of the land to support the building structure. Ground investigations can vary both in cost and extent depending on the size of the proposed scheme and information already known. An investigation will normally include a whole series of bore holes which are taken at strategic locations on the site. Samples taken from the boreholes need to be analysed in a laboratory to establish the nature of the soil, substrata and water table, together with the existence of any contamination (as discussed in the next section).

The results of the investigation will be given to the structural engineer, architect and quantity surveyor. They will need to analyse the results to establish whether any remedial work is necessary to improve the ground conditions or whether any pile foundations are required. An example would be the addition onto the site of external soil as fill material which is then compacted. Both circumstances will have an impact on the cost of the development scheme, which may affect the overall viability of the development.

2.5.3 Contamination

The existence of any contamination on a site has become an issue that developers cannot ignore. The process of property development commonly requires a change of land use – for example from a previous industrial use to a residential use. Industrial processes typically cause land contamination from the chemicals and products used in manufacturing (e.g. oil, acid, metals, etc.). Long-term exposure to these contaminants is harmful to humans. Contaminated land is generally referred to as land which represents a natural or potential hazard to health or to the environment as a result of current or previous uses. When contaminated land is suspected it is important to employ the services of professionals and assess the cost (if any) and timescale for remedial action. It has become common for the financiers and end use purchasers of development schemes to demand evidence from developers that sites are not contaminated and, where they are, that satisfactory remedial action has been taken. In many instances a developer will not obtain finance for a scheme if there is the slightest risk of contamination.

Contamination is typically caused by a previous occupier's use of the land; however, with many sites the use of contaminants is unrecorded, which makes it incumbent on the developer to undertake a thorough investigation especially of brownfield sites. Occasionally contaminants will migrate onto land from other sites and thus pollute the land, although such an occurrence can cause complex issues to arise in terms of getting the polluter to pay costs of remediation. It is worth noting the types of uses that have led to contamination in the forms listed below.

- Manufacturing and industry sites
  - asbestos works and buildings containing asbestos
  - dry cleaners
  - food processing plants
  - metal mines
  - brickworks and potteries
  - chemical works
  - steel works
  - plating works
  - munitions factories and test sites
  - paint works
  - paper and pulping works
  - printing works
  - tanneries
  - textile mills.
- Infrastructure sites
  - cemeteries
  - docks canals and ship yards
  - quarries
  - railway land
  - sewerage works.
- Industrial storage sites
  - landfill sites
  - military airfields
  - oil and petroleum refineries and storage sites
  - scrap yards.
- Power generation sites
  - gas works
  - nuclear power stations.

The cost of ground investigation is usually much higher than normal when any level of contamination exists, representing a potentially substantial upfront cost for the developer. As much information as possible should be obtained on the site's history of uses before any ground investigation is started. This is achieved by looking at ordnance survey maps, local authority records, title deeds and any other likely source of information. In regions where contamination is widespread, the local government authority may have already compiled records of contaminated land. However, information obtained from records may be limited and will always need to be checked.

The ground investigation will usually involve taking soil samples down to the water table level. This may be accompanied by extensive surveys of all underground and surrounding surface water due to the risk of contaminants seeping through water (for example see the UK legal case *Cambridge Water v Eastern Counties Leather plc* [1994] *1 All ER 53* involving the contamination of the local water supply from the operation of a tannery). The results of the ground investigation will enable an assessment to be made of the extent and cost of remedial measures.

There are many different ways of remedying and treating contamination. The main options available are to (a) remove the contaminated soil and replace it, (b) treat the contaminated soil in-situ on-site or (c) contain it under a blanket of clean earth which is often referred to as capping. On-going measures may be required once the development is complete such as venting methane gases to the surface, especially in the case of landfill sites. If contamination is limited to one area of the site it may be possible to design around the problem – for example by locating a car park there. If ground has to be filled with imported material as part of the process then, depending on the standard of treatment required, deep piled foundations may be required.

In general terms when a developer is faced with a contaminated site the remedial measures are often expensive, which in turn rules out all land uses but the higher value uses such as commercial office or retail warehousing. In view of the increasing concern about contaminated land and the debate about who should be financially responsible, it is worthwhile for developers to undertake a thorough due diligence and spend time in the preliminary investigations to thoroughly investigate its existence. The appropriate professionals should undertake a full environmental audit, which can then be presented to potential purchasers, financiers and final owners.

2.5.4 Services

The site survey should establish the existence of services which are available to the site including the provision of water, gas, electricity and drainage. All of the utility companies should be contacted to establish that the services identified in the survey actually correspond with those on record. In addition, the capacity and capability of the existing services to meet the needs of the proposed development should be ascertained. If the existing services are insufficient then the developer will need to negotiate with the company concerned to establish the cost of upgrading or providing new services, for example a new electricity substation to provide the additional electricity needed for the development scheme. Where work needs to be carried out by either an electricity or gas company, the developer will be charged the full cost of the additional service requirements. At times a partial rebate may be available once the development is occupied and the company is receiving a minimum level of income. The route of a particular service may need to be diverted to allow the proposed development to take place, e.g. one lane on a road or a complete road may be closed. The cost of diversion and the time it is likely to take to complete should be established at the earliest possible stage. The legal search of the title deeds will reveal if any adjoining neighbours or occupiers have rights to connect to or enjoy services crossing the site. The developer may need to renegotiate the benefits of these rights if they affect the development scheme. It hardly needs to be stated that access to services is essential to any new development and the developer must be 100% certain that the site will have full access to those services required by the occupier before proceeding further.

2.5.5 Legal title

In most cases a solicitor or legal expert will be appointed by the developer to deduce the legal title to be acquired and to carry out all the necessary enquiries and searches before contracts are entered into with the landowner. The developer's solicitor will apply to the body governing, or in control of, land registration to examine the official register of the title. If the land to be acquired is leasehold it is essential to closely examine brief particulars of the lease and the date it was entered into. The developer will need to establish the length of the lease, the pattern of the rent reviews and the main provisions of the lease. Such provisions need to be checked to ensure the terms are acceptable to the provider of development finance. The solicitor needs to establish that the land will be acquired with vacant possession and that there are no unknown tenancies, licences or unauthorised occupants. The fact that a site or building is unoccupied does not necessarily mean that no legal rights of occupancy exist. The search of the lease and title deeds will also reveal the existence of any conditions or restrictions affecting the rights of the landowner to sell the land. In addition, all rights and interests adversely affecting the title will be established, such as restrictive covenants, easements, mortgages and registered leases.

The existence of an easement might fundamentally affect a development scheme. An easement may be either (a) positive, e.g. a public or private right of way to an adjoining property, or (b) negative, e.g. a right of light access or views for the benefit of an adjoining property. If the easement exists to the detriment of the proposed scheme then the developer may be able to negotiate their removal or modification to allow the scheme to proceed. Rights of natural light might affect the proposed position of the scheme and the amount of floor space available in the property development. If a party (or inter-tenancy) wall exists then it will be necessary to agree a schedule of condition with the adjoining property or make a payment of compensation. A property professional with specialist knowledge on party wall matters may need to be appointed by the developer.

The existence of restrictive covenants may adversely affect the development scheme, for example a covenant restricting or prohibiting a particular use of a site. It is sometimes difficult to establish which landowner has the benefit of a covenant as the covenant might have been entered into some considerable time ago. If the beneficiary can be found then the developer may be able to negotiate the removal of the restriction. If not, then the developer may be able apply to the relevant government authority for its discharge, which is a lengthy process. Another option is for the developer to take out an insurance policy to protect against the beneficiary enforcing it. The insurance cover may be able to compensate against the loss in value caused by any successful enforcement action.

A solicitor will also carry out or request a search of local government authority records (also referred to as the 'Local Land Charges register' in the UK). This will reveal the existence of any planning permissions or whether any building or site is listed as a building of special architectural or historic interest or other charges on the property or land being sold. Enquiries will also be made of the local government authority to establish whether the road providing access to the site is adopted and maintained at public expense. The existence of any proposed road improvement schemes might affect the site, for example a strip may be protected at the front of the site for road widening purposes.

Enquiries should also be made of the existing site landowner (seller/vendor) and will include standard questions on matters such as boundaries and services. Enquiries will also reveal the existence of any overriding interests (e.g. rights and interests which do not appear on the register of the title) or adverse rights (e.g. rights of occupiers of the land). Solicitors may also make additional enquiries of the vendor which are particular to the land being acquired.

The developer must aim to acquire the freehold or leasehold title of the development site free from as many encumbrances as possible by renegotiating or removing these restrictions and easements. However, this not always possible and the developer may need to make some compromises, especially with reference to the retention of heritage buildings and facades on the site. Lenders, particularly the financial institutions, prefer to acquire their legal interest with the minimum of restrictions which might adversely affect the value of their investment in the future. The developer has to be able to 'sell' the title to the final occupants, being either purchasers or tenants, as quickly as possible without complications.

2.5.6 Finance

The provision of finance is fundamental to the success of the development. No prudent developer (unless there are sufficient internal cash resources and equity) would consider entering into a commitment to acquire a site without first having secured the necessary finance or development partner to at least cover the cost of acquisition, including interest on the acquisition cost, while the site is held pending development. The developer should aim to ensure that the financial arrangements are confirmed to coincide with the acquisition of the site. If no financial arrangements are in place then the developer must be satisfied that either the finance will be secured or that the site can be sold on the open market if no funding is forthcoming. The developer must ensure that all investigations have been carried out thoroughly so that any financier or partner has a full and complete picture of the site. Every area of doubt must be removed if at all possible. Downturns in the marketplace such as the global financial crisis create a situation where lenders are more risk averse and typically tighten their lending criteria to reduce their risk. Often this means they will require a pre-let with an existing tenant for all or a substantial proportion of the proposed scheme before they are prepared to finance the venture. The developer needs to understand what the lender is looking for regarding lending criteria and tailor their application accordingly.

Discussion points

- When evaluating a particular site, which factors do developers have to consider?
- What are the elements when undertaking a site survey?

2.6 Site acquisition

The findings from all of the investigations undertaken, as discussed above, need to be reflected in the site acquisition arrangements. The degree to which developers reduce the risk inherent in the property development process depends to some extent on the type of transaction agreed at the site acquisition stage. The prudent developer will always endeavour to reduce the element of risk to a minimum and the site acquisition arrangements are important in this respect. Ideally no acquisition will be made until all the relevant detailed information has been obtained and all problems resolved. Hence a full due diligence has been completed. However, in practice, it is virtually impossible to remove every aspect of uncertainty due to the inherent characteristics of the real estate market. The degree to which the developer can reduce risk to the site acquisition stage is dependent on the landowner's method of disposal, the amount of competition and the tenure. It is possible to pass some of the risk to the landowner but this will largely depend on the developer's negotiating abilities.

The majority of site acquisitions are on a straightforward freehold basis where the developer then owns the site outright. The freehold title transfers from the vendor/landowner to the developer once contracts have been completed and from this moment forward the property developer is responsible for all of the risk. The developer can reduce the risk inherent in the transaction through negotiation of the contract terms. The contract can be conditioned and payments can be phased or delayed. For example the property development may be based on obtaining permission for a change of land use and therefore the developer should negotiate that the contract is subject to a 'satisfactory planning consent' being obtained. The vendor, if such a condition is acceptable, will try to ensure that the term 'satisfactory planning consent' is clearly defined.

The developer may obtain a planning consent which does not reflect the optimum value of the site but satisfies the condition in the contract and then, at a later stage, obtain a better planning consent. It is not uncommon for 'top-up' arrangements to be made whereby the vendor benefits from any improvement created by planning consents obtained by the developer. Developers will carefully weigh the degree of uncertainty in relation to planning and it will be a matter of judgement as to whether the additional exposure risk (i.e. uncertainty) is acceptable. If the vendor is undertaking to sell the site with vacant possession then the contract should be conditional upon this for there could be a delay in the occupants of a building leaving.

While the normal period between signing a contract to purchase a site and then access can be relatively short (e.g. 28 days), the developer may negotiate an extended time delay for the completion, e.g. six months. Any delay in the development process will cost money, therefore the developer should ensure that any potential problems revealed by investigations are dealt with before contracts are completed, or the time needed to resolve them is reflected in the evaluation and hence the price paid for the land. As an alternative, especially if the planning process is perceived to be long and difficult, the developer will often consider it advantageous to pay for an option to reserve the land. An option involves the developer paying a nominal sum to secure the right at a future date to purchase the freehold. There is usually an agreed date (also known as a 'long stop' date) after which the vendor is free to sell the land to anyone if the developer has not taken up the option. The option agreement might specify that after certain conditions have been complied with the developer has to purchase the land. If the developer fails to complete the purchase by the 'long stop' date then the vendor is free to market the site elsewhere. Alternatively, the agreement may allow the developer to call upon the vendor at any time to sell the site after sufficient notice. The developer will aim to fix the value of the site at the time the option agreement is entered into, but in practice this is often difficult to actually achieve. At least in a rising market where values are increasing, the vendor/landowner will usually try to ensure that the open market value is fixed at the time the developer actually purchases the land.

The developer may only be able to acquire a long leasehold interest in the land at a premium with a nominal ground rent – this is sometimes referred to as a 'peppercorn rent'. This happens when the landowner (a) is only able to dispose of a leasehold interest or (b) wishes to retain some control over the development, e.g. the landowner is a local government authority. The developer may be able to take out a lease on a building in the first instance, which will immediately form a legal estate although probably subject to covenants relating to the satisfactory completion of the property development. Alternatively, the transaction might be arranged on the basis of a building agreement and lease which, in the first instance, merely gives to the developer a licence to enter onto the site and construct the building with a commitment by the landowner to grant a lease when the building has been satisfactorily completed.

A similar arrangement is sometimes made by a local government authority in relation to freehold transactions when the developer is able to carry out the development under a building agreement and the freehold is transferred on the satisfactory completion of the property development. Under this type of transaction the local government authority may become an equity partner in the scheme. The value of the scheme is assessed on completion and, therefore, the authority can share in any growth. The developer can use this type of transaction to reduce risk at the outset. The property developer may only be required to pay a nominal premium to enter into a building agreement and the consideration to the local government authority may be only payable if a profit is made on completion. The consideration paid to the local government authority may be in the form of a profit share or it may be the land value on completion. This method of acquisition is advantageous to the property developer where the development scheme is large and likely to take a number of years to complete. The risk to the property developer is substantial; however, the government authority may be willing to be flexible due to their interest in the implementation of the scheme. A prudent developer will not enter such a building agreement on a large property development scheme without making it conditional upon funding.

It is important that the building agreement is carefully negotiated as otherwise long drawn out arguments can take place on completion in relation to the calculation of profit. Larger property development schemes will take a number of years to complete and as a result it is quite likely that the personnel and the political party in overall control at the local government authority will have changed by the time the project is finally completed. This may lead to disputes over matters such as the definition of development costs which adversely affect the authority's profit share. One alternative approach to undertaking endless negotiations about how to calculate the profit is to form a joint company with the authority. With a joint company the profit is clearly shown in the audited accounts of the company and there is no dispute as to the level of acceptable development costs.

For some developments the site is acquired on the basis of a long leasehold interest with an open market ground rent payable, instead of a premium being payable with a nominal ground rent. Some leaseholds can be for an extremely long timeframe such as 99 or 125 years. The actual level of ground rent payable can be reviewed in a number of different ways and might be geared to (a) a percentage of the current market rent of the property, (b) a percentage of the rents received less outgoings, or (c) the rents receivable. A developer should have high regard to the preferences of financial institutions and lenders in the local region in which they operate. As a rule many financial institutions prefer freehold arrangements to long leaseholds and this will be reflected in the yield at which the institution values the completed investment. The reviews might be to cleared-site value with planning permission on the assumption that a similar term of lease will be granted at the time of review. Some financial institutions prefer the revised rent of the building to be fixed at the time the ground lease is granted. This means that the ground rent might cease to rise or may even fall towards the end of the term. Ground leases may have a user covenant which will limit the use of the site to a particular planning use class.

2.7 Government assistance

There can be barriers to development when (a) land is not made available or (b) development is not initiated by private market forces due to the fact that development is not viable. Development is often not viable due to low market rents and lack of occupier demand in a particular area and/or prohibitively high development costs as a result of the physical condition of a particular site. Typically many of these areas are within the inner cities or regions in economic decline burdened with high unemployment. Very often the infrastructure in such areas is extremely congested or non-existent and there may also be widespread contamination from previous use of the land by heavy industries.

To tackle this problem more innovative forms of funding urban regeneration have been encouraged, typically involving more risk sharing by the public sector rather than just handing out grants to developers.

2.7.1 Government agencies

Various government agencies exist to implement and administer urban regeneration policies on behalf of the government. Just to make matters a little confusing, the UK has different agencies operating within England, Scotland, Wales and Northern Ireland. Their roles differ but they all take an active, initiating role in the development process by making land available for development and providing financial assistance and they may even participate directly in development. In the UK, for example,

Urban Development Corporations were established under the UK Local Government Planning and Land Act 1980. UDCs had a broad remit to:

- secure the regeneration of a defined area achieved by bringing land and buildings into effective use;
- encourage the development of existing and new industry and commerce;
- create an attractive environment; and
- ensure housing and social facilities are available to encourage people to live and work in the area.

At one stage there were 14 UDCs where the last one to operate was 'West Northamptonshire'.

2.7.2 Funding and grants

Direct government financial assistance via funding or grants is usually available to local authorities and/or developers proposing urban regeneration schemes in specific areas of the country. Financial grants are available to developers and the public sector in the UK to develop derelict or rundown inner-city sites and buildings from the government and Europe directed towards Urban Priority Areas (UPAs), Assisted Areas and Urban Development Corporations (UDCs).

Discussion points

- List the various considerations when undertaking a site acquisition.
- What are some of the ways in which governments assist the development process?

2.8 Reflective summary

This chapter discussed the framework surrounding the selection of land for development. Even if a developer has undertaken a substantial amount of research prior to implementing a land acquisition strategy, successfully achieving this objective within the stated timeframe and within budget is often beyond the control of the developer. The following preconditions need to be in place for the development process to be initiated through land acquisition:

1 The landowner's willingness to sell the land at a price with associated conditions to enable a viable development to proceed.
2 Planning permission for the proposed development or allocation of the proposed use within the relevant development plan and zoning constraints.
3 The existence of adequate infrastructure and services to support the proposed development.
4 The existence (if necessary after appropriate decontamination at a reasonable cost) of suitable ground conditions to support the development.
5 The necessary development finance at an acceptable cost.

6 An identified end-user or confirmed hypothetical occupier demand for the proposed development.

If any one of these conditions cannot be confirmed then development should not proceed or subsequent development will represent a considerable risk to the developer. Local authority involvement and government assistance may be available in relation to compulsory acquisition, provision of infrastructure, site reclamation, finance or occupier/ investor incentives depending on the nature of the proposed development and its location. Most importantly, the requirements of occupiers is always the central component of a successful developer's land acquisition strategy.

2.9 Case study – redevelopment of a pottery kiln into residential accommodation

This case study examines the successful conversion of a heritage-listed brick manufacturing plant including two kilns into residential accommodation.

Historical site background

(This section is based on information sourced from the relevant government body: Heritage Victoria 2013.)

As early as the 1860s it was recognised that the Brunswick district of Melbourne, Australia contained some of the best clay and stone resources. In addition to brick-making the valuable clay deposits of Brunswick enabled the production of all kinds of pottery to be manufactured in the district. In 1870 the Hoffman Patent Brick and Tile Company was established in Albert Street, Brunswick and this company introduced large-scale brick-making to Victoria. Central to this process was the Hoffman kiln for which the company had patent rights. This kiln, developed by Friedrich Hoffman in Stettin, Prussia in 1859, revolutionised the brick-making process by allowing a continual process of loading 'green' bricks as well as being more economical with fuel. The speeding up of the brick-making process which followed encouraged the mechanisation of the making of 'green' bricks and as a consequence the development of the Bradley–Craven brick press and other brick-making technology and improvements in work processes. The Bradley–Craven principle was employed by the Hoffman Company in 1887 when they accepted a tender by Langland's foundry to manufacture one in Victoria and a year later purchased another at the centennial exhibition. This 'copying' of the Bradley–Craven design by local heavy engineering works would appear to account for the majority of machines which survive.

In 1884 the Hoffman Company purchased 36 acres of the 'Dawson' estate and opened a new yard which boosted employment to over 400 men and production to over 40 million bricks a year. The depression of the 1890s saw the collapse of the building industry, although the company had begun to diversify out of an exclusive dependency on brick-making by the late 1880s with the manufacture of drainage pipes and other domestic items such as urinals and pottery ware. From 1900 the building industry returned to normal and this saw the continued expansion and development of the Dawson Street site. In 1907–1908 the works were 'modernised' and a further Hoffman kiln erected. Gradual expansion by the company appears to have continued after the 1914–1918 war until the depression of the 1930s which again halted works. After this the company concentrated on the Dawson Street site following the realisation that the clay hole at the No 1 Works at Albert Street had reached its limits.

In the post-war period the development of new, and cheaper, kiln technology saw the emphasis shift away from the Hoffman mode of operation; although the Dawson Street complex still continues making bricks and the general superiority of its process, as far as quality is concerned, remains acknowledged. In 1960 Clifton Holdings took over Hoffman's. The drain pipe division was closed in 1962 and the other pottery works in 1969. A large amount of the company's land holdings which had little existing infrastructure and buildings were subdivided and sold. The Dawson street site remains operative as a brickworks utilising the Bradley–Craven brick press principle and Hoffman kiln technology. It is the last collection of Hoffman kilns and associated technology operative in the metropolitan area and the most important in the state, possibly the country. The former Hoffman Brick and Pottery Works are of architectural and historic importance for the following reasons:

- The complex is the sole survivor of the clay manufacturing industry which was central to the history of Brunswick.
- The Dawson Street works of the company were established in 1884 at the beginning of the building boom of that decade. The three Hoffman kilns, brick presses and buildings which date either in whole or in part from this time, or are their successors, are a vital link with the boom decade of the 1880s.

Challenges from an overall development perspective

The redevelopment of every site presents unique challenges, especially when there is a substantial change of land use such as from industrial use to medium density residential land use. This particular development presented many challenges such as working within the strict guidelines of heritage restrictions including the conversion of the purpose-built kilns into residential accommodation. Other challenges relate to addressing obvious contamination issues on the site and ensuring the residential accommodation is both appealing and competitively priced in the marketplace. The building structures such as the kilns were also interesting due to the requirement to provide services (e.g. sewerage, water, electricity) to each unit and also meeting fire safety regulations, itself being an oxymoron in the broader context when considering the original purpose of the kilns. Another challenge related to the provision of useable accommodation space in each kiln, being originally designed with rounded sides and little natural light in order to retain heat.

The end product was a successful development which was very well received in the marketplace. The construction work on the development was undertaken in multiple stages over many years to ensure a high profile in the marketplace was maintained. The kilns are located in the centre of the developed land; however, the kilns were only developed in one of the final stages of the development and released onto the market in 2013. The initial stage of the development, being promoted and released onto the

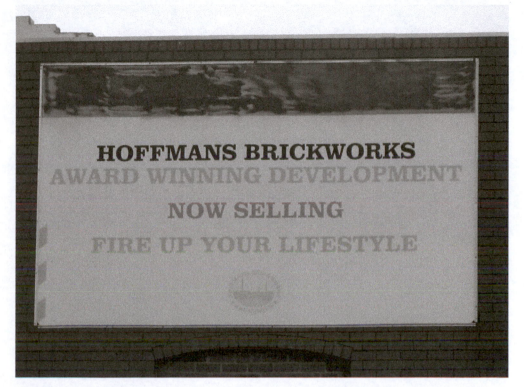

Plate 2.1 Initial promotion of the Hoffman Brickworks development based on the historical signifi-cance of the site (Source: Richard Reed)

market nearly ten years prior (Plate 2.1), was a large number of newly constructed units at the rear of the property, although designed and constructed in character with the development. The construction of this relatively inexpensive accommodation, two or three levels high with secure underground car parking, allowed the developer to benefit from cleared level land, while also promoting the completed units as part of the historical 'Brickworks' development. Only a relatively small proportion of this development included the conversion of a previous industrial building into residential accommodation (Plate 2.2). The marketing of the development was greatly assisted by the high profile chimney stacks (Plate 2.3) associated with the kilns where remain visible from a substantial distance from the site.

Challenges for redevelopment of the kilns

While there are particular challenges for any redevelopment, there were specific challenges for the change of use of the kilns. Plate 2.3 shows one of the two redeveloped kilns on the site shortly after release to the public. Plate 2.4 shows the opposing kiln at the other side of the car park.

The original design and construction of the kiln were based on 'firing' clay bricks on the ground level via the numbered doors which have been bricked in (Plate 2.4). After the redevelopment the internal space on this level remains relatively empty due

Plate 2.2 An early stage of redevelopment (Source: Richard Reed)

Plate 2.3 Converted kiln with the high profile chimney stack being part of the marketing process (Source: Richard Reed)

Plate 2.4 Kiln redevelopment into residential accommodation (Source: Richard Reed)

to the lack of natural light and access only via one external door. The only use for this space is for individual secure storage 'cages' (e.g. for bicycles, boxes) for each occupier in the building which was relatively inefficient but seemingly unavoidable.

The next level (level 1) of each kiln was converted into residential accommodation. There is only one entry/egress point for each building in the form of external steel stairs. In a similar manner to all buildings in the development the entry is via secure password only. On level 1 the units are a combination of one or two bedrooms with each unit having a balcony (see undercover rectangular balconies in Plate 2.4). The next level of each kiln (level 2) is within the cavity of the roof and is restricted to bedroom accommodation predominantly via internal stairs in each unit. In addition each unit has either one or two bathrooms.

The redevelopment of the kilns as per Plates 2.3 and 2.4 highlights the innovative approach of the developer and associated stakeholders including the architect and the heritage body. For example the roof drainage was affixed externally to ensure the original brickwork remained unaltered. The silver corrugated iron sheeting in Plate 2.4 was added to facilitate balcony accommodation. Internally the units are professionally designed and substantial effort was required to ensure the living accommodation was appealing. However, there are obvious constraints which include the sloping ceiling, the relatively narrow width of each kiln and the round chimney in the centre of each building. Another limitation was the absence of undercover parking which is located between the buildings as observed in Plates 2.3 and 2.4.

This development is viewed as successful in light of the substantial heritage-related barriers which were also promoted benefits for marketing purposes. Such restrictions being linked to the historical nature of the buildings ensured the completed units do not appeal broadly to all types of purchasers, especially when considering the pricing of the each unit which was marginally above the median selling value of other units. On the other hand, few occupiers reside in a converted kiln and this aspect has been successfully promoted by the developer as a unique selling point for the development.

Chapter 3
Development appraisal and risk

3.1 Introduction

This chapter examines how property development projects are evaluated from a financial profit and risk perspective, hence focusing upon the process of development appraisal and valuation. Assessing and evaluating the financial viability of a development constantly occurs throughout all stages of the development process. A developer should not just conduct a one-off appraisal prior to the acquisition of a development site, but rather should constantly re-appraise the profitability of the scheme throughout the development process due to the effect of a myriad of influencing factors. Many dynamic factors influence development and their status must be updated and re-evaluated in light of the overall risk attached to the development.

Risk is an inherent and unavoidable part of the property development process and we shall consider how this is assessed as part of the overall evaluation process. We shall also consider the importance of market research (covered in detail in Chapter 8) in the appraisal and analysis of profit and risk. This chapter commences with a discussion on the conventional approach to development appraisal before introducing the various cashflow concepts, including the discounted cashflow approach. Finally we will examine the influence of uncertainly and how this can be contained in order to reduce the effect on risk.

3.2 Financial evaluation

3.2.1 Conventional technique

The financial maths and models for undertaking a property development are not difficult or complex if examined in individual sections. The best approach is to commence with a basic 'no frills' short-term development and then increase the skill level to complex long-term developments. Conventional techniques of identifying the various components of value in a proposed development are relatively straightforward based on using a form of 'residual' valuation. This type of model is designed to isolate an individual component of a development such as the level of risk/return or the land value, and assess its individual 'unknown' value when information about all of the other variables is known. In other words, it starts with the 'known' or certain variables and then moves to the 'unknown' variables to complete the analysis. It commences with the total value of the completed project and then deducts selected components to arrive at the remaining or 'residual' component.

The main variables in the model are:

- land purchase price
- building construction cost
- prevailing market rents/prices
- interest rate level
- investment yields (return on investment), and
- time (months).

Two primary types of residual valuation are undertaken depending on the final outcome sought. The first focuses on calculating the investment risk/return, the second on calculating the remaining (residual) cash available to purchase the land once all other development costs, including the developer's profit, have been paid.

1. Investment risk/return

This model commences with (a) the final estimated value of the completed property development based on estimated final market prices. Then (b) the total development costs (e.g. land, construction cost) can be deducted from (a) to establish whether the project produces (c) an adequate rate of return for the developer or financier, either in terms of a trading profit, an investment yield or return on capital. In this model the financial amounts in both (a) and (b) are generally fixed by the market and outside the control of the developer. Therefore, the result for (c) may be too low at present to undertake the property development, possibly because (a) is too low (e.g. low predicted demand) or (b) is too high (e.g. high borrowing costs at present).

2. Affordable land purchase price

An alternative approach for using a residual valuation is to assess (a) the likely costs of producing a development scheme and by deducting these costs from (b) an estimate of the value of the completed development scheme, to arrive at (c) a land purchase price. The developer will include an allowance for the required return in assessing the total development costs. Once again, the financial amounts for both (a) and (b) are generally fixed by the market and outside the control of the developer. Therefore, the current 'for sale' price of the land (c) may be too high to make the development 'affordable'. The manner in which the above-mentioned variables are brought together in a development appraisal can be shown best via a working example.

Let us assume a 0.8 ha site (2 acres) is on the market and the vendors are seeking a price of £$2,900,000. The site is in a good location in a town with adequate transport access and the vendor has obtained planning consent for 3,908 m² (42,065 ft²) of offices. Through market research, a developer has established that rents are currently £$322.92 per m² (£$30 per ft²) for comparable office space. A bank has agreed to provide short-term finance for the scheme at an interest rate of 2% above the bank's base rate of 6.25%, to be compounded quarterly, i.e. an effective annual rate of 8.47%. The developer's quantity surveyor has advised him that building costs are currently £$1,421 per m² (£$132 per ft²). The agents have advised the developer that the completed scheme should achieve a yield of 8.0% when sold to an investor. The developer will carry out a typical conventional evaluation as shown in Example 3.1. A model showing the Excel formulas is shown in Example 3.2. Note there may be rounding errors in some of the calculations when totalled.

We shall now examine each of the elements of the appraisal model in further detail.

3. Net development value

There are two important variables in establishing a development scheme's value, namely 'rent' and 'investment yield'. These terms vary between regions and countries; however, here the term 'rent' means the annual financial amount a tenant pays for occupying the final product and the term 'investment yield' refers to the income return on an investment which is usually expressed as an annual percentage based on the cost of the investment. Both variables can have an adverse effect on the appraisal model and it is critical that careful attention is paid to accurately defining each variable.

(A) RENT

The anticipated level of rent that a tenant is likely to pay to occupy part of, or all the completed property scheme is usually established in consultation with a real estate agent or a valuer. This amount should reflect the interaction between supply and demand factors in the property market at a certain point in time (this is usually based on a forecast of the property market at the time the development will be completed.). An estimate of the future rent must be as reliable as possible, using as few assumptions as possible. The estimate will be based on a thorough analysis of the present and future market trends and is often referred to as 'fair market rent'. According to the International Valuation Standards Committee (2013), market rent can be defined as 'the estimated amount for which a property, or space within a property, should lease on the date of valuation between a willing lessor and a willing lessee on appropriate lease terms in an arm's-length transaction, after proper marketing, wherein the parties had each acted knowledgeably, prudently, and without compulsion'.

The best source of information about the current market levels of rent is to refer to comparable evidence based on recent lettings of similar schemes in the close vicinity, in the local region or surrounding area in accordance with the definition of fair market rent stated above. It is important that each lease complies with each section of the definition and in all respects is representative of a normal rental agreement. In other words, the tenant involved in the comparable lease must have been fully conversant with the market and hypothetically may have even leased the proposed development in question (if it was completed). As no two properties are exactly identical, it is important that the property involved in each rental lease is adjusted to reflect differences in age, quality and specification. For instance where there are few (if any) recent comparative leases, a thorough market research exercise needs to be undertaken (see Chapter 8). It is essential that the developer bases the estimate on firm reliable evidence and careful

Example 3.1   Residual valuation

Evaluation of profit (risk/return)	£$	£$
(a) Net Development Value		
(i) Estimated Rental Value (ERV)		
Net lettable area 3,908 m^2 (42,065 ft^2) @ £$322.92 per m^2 (£$30per ft^2)	1,261,971	
(ii) Capitalised @ 8.0% YP in perpetuity	12.50	
	15,774,642	
(iii) less purchaser's costs @ 2.75%	433,803	
Net Development Value (NDV)		15,340,839
(b) Development costs		
(c) Land costs		
Land price	2,900,000	
Stamp duty @ 1%	29,000	
Agent's acquisition fees @ 1%	29,000	
Legal fees on acquisition @ 0.5%	14,500	
		2,972,500
(d) Building costs		
Estimated building cost		
Gross area 4,599 m^2 (49,500 ft^2) @ £$1,421 m^2 (£$132 ft^2)	6,535,179	6,535,179
(e) Professional fees		
Architect @ 5%	326,759	
Structural engineer @ 2%	130,704	
Quantity surveyor @ 2%	130,704	
M & E engineer @ 1.5%	98,028	
Project manager @ 2%	130,704	
		816,897
(f) Other costs		
Site investigations – say	19,250	
Planning fees – say	6,600	
Building regulations	33,000	
		58,850

Evaluation of profit (risk/return)	£$	£$
(g) Funding fees		
Bank's legal/professional fees – say	66,000	
Bank's arrangement fee	99,000	
Developer's legal fees – say	55,000	
		220,000
(h) Finance costs		
(i) Interest on land costs (£$3,946,250) over the	673,225	
development period and void of 30 months @ 8.25% compounded quarterly = (1.0206)10		
(ii) Interest on (d) building costs, (e) professional fees, (f) other costs and (g) funding fees divided by a half (£$7.63m/2) over building period of 12 months @ 8.25% compounded quarterly = $(1.0206)^4$	324,649	
(iii) Interest on building costs, professional fees, other costs and funding fees (£$7.63m) over void period of 12 months @ 8.25% compounded quarterly = $(1.0206)^4$	649,297	
		1,647,171
(i) Letting and sale costs		
Letting agents @ 15% ERV	189,296	
Promotion	95,000	
Developer's sale fees @ 1.5% NDV	230,113	
Other costs (see text)		
		514,408
(j) Total Development Costs (TDC)		12,765,006
(k) Developer's profit		
Net Development Value	15,340,839	
Less total development costs	12,765,006	
Developer's profit		2,575,834
Developer's profit as % of total development costs	20.18%	
Yield on development cost	9.89%	

Example 3.2 Excel formulas

	A	B	C
1	Evaluation of profit (risk/return)	£$	£$
2	(a) Net Development Value		
3	(i) Estimated Rental Value (ERV)		
4	Net lettable area 3,908 m² (42,065 ft²) @ £$322.92 per m² (£$30per ft²)	=3908*322.92	
5	(ii) Capitalised @ 8.0% YP in perpetuity	=100/0.08/100	
6		=B3*B4	
7	(iii) less purchaser's costs @ 2.75%	=B7*0.0275	
8	Net Development Value (NDV)		=B6-B7
9			
10	(b) Development costs		
11	(c) Land costs		
12	Land price	2,900,000	
13	Stamp duty @ 1%	=B12*0.01	
14	Agent's acquisition fees @ 1%	=B12*0.01	
15	Legal fees on acquisition @ 0.5%	=B12*0.005	
16			=SUM(B12:B15)
17	(d) Building costs		
18	Estimated building cost		
19	Gross area 4,599 m² (49,500 ft²) @ £$1,421 m² (£$132 ft²)	=4599*1421	=B19
20			
21	(e) Professional fees		
22	Architect @ 5%	=B19*0.05	
23	Structural engineer @ 2%	=B19*0.02	
24	Quantity surveyor @ 2%	=B19*0.02	
25	M & E engineer @ 1.5%	=B19*0.015	
25	Project manager @ 2%	=B19*0.02	
27			=SUM(B22:B26)
28	(f) Other costs		
29	Site investigations - say	19250	
30	Planning fees - say	6600	
31	Building regulations	33000	
32			=SUM(B29:B31)

	A	B	C
33	(g) Funding fees		
34	Bank's legal/professional fees - say	66000	
35	Bank's arrangement fee	99000	
36	Developer's legal fees - say	55000	
37			=SUM(B34:B36)
38	(h) Finance costs		
39	(i) Interest on land costs (£$3,946,250) over the	=(C16*((1+020625)*10))-C16	
40	development period and void of 30 months @ 8.25% compounded quarterly = (1.0206)10		
41			
42	(ii) Interest on (d) building costs, (e) professional fees, (f) other costs and (g) funding fees divided by a half (£$7.63m/2) over building period of 12 months @ 8.25% compounded quarterly = (1.0206)^4	=(((C19+C27+C32+C37)/2)*((1+0.020625)^4))-((C19+C27+C32+C37)/2)	
43			
44	(iii) Interest on building costs, professional fees, other costs and funding fees (£$7.63m) over void period of 12 months @ 8.25% compounded quarterly = (1.0206)^4	=((C19+C27+C32+C37)*((1+0.020625)^4))+(C19+C27+C32+C37)	
45			=SUM(B39:B44)
46	(i) Letting and sale costs		
47	Letting agents @ 15% ERV	=B4*E47	
48	Promotion	95000	
49	Developer's sale fees @ 1.5% NDV	=C8*E49	
50	Other costs (see text)		
51			=SUM(B47:B50)
52			
53	(j) Total Development Costs (TDC)		=SUM(C16:C51)
54			
55	(k) Developer's profit		
56	Net Development Value	=C8	
57	Less total development costs	=C53	
58	Developer's profit		=B56-B57
59			
60	Developer's profit as % of total development costs	=C58/C53	
61	Yield on development cost	=B4/C53	

analysis to establish today's rent level, i.e. at the date of the appraisal. The importance of ensuring this rental estimate cannot be over-emphasised, since an overly optimistic estimation in rental levels can result in a false presentation of potential risk and return at the completion of the development. Note that it is not recommended that a developer should rely heavily on a forecast (based on assumptions) or inflated rent in this type of appraisal. When there is reliance on a forecast rent then this should be explicitly stated in a valuation model (discounted cashflow or capitalisation of income approach) rather than bound up within the discount rate or an 'all risk' yield. The property market is a complex interaction of variables and it is extremely difficult, if not impossible, to forecast future rent levels with a relatively high degree of accuracy. The foundation for this modelling should place the emphasis on the existing market and the current level of agreed rents.

Rents are usually analysed by reference to a rate per square foot or per square metre based on the cost per year. In the case of an office building, the net area of the building (i.e. the internal usable space excluding circulation space and toilets, etc.) needs to be established and is known as the net lettable area (NLA). The tenant is most interested in the NLA available, and tenants seeking a larger area to lease will often be rewarded with a smaller rate per square metre (m^2). Larger rental areas are usually cheaper (on a 'per unit of area' basis) than smaller rental areas due to bulk discount, even though other influencing factors (e.g. use, building, floor level) remain unchanged. In addition, discounts will be applied to the rate used to take account of areas with less demand, such as basements (i.e. due to no or restricted natural light) and floors with restricted headroom. At times it can be difficult to ascertain what components on a floor should be included in NLA and what is excluded, and this can have an adverse effect on the aggregate amount of NLA especially in a multi-level office or retail building. All measurements should be made in accordance with the relevant guidance notes as accepted by lessees and lessors in the local property market which sets out the various definitions of measurement. Often the guidance notes are produced by the industry body which has relevance for that location and market. For example in the UK that body is the Royal Institution of Chartered Surveyors (RICS) and definitions and guidance for measuring is set out in their Code of Measuring Practice. Many countries have a unique approach to measuring areas which are accepted by local surveyors, property owners and tenants, for example in Australia the accepted publication is *Methods of Measurement* as published by the Property Council of Australia (PCA).

With an industrial scheme it is usual for the 'gross internal area' to be measured, which includes all the internal space within the external walls. By contrast the approach to retail property varies. For example in some regions the rent for retail units may be analysed in sections or zones measured from the front boundary of the shop (e.g. zone widths are usually a depth of approximately 20 feet or 6.1 metres in the UK, London has variable widths, some parts of Ireland have zones of approximately 15 feet or 4.5 metres). The first zone is known as Zone A, the second Zone B and the remainder Zone C. Retail units are valued in relation to the Zone A rent which is the rate applied per square metre (or per square foot) to the area of Zone A. Then half the Zone A rate is applied to the area of Zone B and a quarter of the Zone A rate is applied to the area of Zone C. This zoning is to reflect the fact that the most valuable space is at the front of the shop. As most shops do not conform to a standard size and shape, adjustments will be made to the above-described analysis to reflect the varying levels of demand for different frontages and unusual shapes, which largely will be based on the experience of the property professional (e.g. valuer) and their knowledge of the prevailing market. Large retail units such as department stores and variety stores, together with shops in small parades, are usually analysed on an overall rate rather than any Zone A rate.

(B) INVESTMENT YIELD

The return on the total financial capital used to purchase a real estate investment is usually referred to as the 'investment yield'. Used in the process of assessing value, the investment yield is used to discount the future rental income stream in order to calculate the capital value of the development scheme in today's money. From the investment a multiplier is derived that can be applied to the future income stream and this is known as the 'year's purchase' (YP) in perpetuity. The YP is the reciprocal of the investment yield.

This modelling approach is based on taking a snapshot of the current rental income and, where this represents current market value, assumes that it will remain at its present level in perpetuity. The growth in the rental income and the risks associated with it are then reflected in the multiplier used. For example, if the future income from the completed property scheme is estimated to be $500,000 per annum and the investment yield is 5% then 5 is divided into 100 to calculate the YP of 20, which is then multiplied by the rental income to produce the final capital value of the completed development. In this example it is $10,000,000. Therefore, the investment yield (5%) is derived by dividing the rental income into the capital value (0.05) and then multiplying by 100 to express it as a percentage. Table 3.1 lists a range of investment yields and their respective YP.

The actual investment yield for a particular property can be obtained by comparing 'like with like' in the property market. The approach is by way of comparison, based on an analysis of recent sales of properties that are similar to the property development

Table 3.1 Investment yields and respective year's purchase (YP)

Investment return (%)	Years purchase (YP)	Investment return (%)	Years purchase (YP)
30	3.3	10	10.0
27.5	3.6	9	11.1
25	4.0	8	12.5
22.5	4.4	7	14.3
20	5.0	6	16.7
17.5	5.7	5	20.0
14	6.7	4	25.0
14	7.1	3	33.3
13	7.7	2	51
12	8.3	1	100
11	9.1		

scheme being proposed. When undertaking the comparison approach, consideration should be given to a large number of variables, as well as each variable's varying level of contribution to value, which will affect the value of each property. A starting point for undertaking this comparison is to consider the two main variables in the property development: (a) the land component and (b) the building component.

The investment yield is a measure of risk and return and will reflect the property investor's perception of (a) the future rental growth against (b) the risk of future uncertainty. In general terms the faster the level of rent is expected to grow in the future, the lower the yield an investor is prepared to pay at the outset. However, the level of perceived risk in an investment can have an adverse effect on the property itself and therefore risk factors also need to be taken into account. The level of yields tends to change with changes in the patterns of rental growth or investor demand. Generally retail developments tend to attract the lowest yields while industrial developments often attract the highest yields, based on their previous history of rental growth.

4. Purchaser's costs

Regardless of whether the developer intends to retain the completed property development as an investment or sell the property on completion, the final completed development value needs to be expressed as a net development value to allow for purchaser's costs such as stamp duty, agent's fees and legal fees.

5. Development costs

(A) LAND COSTS

The land cost includes the land purchase or acquisition price, which is either the price already negotiated with the landowner or, as in this example, the price being sought by the landowner. All costs and expenses associated with acquiring the site need to be included. Accordingly acquisition costs and fees will include stamp duty (or tax) as a percentage of the land purchase price, legal fees associated with acquiring the site and (if applicable) a real estate agent's introduction fee. In many regions the applicable stamp duty (tax) is not a flat rate irrespective of value but rather can be separated into graduated bands based on the land price – fortunately most government authorities have online tax conversion tools on their websites. For the purposes of this simple example the stamp duty rate is assumed to be a flat 1%. Legal fees are often between 0.25 and 0.5% of the land price, depending on the complexity of the deal. Usually the real estate agent's fees are between 1–2% of the land price, depending on whether the agent is to be retained as the letting and/or funding agent.

Other considerations will affect the modelling process for each unique property development. Complications may arise in certain jurisdictions with regards to stamp duty, property tax and land tax (if applicable). For example, if the project is being forward-funded by a financial institution, the developer may have to allow for double stamp duty on the land cost if the developer purchases the site before completing funding arrangements with the financial institution. This is because stamp duty will be incurred on the initial purchase of the land by the developer and also on the subsequent transfer to the financial institution.

(B) BUILDING COSTS

Building costs are estimated by the developer's quantity surveyor and are usually expressed as a 'per unit' cost, such as the overall rate per square metre (or square foot) – this 'per unit' cost is then multiplied by the gross area of the proposed building. The building costs are estimated at the time of the proposed implementation of the development project. In many cases no allowance is made for cost increases (e.g. due to inflation) during the building contract period, although some developers may inflate building costs in their appraisals based on informed assumptions by third parties, particularly in periods of rapidly rising building costs or if the construction period covers an extended period of time and there is uncertainty surrounding future building costs.

(C) PROFESSIONAL FEES

These fees are normally calculated as a percentage of the total building costs and include all fees for professional services employed in the completion of the development. Often this includes the architect, the quantity surveyor, the structural engineer, the mechanical and electrical engineer, and the project manager. The actual rates per professional can vary considerably in accordance with factors such as the size of the project, the capital outlay and the complexity of the task. Where the total cost of professional fees are not standardised they are often in excess of 10% of the total building costs, say about 12–13% and are either calculated on a 'flat fee' basis or based on (a) the 'scale of fees' or charges for each profession, (b) a negotiated percentage or (c) a fixed fee. The percentage agreed with each member of the professional team depends on factors such as the nature and scale of the property development, as well as the relationship/goodwill between the developer and each professional. Small refurbishment schemes normally attract higher percentages than larger, complex development projects due to the economies of scale. Perceived high profile or 'blue ribbon' developments may cause competition between industry professionals who are more than willing to be a part of an important project to raise their professional profile, which in turn may equate to a lower rate. If a developer is to appoint other professionals, such as a traffic engineer, a landscape architect or a party wall surveyor, then these need to be included in any evaluation of the project.

(D) SITE INVESTIGATION FEES

These include fees for ground investigation and land surveys. Especially for residential developments, any potential land contamination must be identified and rectified. There are many examples of a building being constructed on contaminated land (not identified in the initial site survey), which necessitated the new building being demolished prior to the site being decontaminated. Other developers were possibly aware of this added risk and bypassed the development for this reason. The importance of a thorough site investigation to reduce exposure to risk of the unknown cannot be over-emphasised.

6. Planning fees

These costs relate to government fees required to make a planning application and securing consent for the property development project. Many developments necessitate a change in the use of the property from a previous land use, such as from industrial to residential, where the highest and best use of the property has changed over an extended time period, for example this may be due to a lower demand by light industry and higher demand by new residents. This component normally only includes the fees paid to the relevant government planning authority and is based on the scale and nature of the scheme. A list of the relevant fees and charges can normally be obtained from the government planning authority and will usually vary depending on the size of the development.

In the above example, planning consent has already been obtained. However, in a situation when obtaining planning permission may prove difficult (especially where there is a substantial change in use), a developer has to allow for planning consultant fees and, in the event of an appeal, costs such as fees for solicitors, counsel and expert witnesses. The extra time period involved will definitely need to be reflected in the interest costs associated with the extended holding period which the developer is directly accountable for.

7. Building regulation fees

Usually these costs are on a sliding scale, based on the final building cost. Details of such fees are available from the building control department of the relevant government authority.

8. Funding fees

Most financial institutions and lenders charge fees when arranging development finance. These fees are related to the costs associated with arranging development finance and will vary on the method of finance. To illustrate this point, Example 3.1 is bank financed so the developer will need to pay the bank's arrangement fees, solicitor's fee and surveyor's fee. These fees are a matter of negotiation but usually reflect the size of the required loan and may be anything between 3 and 10% of the value of the loan (see Chapter 4 for details).

If the property development is to be 'forward-funded' by a lender or financial institution, then the developer will pay the fund's agent (if appointed) and associated solicitor's fees, as well as their own agent (if appointed) and their own solicitor's fees. The developer may also have to pay the fund's building surveyor's fees to monitor the construction of the building on behalf of their client.

9. Finance costs/interest

Interest costs for borrowed funds are a critical element of the appraisal and can have an adverse effect on the overall viability of any development proposal. These costs reflect either (a) the actual cost to the developer of borrowing money over time or (b) the implied or notional opportunity cost (reflecting the investment foregone, i.e. the capital could be earning money elsewhere at a comparatively higher return but not necessarily with the added risk). The actual cost of the finance/interest is affected by many factors including the loan-to-value ratio (LVR), the risk in this land use sector and the location (e.g. a new retail development in the city), the established relationship between the borrower and the financier/lender, and the borrower's estimated risk that the borrowed funds will be completely repaid in full by the due date. In Example 3.1 the development company will be borrowing money from the bank and, as a condition of the loan, will be providing some capital from its own resources. It is assumed that the interest rate charged by the bank and the opportunity cost of the developer's own money are identical.

In order to calculate the interest costs, the developer must estimate the total length of the development. Normally the cash inflow will commence when the completed property development is either (a) let to a tenant/s (and therefore becomes income-producing) or (b) sold as is usually the case in residential schemes, where property is often sold before the development is completed, enabling the developer to use this income to fund the rest of the scheme and reduce the cost of borrowing finance to the build the scheme. The decision depends on whether the developer wishes to (or is financially able to) retain the scheme or not. In addition, the developer must allow sufficient time for all the preparation work needed after the site has been acquired but prior to the commencement of the building contract. Also there must be a careful estimation of the time it will take to either let or sell including a vacancy/void period which is based on a judgement of prevailing market conditions at that future point in time.

In Example 3.1 the development timetable is assumed as shown in Table 3.2. These periods can be expressed in a timeline diagram as shown in Figure 3.1.

Table 3.2 Development timetable

Site acquisition, preparation and pre-contract	6 months
Building contract	12 months
Letting period	6 months from completion
Investment sale period	6 months from letting
Total development period	30 months

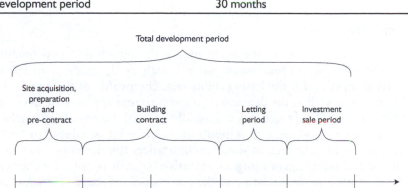

Figure 3.1 Development timetable

The site acquisition is the first commitment and requires a major capital financial outlay. As this payment is at the very beginning of the development, interest is calculated on all site acquisition costs over the entire development period – this is the longest timespan within the development period which runs from the date of acquisition to the eventual letting/sale of the building. In Example 3.1 the total timeframe of the development is thirty months. In some circumstances, additional expenses may be incurred prior to the site acquisition, such as the costs associated with searching for potential sites. However, such costs are usually not considered to be substantial capital outlays that attract a high interest cost. Once the building contract is signed then most of all the other costs will be incurred at various different times over the building contract period (in Example 3.1 this period is twelve months), although the actual timing of these cashflows will vary and are often difficult to accurately quantify. Accordingly, a 'rule of thumb' simplistic method of calculating the interest is adopted for straightforward short-term developments which assumes the costs are incurred evenly over the entire contract period. Therefore all the costs, with the exception of promotion and letting costs, are divided in half and then the interest is calculated on that sum over the whole period. In Example 3.1 this time period is twelve months.

Once the building contract is complete, then the interest costs payable will continue to accrue on all the building and other costs spent (except some of the promotion costs and letting/sale fees) until the date when it is assumed that the building will be eventually let/sold. In Example 3.1 it is assumed that the building is let within six months of final completion and then on-sold to a third party investor within twelve months of completion. It is further assumed that a period of six months rent-free are granted to the tenant. If it were to be assumed that some rental income would be received before the sale of the investment then such income would be included in the appraisal and offset against the interest calculation. In Example 3.1, interest is calculated by using the Amount of £$1 formula for compound interest (Baum and Mackmin 2011). In order to calculate compound interest on a quarterly basis the interest rate of 8.25% is divided by 4 to obtain the quarterly rate of 2.06%, which produces a compound interest formula of $(1.0206)^n$, where n represents the number of quarters over which the interest is calculated. In Example 3.1 this is four quarters.

10. Letting agent's fees

These fees relate to the cost of the agent letting the building to new tenants. The actual amount will vary depending on factors such as the number of letting real estate agents competing for the letting rights (e.g. the profile of the development in the marketplace), as well as the demand by tenants to rent space in the development. If joint agents are involved these fees are usually 15% of the rental value achieved at letting. If only one real estate agent is involved, then the fee is reduced to 10% of the rental value achieved at letting. In some circumstances the developer may negotiate a fee with the real estate agent using an incentive basis. It is usual for the tenant to pay the developer's legal fees relating to the completion of the lease documentation.

11. Promotion costs

The developer has to make an assessment of the likely sum of money that needs to be spent on promoting the project in order to let the property, and very often it is this element of the evaluation that is underestimated at this initial stage (see Chapter 10). This amount will be affected by the perceived level of demand for the development (i.e. a high profile development may be in high demand and require less promotion) and the location of the prospective tenants, e.g. advertising in a national newspaper or via other costly mediums such as television or radio.

12. Sale costs

Costs associated with selling the completed development will need to be included if the developer intends to sell the building once it is fully let. These will include any real estate agent's fees together with those of the developer's solicitor, often equating to between 1% and 2% of the net development value (NDV).

13. Other development costs

The inclusion of other costs within the evaluation will depend on the nature of the development and will be specific to the project (e.g. inter-tenancy party wall agreements, planning agreements with the government planning authority and rights of light agreements).

If the developer considers that there may be a void/vacancy period between completion of the development and letting the property, then costs such as maintenance and insurance will need to be included. If a lengthy void/vacancy period is anticipated, an allowance will need to be made for additional costs such as maintenance and management expenses. If the property development scheme has been forward-funded by a financial institution then, under the terms of the funding agreement, rent may be payable to the fund until a letting is achieved (see Chapter 4).

14. Contingency allowance

In reality there are relatively few property developments, if any, completed on time exactly as originally planned nor are they likely to adhere entirely to the initial budget forecast. This is partly due to assumptions incorporated into the original planning phase including forecasts about future expenses and rental levels. It is therefore essential to include a contingency allowance to cover unexpected costs. However, the actual contingency itself is an assumption and will vary from project to project depending on variables such as the risk profile of the developer, the developer's ability to plan and execute an accurate development plan, the associated time period, the level of risk/return built into the proposal and the level of flexibility. For Example 3.1 a flat contingency allowance of 1% was adopted.

15. Developer's profit/risk allowance

The residual in an appraisal model is often the developer's profit/risk allowance which is usually expressed as a percentage of the total development costs or, alternatively, as a percentage of the net development value (NDV). As per standard economic theory, the level of profit that a developer will require will depend on their exposure to risk for the same property development scheme. A higher level of risk will be commensurate with a higher level of return and vice versa. It is difficult to generalise here but often developers will seek between 15 and 25% of the total cost, the percentage rising with the perceived risk. The profit may also contain an element for contingencies, rather than a separate allowance for contingencies as discussed above.

If the developer is an investor wishing to retain the development, then profit may be assessed by reference to the yield on cost (in Example 3.1 it is 9.89%). The yield or return on cost is the total development cost (excluding profit) divided into the first year's rental income. The resulting yield needs to be higher than the yield applied to obtain the NDV (which is comparable to the yields on other similar standing investments) as the difference between the two yields represents the profit to the investor.

In Example 3.1 the land price used in the evaluation is the asking price and therefore is fixed. However, in most cases the developer has to establish the land price that can be afforded in order to enjoy a fixed target rate of profit. At the same time, the landowner (i.e. vendor) will be seeking to maximise the sale price, and may not even quote an asking price. In Example 3.3 we assume the developer wishes to ensure a rate of profit (risk/return) of 20% on total development costs, so a residual land evaluation is carried out to determine the affordable land price.

Discussion points

- What are the various components of value when undertaking a property development?
- Explain the relevance between the level of profit/risk and the overall development.

3.2.2 Cashflow method

The conventional method of evaluating a proposed property development, as shown in Examples 3.1 and 3.3, has two basic weaknesses which require further discussion. First, it is inflexible in its handling of the timing of when the expenditure and revenue actually occur. As a result, the calculation of interest costs is very inaccurate and may vary substantially – unless the projected time period is the exact same length in reality, then the evaluation is incorrect. Second, by relying on single-figure 'best estimates' it hides the uncertainty and assumptions that lie behind the calculation.

The problem associated with inflexibility in the first point can be overcome by carrying out a cashflow appraisal or evaluation which enables the flow of expenditure and revenue to be spread over the period of the development. This model presents a more realistic and accurate assessment of development costs and income against the variable of time. As commonly accepted, the amount of compound interest accrued over an extended period of time can have an adverse effect due to the time value of money. Therefore, the conventional evaluation shown in Example 3.1 is presented as a cashflow appraisal model in Example 3.4.

Example 3.3   Residual valuation

	£$	£$
(a) Net Development Value (NDV)		
(i) Estimated Rental Value (ERV)		
Net lettable area 42,065 ft² (3,908 m²) @ £30 per ft² (£$322.92 per m²)	1,261,971	
(ii) Capitalised @ 7.0% YP in perpetuity	14.29	
	18,028,162	
(iii) less purchaser's costs @ 2.75%	495,774	
Net Development Value (NDV)		17,532,388
(b) Development costs		
(c) Building costs		
Estimated building cost		
Gross area 49,500 ft² (4,599 m²) @ £$132 ft² (£$1,421 m²)	6,535,179	6,535,179
(d) Professional fees		
Architect @ 5%	326,759	
Structural engineer @ 2%	130,704	
Quantity surveyor @ 2%	98,028	
M & E engineer @ 1.5%	130,704	
Project manager @ 2%	130,704	
		816,897
(e) Other costs		
(iv) Site investigations -- say	17,500	
(v) Planning fees – say	6,000	
(vi) Building regulations	30,000	
		53,500
(f) Funding fees		
Bank's legal/professional fees – say	60,000	
Bank's arrangement fee	90,000	
Developer's legal fees – say	50,000	
		200,000
(g) Finance costs		
(i) Interest on (c) building costs, (d) professional fees, (e) other costs and (f) funding fees divided by a half (= £$6.33m/2) over building period of 12 months @ 8.25% compounded quarterly = $(1.0206)^4$	323,570	

continued...

Example 3.3 continued

	£$	£$
(ii) Interest on (c) building costs, (d) professional fees, (e) other costs and (f) funding fees (£6.33m) over void period of 12 months @ 8.25% compounded quarterly = $(1.0206)^4$	647,140	
		970,711
(h) Letting and sale costs		
Letting agents @ 15% ERV	189,296	
Promotion	95,000	
Developer's sale fees @ 1.5% NDV	262,986	
		547,282
Net total development costs excluding land costs and interest on land costs		9,123,568
(i) Developer's profit		
@ 20% on net total development costs (£$9,123,568) excluding land costs and interest on land costs		1,824,714
(j) Net Total Development Costs (NTDC)		10,948,282
(k) Residue i.e. NDV less NTDC		6,584,106

This residue is made up of the following elements:

Land price =	1
plus cost of acquisition @2.5%	0.025
	1.025
multiplied by cost of interest of holding land for development period and void (30 months) @ 8.25% compounded quarterly $(1.0206)^{10}$ =	1.226
Total land cost	1.257
multiplied by profit on total land cost @ target rate of 20%	1.2
	1.508

The residual land value i.e. the price the developer can afford to pay to ensure the target rate of profit, is therefore derived as follows:

	£$	£$
Residue	6,584,106	

Divided by factor (calculated above) to take account of land price, acquisition costs, interest and profit as calculated above	1.508	
Residual land value	4,366,176	
Say	4,365,000	

This calculation can be checked as follows:

Land price	4,365,000	
plus cost of acquisition @ 2.5%	109,125	
Total land costs	4,474,125	
multiplied by interest for 30 months @ 8.25% compounded quarterly $= (1.0206)^{10} = 1.226$	1,013,319	
Total Land Cost (TLC)	5,487,444	
plus Net Total Development Cost excluding profit	9,123,568	
Total Development Cost (TDC)	14,611,013	
Net Development Value, as above	17,532,388	
Less Total Development Cost (TDC)	14,611,013	
Developer's profit	2,921,375	
Developer's profit on cost	20.0%	

Note: this result confirms that at a land price of £$4,365,000 the target level of profit of 20% can reasonably be expected to be achieved.

Example 3.4 Cashflow approach (with rounding)

Monthly Interest Rate	0.680%																														
Months	1	2	3	4	5	6	7	8	9	10	11	12	13	14	15	16	17	18	19	20	21	22	23	24	25	26	27	28	29	30	Total
Cost (£$000)																															
Land cost	2,972																														2,972
Building cost							182	205	305	492	622	505	512	686	655	790	545	440				184	197	215							6,535
Professional fees				48		48		72	45	46	60	95	92	45	46	42	44	41	31	30				31							816
Other fees	9	8		8			8		12		8	6																			58
Funding fees																						155	65								220
Letting fees																								189							189
Promotion																				25	25	30		15							95
Sale fees																														230	230
Tax paid	620	2	0	8	0	42	41	77	45	81	98	99	89	120	129	120	83	44	6	11	6	6	0	69	0	0	0	0	0	40.5	1,833
Tax reclaimed	0	0	0	-620	-1.5	0	-7.5	0	-42	-41	-76.5	-45	-81	-97.5	-99	-88.5	-120	-129	-120	-82.5	-43.5	-6	-10.5	-6	-6	0	-69	0	0	-40.5	-1,833
(a) Sub-total (Month)	3601	9.5	0		-1.5	90	223	353.5	365	578	710.5	660	611.5	753.5	731	863.5	551.5	395.5	-83	-17	-12.5	369	251.5	513	-6	0	-69	0	0	230	11,115
(b) Balance B/F	0	3,625	3,660	3,685	3,149	3,169	3,281	3,528	3,908	4,302	4,913	5,662	6,365	7,024	7,830	8,620	9,548	10,168	10,635	10,624	10,679	10,739	11,184	11,513	12,108	12,184	12,267	12,281	12,364	12,448	12,765
(c) Total (a + b)	3,601	3,635	3,660	3,128	3,148	3,259	3,504	3,882	4,273	4,880	5,624	6,322	6,977	7,778	8,561	9,483	10,099	10,563	10,552	10,607	10,667	11,108	11,435	12,026	12,102	12,184	12,198	12,281	12,364	12,678	
(d) Interest	24	25	25	21	21	22	24	26	29	33	38	43	47	53	58	64	69	72	72	72	73	76	78	82	82	83	83	84	84	86	1,650
Balance C/F (c + d)	3,625	3,660	3,685	3,149	3,169	3,281	3,528	3,908	4,302	4,913	5,662	6,365	7,024	7,830	8,620	9,548	10,168	10,635	10,624	10,679	10,739	11,184	11,513	12,108	12,184	12,267	12,281	12,364	12,448	12,765	

Total development cost (£$)	12,765,006
Net development value (£$)	15,340,839
Developer's profit (£$)	2,575,834
Developer's profit as % of cost	20.18%

Table 3.3 Normal S-curve irregular pattern of expenditure

Months	1	2	3	4	5	6	7	8	9	10	11	12
% total costs	3	10	14	22	31	40	48	60	73	85	93	97

As this example shows, by enabling the expenditure to be allocated more accurately over varying timelines a better assessment can be made of interest costs. The 'rule-of-thumb' conventional evaluation, described above, assumed that building costs would be spread in this way. In practice, building and other development costs are seldom spread evenly over the period. In Example 3.4 some of the development costs are incurred before or at the start of the building contract period, e.g. funding fees and some of the professional fees. Often the majority of professional fees are incurred during the pre-contract stage and early in the building contract period, as most of the design and costing work is carried out then. In Example 3.4 only 40% of the building cost has been incurred after six months of the contract which is the half-way point. Often the pattern associated with building costs follow a normal S-curve irregular pattern of expenditure – for example it may be as shown in Table 3.3.

In this example the remaining 3% of the costs represents the standard practice of holding a retention sum under the building contract usually for a period of six months. The retention sum and period may vary and are often perceived as 'insurance' on the overall property development process.

In practice, the quantity surveyor should be consulted to assess the timing of building costs. Computer programs are available to calculate the S-curve for a particular project and convert that into the expenditure flow. The project manager can assist in assessing the flow of other costs directly related to the building costs. The timing of all other costs should be capable of assessment by the developer based on experience.

The cashflow method enables the developer to allow for such an irregular pattern of cost, giving a more explicit presentation of the flow of expenditure and a more accurate assessment of the cost of interest. It is the nature of property development that the timing of cashflows is irregular and uneven – for example, a capital outlay for a parcel of land is usually unavoidable as, in most cases, a building cannot be constructed unless the land is owned outright in the first place. In Example 3.4 the total interest figure ($£1,650,000) is higher than calculated in the conventional evaluation ($£1,647,171) in Example 3.1. However, given a different pattern of expenditure, with professional fees and funding fees being incurred later on, then the total interest figure may well have been lower than in Example 3.1. It is impossible to generalise or include additional assumptions, which is why the conventional 'rule of thumb' method is simple but at the same time relatively inaccurate. In the cashflow example, interest is calculated on the outstanding balance (including interest) at the end of each month at the rate of 0.68% per month calculated as in order to equate to the effective annual rate in Example 3.1 of 8.47% per annum.

In the above example the project is a single office development and therefore cash inflows (in the form of final sale or rent to a tenant) would not generally occur until the entire building is complete. The advantages of the cashflow method are more clearly demonstrated in relation to developments where receipts (or cash inflows) occur during the development period prior to final completion of the entire scheme, e.g. a development of phased industrial units, a major retail scheme and a residential scheme. Another example would be a large and complex mixed-use development scheme which will take a number of years to complete fully and therefore will be developed in phases. In this case, it may be possible to let or sell the early phases while construction continues.

As in most businesses, cashflow is critical due to the cost of repaying borrowing funds and the effect of compound interest over an extended period of time. Consequently the potential to develop a property in phases can be a major advantage for the overall viability of the project. Recent developments in construction technology have assisted more building types to be developed and released in phases. For example, some offices in high-rise buildings can be let or even sold-off and allow the new owners to occupy the lower floors, even though the upper floors or other sections of the building are still under construction. This example applies to large-scale projects that take years to complete where the developer has been creative in their project management with a desire to commence cash in-flows at the earliest available opportunity.

There are other advantages with adopting the cashflow method. For example the model enables the developer to adjust for changes in interest rates easily over the development period or for different sources of finance within the appraisal. In addition, this method disciplines the developer to think hard about the nature of the cashflow of the project. It highlights, where possible, the need to delay outgoing payments and bring forward receipts (cash income). It shows the developer that cashflow is an important tool in identifying a competitive advantage over competitors, which can be achieved by maximising profitability and reducing the cost of borrowing. A developer will certainly have to produce a cashflow appraisal to satisfy potential sources of finance, when a detailed 'business case' is a standard request prior to funds being advanced. In reality, many developers use both conventional and cashflow techniques. They will use the cashflow method to calculate the interest cost and input the resultant figure into a conventional evaluation for presentational purposes. In addition, the cashflow method will be used throughout the development period to constantly evaluate the project as costs are incurred and influencing variables change, such as changes in the level of interest rates in the broader economy.

It is important for the developer to assess the impact of taxation on the project, which may involve many different charges for different government authorities and is constantly subject to change. Rates of taxation are also likely to vary depending on the type of developer and the type of scheme under consideration. Different developers will adopt different vehicles for operating their business – for example, one developer may be operating as a private operator under their own personal name, rather than forming a private company with limited shareholders. Larger development companies might list on the equities market in which shareholders have an option to buy shares in the company. Despite the complexity of taxation a developer must be fluent and up to date with any taxation implications or government restrictions relating to money. The property developer should enlist the services of an accountant, financial adviser and/or legal adviser to ensure they avoid paying additional taxation costs which could otherwise be minimised. A developer who is paying more tax than a competitor, when other variables (e.g. land cost, building costs, interest rates) remain unchanged will be unable to compete in the marketplace or generate a healthy profit.

Tax can have a varying impact on the cashflow in different stages of a development project. In some regions it will be possible to recover the tax paid on land transactions and construction costs if the developer (subject to approval) elects to then pass on the tax assessment with the sale of the completed building or on rents from the letting of the completed building. However, a cashflow implication arises as there may be a delay between the initial payment of tax and its subsequent recovery – in Example 3.4 there is a three-month delay in recovering the tax from the final occupier. In this model the delays have no effect on the overall interest figure but, on larger schemes, the delay in repayments will almost certainly impact on the interest calculation due to the leverage involved. Tax legislation is very complex and it is beyond the scope of this book to examine all of the implications for each region and country in detail. It is important to stress that a developer must fully assess all tax implications and the direct or indirect effect on a particular development project when carrying out an appraisal.

3.2.3 Discounted cashflow method

Although a discounted cashflow (DCF) can examine different cashflow models, they are all discounted back (i.e. using a present value formula) to a common point in time to facilitate an even comparison or analysis. The discounting component acknowledges the relationship between time and money which is especially relevant in property development. For example there is usually an extended period of time between when the land is purchased and when the building is completed and cash inflow commences. Example 3.4 calculates interest on a month-by-month basis to reflect a normal development pattern, so that at any point in the development programme the developer can establish the outstanding debt at that particular time. The time periods can be modified to any time period, such as days or years depending on the intended complexity of the DCF. Alternatively, a combination of two cashflow techniques can be used, being the 'net terminal approach' and 'discounted cashflow' (DCF) methods. As Example 3.5 shows, the 'net terminal approach' simply calculates the interest in a different way but produces exactly the same result as the normal cashflow method. The interest is calculated on each month's total expenditure until the end of the development period, i.e. when the development is let (or sold) and the debt is fully repaid (plus profit/risk allowance). Note that the 'net terminal approach' will overstate the amount of debt outstanding at the end of each month and has no advantage to the developer over the normal cashflow in Example 3.4. The model displayed in Example 3.4 is in the format of a traditional cashflow where the time periods are on the x-axis and the variables for each time period are listed on the y-axis.

The DCF method is distinctly different as it does not calculate interest on the monthly expenditure. It sums the income and expenses for every month and then discounts the amount for each month back to present-day equivalents to establish the value of the profit in today's money, rather than at the end of the development. The discount rate used is the cost of borrowing the money and the formula used to convert costs and values to the present day is the 'Present Value of $£1$', which is $1/(1 + i)^n$ (or alternatively $(1 + i)^{-n}$), where i represents the prevailing interest rate (e.g. 0.075 for 7.5%) and n represents the number of periods (e.g. in months). This formula is the reciprocal of the amount of £1 used for compound interest.

Using the same figures as those contained in Example 3.4, a DCF is calculated in Example 3.6.

Example 3.5   Net terminal approach

Months	Cashflow (£$000)	Interest until completion at 0.68%	Total (£$000)
1	3601	1.2255	4413
2	9.5	1.2172	12
3	0	1.2090	0
4	−556.5	1.2008	−668
5	−1.5	1.1927	−2
6	90	1.1846	107
7	223	1.1766	262
8	353.5	1.1687	413
9	365	1.1608	424
10	578	1.1529	666
11	710.5	1.1452	814
12	660	1.1374	751
13	611.5	1.1297	691
14	753.5	1.1221	846
15	731	1.1145	815
16	863.5	1.1070	956
17	551.5	1.0995	606
18	395.5	1.0921	432
19	−83	1.0847	−90
20	−17	1.0774	−18
21	−12.5	1.0701	−13
22	369	1.0629	392
23	251.5	1.0557	266
24	513	1.0486	538
25	−6	1.0415	−6
26	0	1.0345	0
27	−69	1.0275	−71
28	0	1.0205	0
29	0	1.0136	0
30	230	1.0068	232

Total development cost		12,764,539
Net development value		15,340,839
Developer's profit		2,576,300

Example 3.6 Discounted cashflow approach

Months	Cashflow (£$000)	Interest until completion at 0.68%	Total (£$000)
1	3,601	0.9932	3,577
2	9.5	0.9865	9
3	0	0.9799	0
4	−556.5	0.9733	−542
5	−1.5	0.9667	−1
6	90	0.9602	86
7	223	0.9537	213
8	353.5	0.9472	335
9	365	0.9408	343
10	578	0.9345	540
11	710.5	0.9282	659
12	660	0.9219	608
13	611.5	0.9157	560
14	753.5	0.9095	685
15	731	0.9033	660
16	863.5	0.8972	775
17	551.5	0.8912	491
18	395.5	0.8852	350
19	−83	0.8792	−73
20	−17	0.8732	−15
21	−12.5	0.8673	−11
22	369	0.8615	318
23	251.5	0.8557	215
24	513	0.8499	436
25	−6	0.8442	−5
26	0	0.8384	0
27	−69	0.8328	−57
28	0	0.8272	0
29	0	0.8216	0
30	230	0.8160	188
31	−15341	0.8105	−12,434
Net present value (profit)			−2,088
Net present value with interest @ 0.068% for 31 months =			2,576,300

The main advantage of this approach to the developer is that it allows a subsequent calculation of the 'internal rate of return' (IRR), which is the measure used by some developers to assess the profitability of a scheme since IRR considers both the timing of the cashflows and the magnitude of each cashflow. This is as opposed to examining only a percentage return on cost (without consideration to the actual timing of the cashflows) or the present value of the profit (which doesn't fully consider the initial financial outlay and the degree of risk the developer is exposed to). Therefore, the DCF method is more likely to be used by investors who wish to retain the development in their portfolio and also seek to analyse the return on their investment. In order to calculate the IRR, the discount rate is varied by trial and error to the rate which will discount all the future costs and income back to a present value of zero. In other words, this is the percentage return when the project does not make or lose any money from the initial outlay. The IRR is also ideal for comparing different potential property developments with their own variations in the timing and size of the cashflows, e.g. comparing a small residential development with a large multi-storey office building. However, the disadvantages of this method are that the DCF method does not show the outstanding debt at a particular time and the profit in today's value rather than the actual sum that will be received at the end of the development.

3.3 The role of uncertainty and risk

Although the relatively simple cashflow method allows a somewhat accurate and explicit form of calculation, it still relies upon a set of fixed variables. The variables that are important components in the calculations, such as building cost and final rent (or sale price), are presented as 'best estimates' without giving a true impression of the range or variance from which they have been selected. If we look more closely at the basic example of a conventional evaluation set out in Example 3.1 then we can see the model is based on a considerable number of variable factors, as follows:

1 land costs
2 rental value
3 square footage (or metres) of building
4 investment yield
5 building cost
6 professional fees
7 time including pre-building contract, building and letting/sale periods
8 short-term rates of interest
9 real estate agents' fees
10 promotion costs
11 other development costs.

In this example the land purchase price is fixed and so these eleven variables can be reduced to the four main groupings listed below which will mainly affect the overall profitability of a development project:

1 short-term rates of interest
2 building cost

3 final rental value or sale value
4 investment yield.

It is important that the financial information input into the cashflow model is as reliable as possible. The level of reliability depends on the developer's experience and assumptions behind sources of information the developer uses. Recent developments completed by a property developer are often a good starting point, although allowances must be made for changes in supply levels and cost that have occurred over time since the development was completed.

Property developers also usually rely to a large extent on the professional advice of the development team to estimate the cost of the main variable groupings outlined above. For example the quantity surveyor and project manager have the expertise to advise on building costs and related costs. The agent will advise on current market rental value and investment yield, and hence the likely development value. However, in the end, developers must form their own judgement about the estimates of each variable factor. They have to assess the likely risk of the main variables changing when deciding on the required level of return in the evaluation process. Thus it is the property developer who has the largest exposure to risk. It is important that the developer uses current and up-to-date rental values and accurate building costs to reflect income and expenses in every development appraisal. While it is possible to value the present or historical value of an asset, future changes in property and rental values will always remain uncertain due to the complexity of the property market and the interaction of many variables and influencing factors. Accordingly it would not be advisable for a developer to predict future rental values, even when building costs in the appraisal are inflated at current inflation rates, since this would expose the developer to a higher level of risk. It cannot always be assumed that increases in building costs during the period of a development, which is a standard practice approach due to rises in inflation levels, will be equally met by increases in the final value of the property.

The rental income, investment yield, and building costs are usually the most sensitive variables and commonly are subject to external fluctuations outside the control of the property developer. In order to fix the level of rent for the development upon completion, the developer may be able to secure a pre-let commitment with a tenant. Alternatively, to fix the investment yield if the property development is to be sold on completion, it may be possible to pre-fund the scheme with an appropriate institutional investor. Either one or both of these options may be achieved before or during the period of the development project. With this approach, rather than the developer placing the emphasis on the financial risk with the property development, they can now focus on the project management aspects such as ensuring the project is built within budget and on time. Much of this will depend upon the quality of project management, but in some cases a fixed price building contract may be secured. The downside to adopting either one or both approaches is that by reducing or effectively sharing the risk, the developer must also expect to share the profit, thus limiting their potential reward.

An understanding of the complexities of risk is essential for a successful developer. Risk is embedded throughout the property market and is the starting point for every analysis involving property. The two major types of risk that affect a property are broadly referred to as systematic (i.e. market) risk or unsystematic (i.e. property-specific) risk. Importantly, a developer should never underestimate the effect of risk and the level of risk in every development scheme should be identified and, if possible, contained or reduced. It is important to remember that as the development process progresses, the developer's commitment increases and the possibility of variation decreases, and these both equate to a higher degree of uncertainty and associated risk.

Land cost

As discussed in Chapter 2, the purchase price of the land (either vacant or partially improved with an existing old structure) is usually the first major financial commitment. In order to reduce exposure to risk, a site should not be purchased until the appropriate planning permission has been obtained and the detailed building cost confirmed. If this is not possible, the developer should try to negotiate a conditional contract which is subject to the obtaining of a satisfactory planning consent and this is standard procedure for many property sales. If the outcome of the planning application is uncertain at the date of agreement, then it may be possible to negotiate an option to purchase the land by a future date once planning permission has been obtained. Alternatively, a joint venture arrangement might be entered into with the landowner whereby the land value plus any accumulated additional 'notional' interest might be calculated at a future date during the development period.

Once the land is purchased the developer is fully committed to a particular location which cannot be changed, which in turn has a major influence on the highest and best use of the land. In other words, the value of the land and any new property development scheme constructed upon it might be affected by external physical factors such a new road or rail network. Depending on market conditions, the developer may be able to make a profit by simply selling the land prior to the commencement of the development scheme. Once planning consent has been obtained the value of the scheme can be established, although further applications may be made to improve the value of the site. Planning applications take time and any improvement in value that might be obtained needs to be balanced against the costs of holding the site over an extended period of time.

Building cost

The building construction cost is the second major financial commitment or capital outlay in combination with a number of other costs (e.g. professional fees) relating directly to the final sum. After signing the building contract the developer is committed to certain construction costs which invariably will move upwards (and rarely downwards), partly due to the effect of inflation over the construction time period. In addition, many of the cost increases incurred during the development period result from the developer's variations or late production of information by the professionals responsible for design. These are matters over which the developer must exercise tight control. There are ways of making the building cost more certain by passing all or some of the risk and design responsibility onto the building contractor, although greater certainty of cost usually means a higher building cost. These are described in greater detail in Chapter 7.

Project management skills are of vital importance in preventing increases in building cost and time delays, as well as decreasing the risk that the builder will pay penalty rates for a late handover (i.e. substantially longer than the agreed contract date). Therefore, in many instances the employment of an experienced project manager is strongly advisable. Furthermore, it is important that the developer and/or project manager constantly monitor every aspect of the building contract in order to contain any problems as they arise.

Rental value

It is essential to obtain the most reliable up-to-date estimate of rental value. Due to the relatively large space and therefore lettable area of some property developments, any errors in the rental estimates on a rate per m^2 basis can have an adverse effect on the final estimated aggregate income. A reliable estimate of rental value must be undertaken via a thorough analysis of the prevailing market as well as in consultation with the letting real estate agents, if they are to be appointed. However, the level of uncertainty associated with achieving an estimated level of rent can be removed if a pre-letting by tenants can be achieved. Due to nature of the property market, when the development is actually completed there may be an over-supply and the property may be difficult to let. Some developers might not proceed with a particular development until a pre-letting is achieved to reduce a considerable element of the risk involved. For example, a developer of an office business park scheme or a large industrial scheme may initially provide all of the necessary infrastructure and landscaping and then build each additional element as required on a pre-let or a 'design and build' basis. At times a property developer may build one or two speculative units to show potential occupiers the type of building that could be provided and adapted to suit their individual requirements. Quite often a developer of major shopping schemes needs to secure the major tenants (i.e. 'anchor' tenants) for larger units at an early stage in order to then attract other retailers to invest in the smaller 'unit' shops. Financiers and lenders are often reluctant to commit to lending money unless there has been a substantial level of pre-let or pre-commitment in order to reduce leasing risk.

The developer needs to evaluate the benefits of achieving a pre-letting, and thereby reducing risk, against the opportunity costs of achieving a potentially higher profit in a rising market. For example in the time it takes to complete the development scheme the level of market rents might rise, or alternatively there may be more tenant demand when the development is nearing completion and a prospective lessee can actually visualise the lettable area, rather than just looking at architectural drawings. In the case of anchor tenants in a shopping scheme, the developer may have to pay what is called a 'reverse premium' to the retailer to secure a pre-letting and ensure the overall property development will receive approval from the financier. Clearly the cost of such a premium must be accounted for in the development appraisal. An additional advantage of securing a pre-letting is to reduce the overall development timetable before any income is received, as the building will be handed over on completion without the risk and uncertainty of a void/vacancy period and unknown additional interest payments.

Short-term interest rates

Unless the property development scheme is being entirely financed by the developer, funding arrangements need to be in place before any major commitment is made. In obtaining the essential finance to acquire the land and build the scheme, the developer will be exposed to any fluctuations in short-term interest rates. However, at a cost, the developer has the option to either fix or restrict the level of interest rates (see Chapter 4). If the developer agrees to an agreement with a lender based on a forward-funding of the property development scheme, then the interest rate agreed with them may be fixed.

Investment yield

Investment yields are dictated by decisions of stakeholders in the property investment market, being the relationship between the total value of the completed property developments (including improvements) and the total annual rent received. This relationship may vary at any particular point in time according to market factors such as the supply of competing developments, investor demand/sentiment and rental growth. However, the uncertainty of the yield changing over the period of the development can be removed if the scheme is pre-sold or pre-funded. If a development scheme is pre-sold to an owner-occupier then the developer is actually performing the role of project manager. With a pre-funding the developer secures both short- and long-term finance by agreeing to sell the completed and let development scheme to the financial institution/lender. Although the developer still bears the risk of securing an acceptable tenant on satisfactory terms and controlling building costs, as with pre-letting, the terms negotiated prior to the commencement of the scheme are likely to be less favourable to the developer than those that can be negotiated at the end of the project. The developer's chances of securing pre-funding significantly improve if a pre-letting is in place.

When a developer is evaluating how to reduce their exposure to risk, a balance needs to be struck between profit and certainty which is referred to as the risk/return ratio. In general terms the greater the certainty then the lower the potential profit. The level of risk a developer is prepared to accept will depend largely on their motivation at that point in time. Occupiers, contractors, financial investors and the public sector involved in the property development process will each be seeking to reduce risk as much as possible, although property development companies typically will often be willing to accept a much greater degree of risk in return for higher rewards. The degree of risk is usually directly related to the complexity and scale of the proposed development. For example, at one extreme a small self-contained office block pre-let to a major corporation represents a very limited degree of exposure to risk by the developer; at the other extreme a substantial degree of risk is involved in assembling, over a long period of time, a large town centre site suitable for a comprehensive mix of uses including shops, offices and residential. Where a high degree of exposure to risk is perceived, it is usual for developers to seek development partners in order to share both the risks and the rewards.

Uncertainty is unavoidable in the process of appraising development opportunities and substantial attention needs to be given to pre-project evaluation to identify and evaluate the optimal balance between risk and reward. The fewer assumptions that can be relied upon will directly reduce the overall risk associated with that development scheme, which in turn will increase the likelihood of a completing a successful project. The cost of a detailed evaluation and the additional time required usually leads to greater savings of cost and time later on in the project. However, the additional time available to the developer at the pre-project evaluation stage is usually very restricted, especially in a competitive tender situation. In this situation, the developer's judgement and expertise is critical.

Establishing the economic viability of a proposed development scheme and the particular characteristics of the marketplace, prior to committing to the major financial burdens associated with land and building costs, is essential. Only when this evaluation has been prepared and closely examined by the development team, should a decision be made as to whether or not it is prudent to purchase or lease a particular site and, if so, under what terms and conditions. Often a financial component of a property development, such as the initial land purchase price, may be higher than initial market expectations. Any additional money outlaid for the land purchase must be deducted directly from the developer's profit, which may result in the project not being viable. It is essential that individual variables in the assessment are accurate and that each variable reflects current market value, as the income components will be based on current market value.

Discussion point

- Why are uncertainly and risk major considerations when using the cashflow method?

Sensitivity analysis

A critical question raised in the evaluation process relates to how a developer measures the level of uncertainty involved in a particular property development scheme and therefore how much profit is required to balance the resultant risk. Developers are often criticised for not sufficiently understanding and analysing the level of risk. This is a valid criticism as property developers can underestimate the level of risk and a project may not reach completion due to unforeseen problems. On the other hand a developer cannot afford to be too conservative as they may never be successful in securing a site. A careful balance has to be struck which relies entirely on the developer's judgement and experience. It is important to identify and examine possible methods of analysis available to assist the developer to examine the level of risk.

Once the land price is known and fixed, the main variables of the evaluation can be identified as being short-term rates of interest, building cost, rental value and investment yield. In most cases, the financial outcome of the development is more sensitive to their variability than to the variability of the other factors previously mentioned because they are the highest proportional values/costs in the evaluation. For example, a 10% increase in the interest rate is likely to have a more significant overall impact on profitability than a similar increase in building costs. The name given to the procedure for testing the effect of variability is 'sensitivity analysis' and, given the nature of the property market with many variables constantly in a stage of

Table 3.4 Sensitivity analysis and effect on adjusted developer's profit

Variable (original value)	Original value less 10%	Original value plus 10%
Land price (2,900,000)	23.71%	16.84%
Interest rate (8.25%)	21.83%	18.55%
Building costs (£$1,421 per m^2)	28.52%	12.85%
Rent (£$322.92 per m^2)	8.52%	31.76%
Gross area and net lettable area (4,599 m^2 / 49,500 ft^2 and 3,908 m^2 / 42,065 ft^2)	16.08%	23.73%
Professional fees (12.5%)	21.05%	19.32%
Capitalisation rate (8%)	31.10%	9.43%
Funding fees	20.41%	19.95%
Letting and sale costs	20.67%	19.70%
Developers sales fees (1.5%)	20.30%	19.94%

change, the assessment of a potential project must acknowledge this risk and have an in-built capacity to adapt to suit. Accordingly, one or more factors in the evaluation or appraisal model can be varied and the effect on viability measured and recorded. The procedure can then be repeated and the different results compared. If, for example, we take the appraisal model set out in Example 3.1, we can carry out the sensitivity analysis shown in Table 3.4.

This analysis shows that the outcome of the appraisal model is most sensitive to changes in investment yield, rent and building cost. If, for example purposes, it is assumed a change in investment yield is unlikely over the total development time period it is possible to concentrate on the effect of possible variations in rent and construction costs. Suppose that the range of possibilities that the development team consider appropriate is a range of rents of £$290–350 per m^2 per annum and a range of building costs of £$1,200–1,700 per m^2. At this stage the discussion is on 'possibilities' and not 'probabilities' where the range of each is likely to be rather wide. The matrix in Table 3.5 shows the level of developer's profit expressed as a percentage (i.e. profit as a percentage of total development value).

The total range of possible outcomes for the developer's profit is extremely wide and varies between −3.99% to +44.45%. The next step for the developer is to narrow the focus of attention by concentrating on the most likely or probable outcomes. For example as a result of discussion among the development team, the outer limits of the ranges of rent and building cost are excluded as being possible but unlikely. The developer can now concentrate on a narrower range of outcomes in Table 3.5.

Although this attempts to narrow down the focus to what is more probable, the range of possible outcomes still remains wide. A developer's profit that ranges from +11.49% to +14.65% gives an indication of the real uncertainty that lies behind the appraisal model. The developer must now try to weigh up the possible outcomes, assigning either objectively or subjectively some level of probability to each estimate of rent and building cost. In the end, the original 'best estimate' of £$322.92 per m^2 per annum rent and £$1,421 per m^2 building cost may be selected, but the context of

Table 3.5 Level of developer's profit expressed as a percentage (i.e. profit as a percentage of total development value)

| | | Rent (£$) per square metre | | | | | | |
		290	300	310	322.92	330	340	350
Building cost (£$) per square metre	1,200	20.50%	24.52%	28.52%	33.68%	36.50%	40.48%	44.45%
	1,300	14.65%	18.48%	22.29%	27.21%	29.90%	33.70%	37.48%
	1,400	9.34%	13.00%	16.64%	21.34%	23.91%	27.54%	31.15%
	1,421	8.29%	11.91%	15.52%	20.18%	22.73%	26.32%	29.90%
	1,500	4.50%	8.00%	11.49%	15.99%	18.45%	21.92%	25.38%
	1,600	0.08%	3.43%	6.78%	11.09%	13.45%	16.78%	20.10%
	1,700	−3.99%	−0.77%	2.44%	6.59%	8.85%	12.05%	15.24%

possibility and uncertainty in which it lies can now be better understood. On the other hand, an attempt may be made to fix one of the variables in one of the ways discussed above. For example, if a pre-letting contract is agreed at £322.92 per m² per annum this will narrow the range of likely outcomes to between +15.99% and +27.21%. On these figures, a maximum profit of +29.90% has fallen to +27.21%, but as a trade-off the minimum level of profit has risen from +11.49% to +15.99% and the degree of uncertainty has been reduced. It is just this kind of trade-off that is made possible by sensitivity analysis and by the understanding of probabilities, particularly when they are matched to the use of cashflow appraisal models.

This is a brief introduction to the concept of sensitivity analysis based on relatively simple examples. In carrying out sensitivity analysis at the initial evaluation stage, the developer is weighing up the balance between risk and reward. The level of uncertainty in the project is therefore a most important factor. Uncertainty can be reduced by fixing any of the four identified variables. Conventional methods of evaluation do not provide any indication of the uncertainty that is an inherent part of the development process. While cashflow methods of appraisal overcome the inaccuracies of the conventional approach, they still only represent a 'snapshot' of the viability of the scheme. Sensitivity analysis is a tool which the developer can utilise in the decision-making process to provide a measurement of the risk of the development scheme. It forces the developer to be more specific about the assumptions and estimates made. It assists but does not replace a balanced and informed decision-making process.

However, there is a danger in relying too heavily on the figures produced in the financial evaluation of a scheme. A developer must avoid the danger of using the evaluation process to justify a development project which may looks good, often referred to as a 'gut' feeling. Although the evaluation modelling process must be thorough and based on the best possible information, it should be approached from the point of view of downside risk, i.e. what can go wrong, may go wrong. Even if the figures indicate a viable scheme the developer should always research the market for other proposed competitive developments in the particular location (see Chapter 8).

3.4 Reflective summary

This chapter examined conventional methods of evaluation used by developers to assess the profitability of a development scheme, or the land price which can be afforded, given a required return, for any particular site. It has been acknowledged that some aspects of the conventional method of evaluation are rather basic and inaccurate. The inaccuracy in the calculation of interest costs can be overcome by using any of the cashflow modelling techniques including the net terminal approach and the DCF. All methods discussed only produce a residual figure based on best estimates at the date of the evaluation, which hides the true level of uncertainty regarding the final outcome of the completed development. Sensitivity analysis and comprehensive modelling of underlying market conditions can improve the developer's understanding and exposure to uncertainty and risk.

Chapter 4

Development finance

4.1 Introduction

The majority of property developments can only be undertaken with the assistance of funding from an external third party source as a developer does not normally have 100% of the cash required for all development costs incurred during the lifetime of the development. This third party is normally a lender/financier, institution or syndicate which has funds to lend in return for an additional interest cost on the capital plus administration costs. This loan fills the financial difference between the developer's available equity, or cash equivalent, and the total cost of the project including all associated expenses over the development period until completion.

There are two forms of finance that are required for property development:

- *short-term finance* to pay for the initial development costs (i.e. purchase of land, construction costs, professional fees and promotion costs), and
- *long-term finance* to enable developers to repay their short-term borrowing/ loan and either realise their profit via selling or retain the property as an investment with tenants.

Either option depends on the developer's motivation, their financial situation and the prevailing market conditions. In this chapter we examine the different sources of finance available to developers and illustrate the methods of financing a development scheme by using worked examples.

4.2 Sources of finance

Most developments are funded by a combination of equity (from the developer) and finance (from the lender). In these instances the lending institution has exposed their funds to part of the risk associated with the development, and at the same time it also charges the developer an interest and service charge designed to be commensurate with the level of risk exposure, inflation and an allowance for profit. As with other service providers there is a diverse range of financial lenders who themselves are of varying size and expertise. Each financial lender will have its own acceptable level for exposure to risk and may specialise in certain lending projects over varying timeframes. A developer must consider his choice of financier carefully to ensure the type of funding is best suited to an individual project. For example after the global financial crisis the lending restrictions were further tightened due to the increased number of bad debts. Therefore, borrowed funds were harder to obtain, were more costly and also required a larger proportion of equity from the developer.

In many countries the clearing and merchant banks have been providers of short-term development finance with long-term investment finance being provided by the financial institutions (insurance companies and pension/superannuation funds) and property investment companies. Financial institutions also take on the role of short-term financier by forward-funding development schemes, such as where they provide the necessary interim development finance to a developer and agree to purchase the property on completion of the scheme. More recently, real estate investment trusts (REITs) have been an additional and alternative source of funding for property. To provide a starting point to understanding development finance we will now briefly review the history of property financing.

4.2.1 Historical perspective

The role of the development financier varies depending on factors such as the position of both the business and property development cycle at any particular time in relation to the credit cycle. It is important to appreciate that financiers are in the business of making money, and real estate is only one of a number of assets they can invest in and also lend money against, with other competing assets including equities/shares and cash. However, real estate generally offers the financier a relatively secure form of investment since the financier can hold a binding mortgage or first claim over the land component including any improvements affixed to the land. The security is linked to restrictions placed on the property title, where the property owner is unable to transfer or sell the land without first clearing the mortgage on the title. In other words the first step a prospective purchaser does is to undertake a search and ensure the property's ownership title is clear and mortgage-free which ensures a first mortgage has a priority claim to be repaid, ever before the existing property owner.

Each of the various financier groups will have different motivations and liabilities, influencing their policy towards property as either an investment or as a security against a loan. Developers in most countries have the ability to move from one financial source to another, depending on the investor/lender's attitude and lending policy at any particular time. Many decades ago the roles of the short-term financiers (i.e. predominantly banks) and the long-term financiers (mainly insurance companies then) were quite separate and developers usually retained their completed developments as long-term property investments. Short-term finance was typically provided by the clearing or merchant banks in the form of loans secured against the site and sometimes the buildings. Long-term funding, often pre-arranged, was generally provided by insurance companies by way of fixed-interest mortgages. On some occasions, the development would not be retained by the developer but sold as an investment to an insurance company or directly to an occupier. With the exception of some merchant banks, it was unusual for the financiers to participate in the profit or risk of the development.

As inflation became a permanent feature of the economy in many western countries, the insurance companies identified the disadvantages of granting fixed-interest mortgages and sought to benefit from the rental growth. At the same time, long-term interest rates rose and developers were faced with an initial shortfall of income over mortgage interest and capital repayments, often referred to as the 'reverse yield gap', which has remained an almost permanent feature of property financing. Therefore, insurance companies were less inclined to grant mortgages which forced property developers to give away some share of future rental growth in order to close the 'gap' and insurance companies became more directly involved with the direct ownership of property. Subsequently an increasingly active property investment market emerged and the traditional division of the roles began to blur.

Initially, to attract the best investments, long-term investors competed with and took on the additional role of the short-term financiers. Simultaneously, some of the traditionally short-term financiers, such as the clearing banks and the merchant banks, began to seek a share in the equity of the development itself. As the competition for the best (i.e. 'prime') investments increased some of the insurance companies and pension funds – either on a project basis or by the acquisition of property companies – began to take on the role of the developer, accepting the additional element of risk in return for a marginally better long-term yield. The funding of developments on a long-term basis became dependent on the property satisfying the criteria of investors and developers initially had a much wider choice of financial sources.

Many property markets are generally cyclical in nature, also being increasingly globally interconnected partly due to technological advances and the instant availability of information about other markets including financial details. For example by 2005 property development was booming, allowing significant regeneration in many cities around the world. However, the start of the global financial crisis in 2007 threatened the collapse of large financial institutions and resulted in government bailouts of banks and a downturn in equity markets around the world. Many key businesses failed and there was a decline in consumer wealth (estimated in trillions of US dollars). This led to a downturn in economic activity contributing to the 'global financial crisis' (GFC) in 2007.

Since the start of 2008 and the subsequent downturn in the cycle, funding property development has become more challenging because banks and other financial institutions have been unwilling to lend money. The lack of development finance, among other factors (e.g. the fall in property values) brought much of the proposed development to an abrupt halt and only those developments already in progress or with funding arranged continued. The reluctance on the part of investors to fund schemes continues to prevent large-scale development and governments may also be less willing to invest in infrastructure to support new developments. Some developers will also face higher risk premiums and greater constraints on proposed schemes. In the UK, public sector spending has been cut which means that developers can no longer expect financial support from the public sector, forcing developers to look elsewhere for development finance.

Despite institutional lenders' adversity to risk and reluctance to fund new development, the 21st century saw a variety of new financiers entering the marketplace who were keen to increase their market share, coupled with the availability of immediate up-to-date information from the internet. Now there is an avalanche of information available about different funding sources, as well as third party organisations that review the attributes of each option. While the global information age may offer an enormous selection of financiers who are not necessarily even based in the same country, it is important for the borrower to be fully aware of the conditions attached to any loan. For example, fee structures can vary considerably between financiers and the developer should pay attention to the detail in the loan documents.

The increased competition has also opened up the lending market to a myriad of new products, which in turn have associated lending fee structures and loan lengths. While the borrower has benefited from the wider choice and availability of financial products, such as the newly introduced REITs, it still remains fundamentally critical for the developer to spend adequate time reviewing the risk profile and suitability of the financier. It is anticipated that the rate of expansion in the lending market will slow and consolidation will occur, especially if there is an eventual downturn in the property market.

Over time there have been variations in different sources of finance for real estate developments and it is worthwhile to reflect on historical trends. For example in many instances the traditional banking sector was the dominant source of finance during the late 1980s development boom which was further encouraged by the rapid increase in rents and capital values caused by occupier demand. This was during a period (the mid-1980s) when institutions reduced their lending for property investment in exchange for the better performance of other forms of investment such as equities (stocks and shares) compared with the poor performance of property in the early 1980s. At the same time many real estate developers preferred to obtain short-term 'debt' finance from the banks to enable them to sell their completed development into a rising market. Alternatively some real estate developers were able to secure medium-term loans or refinance initial short-term loans to enable them to retain their developments as investments. It is commonplace, in accordance to economic supply and demand principles, for new financiers to enter the commercial property market including foreign banks, foreign investors and to a lesser extent building societies.

With the regular cyclical downturns in the property market, reference is usually made to the combination of the economic recession, high interest rates and an oversupply of new buildings. There was a similar scenario in the global financial crisis. In such scenarios the banks were exposed to excessive 'bad' property debt and many development companies went into receivership or concentrated on their levels of debt. In turn this has had the effect of driving yields down, although Figure 4.1 emphasises the relatively stable level of yields in the next phase of the cycle. This stability is due to numerous factors including increased competition for and lack of availability of 'prime' stock, a tighter monetary policy and enhanced research information available to lessen investment risk.

Discussion point

• What is the effect of a market upturn or downturn on the amount of finance available to real estate developers?

4.2.2 Financial institutions

A financial institution or financier are general terms used in the property and real estate industry to describe pension or superannuation funds, insurance companies, life assurance companies, investment trusts and unit trusts. They invest in property directly and indirectly through the ownership of shares in property investment companies and property development companies, which differ from investment in REITs that are traded on the stock/share market. Direct property investment includes the owning of completed and tenanted/let developments, the forward-funding of development schemes and the direct development of sites and existing properties.

Pension funds vary considerably in size and include the individual occupational funds managed exclusively for the employees of former/present nationalised industries and large publicly quoted companies, together with company and personal pension schemes managed by insurance and life assurance companies. They invest the premiums paid by the clients to achieve income and capital growth in real terms in order to meet the future payment obligations of pensioners on retirement – in many ways real estate is ideally suited as it is a secure and long-term investment. Many pension funds and schemes are under pressure due to the approaching maturity of their schemes (i.e. the ratio of expenditure on pensioners to income received from premiums is increasing), which has placed the emphasis on cashflow and real income growth. In addition, many pension schemes are linked to the final salaries of employees which add to this pressure as incomes have generally continued to rise above the rate of inflation. Listed in the Appendix is the level of CPI by individual country between 2004 and 2012. This table highlights the variations in CPI between each country which ranged in 2012 from a high of 59.22% (Belarus) down to −0.94% (Georgia). Reference should also be made to the volatility over this period for each country and the effect this has on overall investment sentiment. Life assurance companies and insurance companies invest the premiums they receive on life and general insurance policies, respectively, to ensure long-term income growth to meet payment obligations when they occur.

Unit trusts are typically managed by financial institutions who offer investment management services, e.g. merchant banks. The unit trust will comprise of unit-holders, such as small investors and institutions who are unable to take on the risk of direct investment. The trust will manage a portfolio of shares on behalf of the unit-holders in order to benefit from diversification and to obtain a reasonable spread of risk. There are two types of unit trust which specifically invest in property:

- authorised property unit trusts – unit trusts which are strictly regulated to ensure they invest in a diversified portfolio of low-risk prime income producing property as part of a balanced portfolio, as their investors include private individuals; and
- unauthorised unit trusts – unit trusts that are unregulated, investing directly in a portfolio of properties and are attractive to tax-exempt financial institutions such as charities and small pension funds.

The underlying primary goal of these financial institutions is to maximise returns to their shareholders at the same time as minimising exposure to risk and adopting a conservative approach with every investment, especially since they are trustees of other people's money and are therefore under constant pressure to perform. In other words, the emphasis is on providing a return commensurate with a relatively low degree of risk. They invest in property as an alternative to other forms of investment, such as stocks and shares (equities), together with bonds and gilts (fixed-interest income). The extent to which a financial institution or lender will invest in property largely depends on the size of the fund and the nature of liabilities in their overall portfolio. It will vary according to the state of the economy and the performance for property investments relative to other investments. Even though they are long-term investors they take a short-term view of performance, being strongly influenced by the recent performance of each type of asset, although they do forecast future trends.

It is essential to examine the advantages and disadvantages of investing in property from a financial institution's point of view. It is important to identify and appreciate factors that influence the investment decisions of financial institutions as this affects the funding and sale of completed development schemes.

Hedge against inflation

One of the main reasons why the institutions first entered the property investment market was the fact that property represented a 'hedge against inflation'. In other words rental growth outstripped inflation and therefore represented an opportunity to achieve income gains in real terms. During periods where the consumer price index is relatively volatile, price levels for goods and services grow rapidly and the direct costs associated with labour and materials for constructing a new building also increased over the same period, which in turn adds to the total cost of a new property development. As the level of the interest rate is linked to the prevailing inflation rate at a given point in time, this also adds further costs to the project. However, in the last couple of decades and with the arrival of global technology (i.e. the internet) the level of inflation has been kept relatively under control, being consistently below 5% in many western countries. This is due to a number of reasons explained previously, but the tighter rein on inflation was also an indication that governments recognised that sharp increases (i.e. volatility) could have an adverse effect on the economy. Accordingly a property developer in a low inflation environment is more likely to have less volatility in future building costs as well as lower interest rates.

Institutional lease

A lease is a binding agreement enforceable by law where the property owner is guaranteed a future income at an agreed rate, usually with the ability to change the rent to reflect current market conditions. Furthermore one of the principal advantages of real estate, as opposed to receiving cashflow from an alternative investment, is the existence of a long-term lease, usually with upwards-only rent reviews, guaranteeing annual returns with increments. However, this assumes that the tenant continues to pay the rent. The legal doctrine of 'privity of contract' ensures that in the event of a default by any tenant (unless it is the original tenant) the landlord will normally be able to require the original tenant, followed by any assignees in turn, to pay the outstanding rent. However, some countries have taken steps to alleviate the heavy and somewhat unfair financial burden placed on the original tenant.

In many regions a tenant may assign their interest in the lease to another party, provided that they have the prior approval of the landlord, which cannot be unreasonably withheld. Usually the test for such approval is the financial standing (covenant) of the tenant and typically institutions require the potential assignee to demonstrate that their last three years' trading profits exceed three times the rent payable. Accordingly, an investor in property is guaranteed a secure stream of income over the long term with very limited risk of voids due to tenant default.

Another reason why property has traditionally formed part of an institution's investment portfolio is diversity of performance risk. If you examine total returns from property compared with equities and gilts it shows that property investment is not prone to short-term fluctuations. Investors are fully aware of the returns from each investment category, as well as the corresponding levels of risk.

Illiquidity and indivisibility

Compared with other investment options there are always other factors to be considered when choosing to invest in property, the most significant being the illiquidity of property. A property represents a large single investment in financial terms and it cannot be sold quickly (if required) in response to market trends, unlike equities or stocks and shares. It is common for the sale of a property to take months and it may not be possible to sell it at all if the market conditions are not favourable. In addition, selling a property usually involves high transaction costs such as an agent's fees (often in the form of commission or a percentage of the total property), solicitors' fees and associated government charges/taxes.

Another significant barrier in the real estate market which is linked to illiquidity is the indivisibility or 'lumpiness' of a property. Many lots, especially small allotments, are unable to be divided into smaller portions and sold individually. Therefore, it is usually relatively impossible to sell off a small proportion of the property to increase cash flow, especially if the property development has not yet been completed. This factor alone reduces the involvement of the smaller funds in the property investment market directly. The indivisibility and illiquidity of property has become a major challenge for stakeholders in the industry who are trying to improve the attractiveness of property as an alternative investment to equities which are both singular and divisible. Therefore, direct property investment of high value real estate, such as a multi-storey office building, is usually limited to large pension funds or syndicates because of their illiquid and indivisible inherent characteristics.

No centralised marketplace

In contrast to equities/stocks/shares which are traded on a centralised stock market where all buyers and sellers meet in one central location to carry out their transactions, there is no common meeting place for buyers and sellers to transfer property. Buyers and sellers generally meet via an agent and agree to transfer ownership privately. Further complications then arise as there is limited knowledge about the current state of the market, and also what the volume of trading is or the current level of prices being achieved. In addition, the lack of a centralised marketplace also places pressure on the seller to pay high marketing expenses, to advertise their property and ensure it reaches the attention of prospective buyers.

Management

Property investment is relatively labour intensive as it also involves a high degree of management expertise measured in both time and cost. Advances in technology, access to information via the internet and specialised computer programs have greatly enhanced the role of the property manager, especially with regard to monitoring payments and keeping up to date with the current market rent or capital values of comparable properties. Although most management costs can be recovered from tenants under the terms of leases, they are still perceived as a disadvantage to the investor. It is generally considered that active management of a property asset will improve the overall level of return received, at least in relation to reducing vacancies and minimising expense costs.

Research and performance measurement

Research into property markets has grown rapidly in the 21st century, mainly because this specialised information is highly sought after by developers, investors, tenants and stakeholders. One of the main differences between property and other assets is the way in which returns are measured, making it difficult to compare property with other alternative assets (e.g. equities) on a 'like for like' or 'direct comparison' basis. Two primary indicators of the current state of the property market are regular returns in the form of rents and yields. The yield of a property investment is the relationship between the total value and the rent, which is generally defined as the annual rental income received from a property expressed as a percentage of its purchase price or capital value. In many ways the yield is a measure of the property investor's perception of the future rental growth and capital growth against future risk, management expenses and illiquidity. Figure 4.1 and Table 4.1 shows the returns for different land uses on an annual basis over a 30-year period with a clear distinction between returns in retail, office, industrial, residential and other properties as well as a metric for all properties. Based on annualised returns over a 10-year period, it can be observed that both residential and other property had a higher return than all property; however, this information refers to the UK only as an example.

Prime yields are calculated by analysing market transactions in 'prime' properties – specific properties which conform to the following narrow criteria:

1 modern freehold or long leasehold property
2 good location and access to services and amenities (including transport)
3 highest quality and specification, and
4 fully let and income producing to tenants with good covenants.

It should be noted that prime properties usually represent only a very small proportion of the entire property investment market. The movement of 'prime' yields represents a benchmark against which the yields of all properties can be measured and compared. The movement of yields generally represents market sentiment about a specific type of property, where the better the perceived prospects for either capital

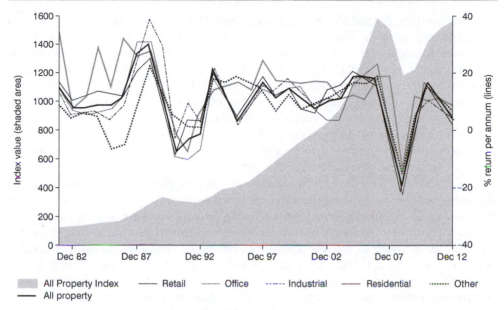

Figure 4.1 Returns by land use (UK) 1982–2012 (Source: IPD 2013)

Table 4.1 Returns by land use (UK) 1982–2012 (Source: IPD 2013)

	Total return index Dec 2011 (Dec 1980 = 100)	Total return index Dec 2012 (Dec 1980 = 100)	Total return (%) 1 year	Income return (%) 1 year	Capital growth (%) 1 year	Annual-ised total returns (%) 3 year	Annual-ised total returns (%) 5 year	Annual-ised total returns (%) 10 year
All property	1513.4	1565.1	3.4	5.8	−2.2	8.7	0.7	6.3
Retail	1873.4	1918.4	2.4	5.8	−3.2	8.4	0.6	6.3
Office	1182.6	1238.3	4.7	5.5	−0.7	9.7	0.7	6.1
Industrial	1991.7	2051.3	3.0	6.8	−3.6	7.0	0.1	5.8
Residential	5930.3	6459.2	8.9	2.4	6.4	10.1	4.9	9.5
Other	1261.3	1323	4.9	6.1	−1.2	10.0	3.4	8.4

or rental growth, the lower the yield (higher capital values relative to income). When making the actual decision about how much money to allocate to real estate in the overall portfolio, financial institutions will place close attention to the performance of both the property investment market and their current proportion of property in their portfolio. There are several established performance measurement indices which measure the performance of institutional property portfolios.

With reference to expectations about the levels of future performance, institutions will typically hold regular meetings to review forecasts of performance and reallocate funds to the different and alternative asset classes. However, investment in property is long term and not as flexible, and therefore does not sit easily with such a rapid review timetable. In addition the integration of property into a wider, multi-asset, context is generally made more difficult by the differences in terminology and valuation practice between property and other assets. This has partially been overcome by the increasing acceptance of accounting based valuation techniques, such as the discounted cashflow (DCF) approach in order to predict the investment's internal rate of return (IRR). Previously the value of property or real estate was viewed using a relatively static approach, which in turn may affect a given level of a fund's assets held in property. It has been argued that issues related to the different terminology and valuation practices may still result in there being lower exposures to property by some fund managers than otherwise would be the case, although this resistance has gradually changed over time and many investment firms now have a department focusing on real estate.

The integration of property into wider multi-asset investment policy is an accepted means of diversifying to reduce exposure to risk and a typical portfolio may include equities, cash and a substantial property holding due to the inherent stability of property as a long-term asset and a 'hedge against inflation'. This also permits a direct comparison between property and returns from other alternative asset classes.

When a decision has been made about the proportion of money to allocate to property investment, it is then the responsibility of the property fund manager to make the decision on what type of property to purchase and, importantly, in what location. Property investment policies are usually based on an analysis by property type and region focusing on recent performance and future forecasts of growth. More importantly, a major consideration is the different policies and investment criteria that each institution uses. However, they all tend to seek a balanced portfolio of property types in order to diversify investment and reduce risk, although the portfolio is usually weighted towards the property type which is performing well at a particular point in time.

Most financiers try to spread their investments geographically. For example a retail investor will often invest in shopping centres in different countries rather put all their 'eggs in one basket' in a single country. Most institutions tend to adopt very rigid selection criteria when making decisions on exactly which property investments to purchase. Many will only search for 'prime properties' which are in demand and accordingly priced at the upper end of the scale. They will look for the best located properties of the highest quality which are fully let on institutionally acceptable lease terms to tenants with good covenants. However, as properties falling within the definition of 'prime property' usually account for less than 10% of all properties at any one time, institutions may have to compromise on some of the following factors:

- location of the property
- quality of specification and design
- lease terms including length of lease and security, and
- tenant quality.

It is not always possible to obtain 'prime property' because either there is an under-supply or the asking ('for sale') price is too high (i.e. market yield is too low) relative to the perceived levels of potential future rental growth. Some investors may be willing to take a balanced view on a specific property by analysing it on its own merits and then adjusting the yield they are prepared to accept to reflect any additional risk. Other investors may give different weights to each of the above factors depending on their individual investment requirements. For example the fund investors who are concerned with the security of income rather than capital growth prospects will put a higher emphasis on the quality of the tenant covenant. The characteristics of a good location in relation to the various property types have already been examined in Chapter 2. Some institutions may be prepared to consider other options based on what may be perceived in the market as 'secondary locations', where there is an under-supply of quality stock although accompanied by strong prospects of future potential rental growth.

The risk associated with the lease needs to be fully evaluated. For example a higher level of flexibility given to a tenant will equate to added risk for the investor. However, many tenants have demanded flexibility in recent times, such as shorter lease lengths as opposed to the extremely long lease lengths (e.g. twenty years available in some countries) and are resisting traditional upwards-only rent reviews. In reality tenants would generally have to pay higher rents for such flexible terms and yields would rise to reflect the risk of fluctuating incomes.

As well as purchasing completed and partially/fully let developments as an investment, many institutions also carry out their own developments or provide development finance. Some institutions primarily restrict their development activity to the redevelopment of properties in their own portfolio. Involvement in development, whether directly or indirectly, will depend largely on a particular institution's attitude to risk and their perception of the development cycle at any one point in time. Once again research into the property market is a critical and essential task for a property developer and investor (see Chapter 8), especially when the core goal should always be to decrease unnecessary exposure to risk where possible. It should be noted that undertaking a proposed property development is generally a riskier proposition than buying an existing completed property as an investment. Building costs and land values are comparatively low when compared with the prices being currently sought for prime standing investments. In addition, undertaking a new development provides the institution with an opportunity to specifically tailor a property to fill a gap in the market for suitable property 'for sale' either now or in the near future. However, it is important to bear in mind that development only represents a small proportion of all institutional property investment held in their portfolio.

4.2.3 Banks and building societies

Banks participate in the funding of property developments due to the potential for growth in capital and rental values and the fact that property offers a relatively secure and low risk investment, especially when it is 'prime' property (i.e. limited supply, high level of demand and top end capital and rental values). Initially most bank loans were for short-term finance; however, encouraged by the property boom of the late 1980s and early 1990s, many banks became involved in longer medium-term loans. Due to exposure to bad debts in market downturns they became understandably cautious about investing in speculative developments and the trend is for most banks to restrict lending to high risk borrowers and reduce their overall level of bad debts.

After a market downturn, many property developments remain vacant or unsold for an extended period of time, or eventually sell at a loss (and are declared a bad debt) or there is a delay until the market picks up again and the property can be sold, less interest and holding costs.

After a downturn, property developers may need time to re-establish trust among lenders, who also conduct due diligence on a property before advancing funds. In addition, most banks now adopt a 'hands on' approach to understanding the property development industry and are assisted by their own in-house valuers and research teams. Also banks are less likely to lend on purely speculative development without the provision of a sound business case with comprehensive and industry-supported market data and reliable market forecasts. Since the recent global financial crisis many banks will only lend on low risk developments or where there is a significant pre-let (around 70%). If the developer is unable to reach the agreed proportion of pre-let or pre-sold, the development may be reconfigured to meet market demand (e.g. modify the quality or scale of development).

Obviously banks are in business to make a direct financial gain from lending money and lending to property companies has always been viewed as profitable, although at times subject to market downturns. Bank lending may take the form of 'corporate' lending to a company by means of overdraft facilities or short-term loans. Alternatively, a loan may be made to enable a specific development project to proceed or for a developer to retain a development as an investment. To reduce exposure to risk, the banks will use the development or the investment property and/or the assets of the company as security for loans. Property is attractive as security for banks as it is a large identifiable asset with a resale value, but importantly it cannot be sold unless it has a clear unencumbered title of ownership. Due to the cyclical nature of property markets and the interaction of supply and demand, there will always be periods where property values will increase and decrease. Accordingly, the banks' willingness to lend money to developers is determined by factors such as their confidence in the property market and the underlying economy at any particular time. Another consideration is a bank's exposure to property as a proportion of their overall portfolio at any given point in time.

In specific property markets it is important for the real estate developer to understand how the market operates and what the process is for obtaining reliable access to competitive funding. Based on a UK example it has been observed that clearing banks, due to their large deposit base, were historically the major providers of corporate finance loans to development and property companies. At the same time a limited amount of project finance was provided by the clearing banks on small schemes being developed by established customers. The merchant banks, with some being subsidiaries of clearing banks, and specialist property lenders have smaller funds but have more property expertise. Accordingly they are more inclined to provide project loans and because of their expertise will take on high risk loans in return for an equity stake in a project. Merchant banks on large development projects have in the past assumed the role of the 'lead' bank and assembled a syndicate of banks to provide finance, and in doing so may underwrite the loan. At times merchant banks also undertook investment management on behalf of institutional investors through investment funds and unit trusts. In this example it should be noted the respective roles of both the clearing banks and merchant banks may overlap, particularly if they are associated. Many foreign banks are represented in major global cities throughout the world and also operate in a similar way to clearing banks. There are also financial intermediaries who act as agents or financial advisers structuring development finance with banks and other sources for a developer in return for a fee, for example say 1% of the value of the loan sought. Several of the large global property and real estate firms offer different services to clients in the form of consultancies as well as having financial service arms.

Building societies in certain regions have been allowed to provide corporate loans and loans secured on commercial rented property; however, this is usually based on the condition that such loans comprise a relatively small proportion of the total loan. In the past some building societies have been exposed to risk due to the ups and downs in property cycles, particularly with residential developers, and as such have withdrawn from development funding. Other societies are still willing to fund commercial property investments. Typically they restrict themselves to smaller loans and tend to be less competitive than banks in relation to the interest rates they charge, although this varies between regions and different societies or banks. The lender's primary task is to ensure a return on their loan which compensates them for their exposure to risk. This risk evaluation will be based primarily on the possibility that the borrower/s will (a) not be able to meet agreed regular payments, (b) not be able to pay back any money whatsoever, and/or (c) the minimum amount the property could be sold if the borrower defaults on (a) or (b). The banks' criteria for making loans will vary depending on their own operating policy and a wide range of unique factors including the size, track record and history of the development company, the nature and size of the development, the total length of the loan, the amount of deposit paid up-front by the developer and the strength of the security being offered (often referred to as the loan-to-value ratio or LVR). In assessing the risk of corporate loans the bank will be concerned with the financial strength, property assets, track record, profits and cashflow of the development company. In relation to loans on specific developments the banks will also be concerned with the security of the development project. The banks need to be convinced that the property is well located, the developer has the ability to complete the project and the overall scheme is viable. The in-house team of property experts with additional external advisers, if necessary, will carry out an assessment and valuation of the project. In the case of medium-term loans, where the developer wishes to retain the completed property until the first rent review, or loans on investment properties, the banks will also be concerned as to whether the rental income covers the interest payments. Previously in periods of market upturns as part of the property cycle, the banks were prepared to provide loans where the rental income did not cover the interest payments (referred to as deficit financing), since they were content that both rental and capital values were rising rapidly. However, following the inevitable downturn, this policy changed to reduce risk and the banks are seeking to quantify their exposure to risk in both a buoyant and a depressed market.

In general, banks often tended to take a short-term view in relation to their lending policies, being concerned primarily with the underlying value of the development company and/or development project (of course they are focused on maximising the return to their shareholders which occurs on an annual basis, rather than over five or ten years as per a property development/investment timeframe). In some countries there was substantial criticism of the banks' lending policies during the late 1980s with many laying some of the blame for the boom–bust situation at their door. Some LTV ratios were exceptionally high (e.g. over 100%) others did not pay enough attention to assessing the risk attached to the borrower and the proposed property development itself. Some banks were quick to blame the valuers whose opinions formed the basis for assessing loans, evidenced by the many negligence cases against valuers, although on occasion the courts felt that lenders had also contributed to their financial loss through their own improvident lending policies. To reduce expose to risk in an ever-changing property market, the banks and their advisers constantly review their policies to reflect the current level of risk and prevailing market conditions, as well as being supported by high quality research and forecasting methods, such as detailed discounted cashflows (DCFs). The role of the valuation industry is also included in the debate from a risk perspective, where the valuer is supposedly accountable for any difference in price between the final sale price and the valuation amount stated when the valuation was carried out.

Property loans usually account for a relatively small proportion, as low as 10%, of all commercial lending by the banks. Merchant banks are also conscious of their exposure to risk from property loans. During a downturn in the property market some development companies will default on loans and the value of their developments/properties can be less than their outstanding debts. In this situation banks obviously want to mitigate their losses however, rather than force the borrower to default, banks are often willing to restructure loans by a combination of measures such as renegotiating loan terms, refinancing loans, swapping 'debt' for 'equity' (i.e. converting part of the debt into mortgages) and also forcing sales of assets in worst-case scenarios. In the last market downturn many companies, particularly those with large development programmes and specific projects, went into receivership. Trader developers were particularly vulnerable and of those who remained many were tightly controlled by their banks. In a depressed market where the banks have decided to stand by and support a developer or project, the problem of vacant or over-rented property continues until the next recovery occurs. An additional complication is that institutional investors are not interested in purchasing over-rented or secondary un-let property. Accordingly, the banks often have to make major write-downs after a downturn and have become increasingly reluctant to be exposed to this type of risk in the future. More recently the banks have also developed a flexible range of different financing products in order to meet the changing demands of the market and in response to an increasingly competitive financing environment.

4.2.4 Property companies and the stock market

There are two broad categories of companies which participate in real estate development: investors and traders. The investor type of company, usually referred to as a property company, is also a source of long-term finance as some purchase property investments for their portfolio as well as retaining their own developments. Their capacity to purchase property depends largely on their ability to raise finance. Property companies and development companies alike are partly financed by their own capital and that of their shareholders, and partly financed by borrowing money either short or long term. The level of 'gearing' (i.e. relationship between borrowed money and the company's own money) will vary between companies. Property companies as opposed to 'trader' development companies tend to have a lower level of gearing due to the strength of their asset base.

Property companies vary from small private firms to large publicly quoted companies. Some specialise in a particular geographical location such as a quadrant in a city, while others hold large portfolios of a cross-section of property types in international markets. Their prime objective is to make a direct profit from their investment and development activities, while some will take a long-term view in relation to the extent to which they 'trade on' their investments and completed developments. Most importantly they always have a responsibility to their shareholders to 'increase shareholders' wealth' by maintaining the share price and providing dividends. Property companies view property investment as both a source of income and as an asset providing security for borrowed money. Property companies, particularly the larger ones quoted on the stock market, will tend to concentrate their investment activities on prime and good secondary properties. However, unlike the institutions, they see the management aspects of property investment as an advantage. They have both the management and development skills in-house to improve the value of properties. Also they are not adverse to multi-let properties provided they are in a good location and of high quality. They may purchase investment properties which are not fully let or are nearing the end of the lease with redevelopment potential.

The shareholders of property companies are a combination of financial institutions and private individuals. Financial institutions invest in property company shares instead of or in addition to their direct property investments. However, it is not tax efficient for pension funds (who are non-taxpayers) to invest in property company shares when compared with investing in property directly. This is because corporation tax is paid on the company's profits before dividends on the shares are paid and capital gains tax is paid on property sales. Taxpaying shareholders will be taxed on the dividends and on any capital gains from selling the shares.

With a quoted property investment company, the shares are valued by the stock market below the value of the assets of the company attributable to the shares, which is known as the NAV or 'net asset value' per share. This discount to asset value is to the tax disadvantage of the company as capital gains tax might be payable on the sale of their assets. Importantly the amount to which shares are discounted varies with stock market conditions and the state of the property market. However, the value of property company shares fluctuates more widely than the value of property, regardless of the state of the property market. This was shown to be the case on 'Black Monday' in October 1987, when stock markets around the world crashed. The crash began in Hong Kong, quickly spreading to Europe before moving on to the USA. At the time the property market was booming but the shares of property companies fell dramatically along with other equity. Other factors affecting the value of the share price of property companies are the financial strength of the company including its level of gearing as evidenced by their balance sheet, as well as the perceived strength of the management team. In a similar manner to other companies listed on the stock market, the 'price-earnings' ratio (P/E) is the main yardstick used to assess the market's perception of the future earning potential of property trading companies.

Equity finance can be raised by issuing various forms of shares in a company, with investors directly participating in the profits and risks of the company. New property companies may float on the stock/share market and raise money by selling shares, often referred to as an IPO or 'initial purchase offering'. Quoted companies can issue new ordinary shares or preference shares to raise equity finance for their development activities, depending on stock market conditions, the overall performance of property company shares and the NAV per share of a particular company. Such finance may also be used to repay bank borrowings and other debts, or alternatively retain strategic developments in an investment portfolio. In addition, companies can raise debt finance via various methods on the stock market. Long-term debt finance is capital borrowed from investors which usually involves fixed interest, and may be secured on the company or unsecured. Debt finance usually has to be repaid by a certain date or converted into shares (i.e. equity). Debt finance instruments became popular as an alternative means of providing long-term finance to hold developments as investments.

4.2.5 Real estate investment trusts (REITs)

Real estate investment trusts, commonly referred to as REITs, have been a successful vehicle for the securitisation of property or real estate in many countries including the USA, the UK, Australia and Singapore. The increased popularity of REITs is linked to many advantages including taxation incentives, availability of up-to-date information about the REIT and being trading on the central stock market. It has been advocated that REITs help to facilitate the development of a high-quality residential letting market and offer the prospect of boosting house building over the medium to long term (RICS 2007; Stockton 2012).

REITs were introduced into the UK on 1 January 2007 in accordance with the Finance Act 2006. The UK REITs have many of the benefits of other REITs including greater flexibility and liquidity. However, one of the most sought-after benefits was from a taxation perspective – for example, UK REITs are treated as normal corporate vehicles which make an election to confer exemption on taxation from relevant company profits, and in return the REIT must withhold tax from distributions paid to shareholders out of these profits. The requirements to qualify for a UK REIT are as follows:

- UK tax resident (and not dual resident);
- listed on a recognised stock exchange;
- not an open-ended investment company;
- only classes of shares allowed are ordinary shares (one class only) and not participating preference shares;
- distribute 90% of its net taxable rental profits (not capital gains) during the relevant accounting period or within twelve months of its end;
- derive at least 75% of its total profits from its tax-exempt property letting business;
- at least 75% of the total value of assets held by the REIT must be held for the tax-exempt property letting business;
- other tax charges exist although no loss of REIT status; and
- additional conditions also apply (KPMG 2007; London Stock Exchange 2008).

To qualify as a REIT in the USA, according to the US Securities and Exchange Commission (2013) a company with the bulk of its assets and income connected to real estate investment must:

- be an entity that would be taxable as a corporation but for its REIT status;
- be managed by a board of directors or trustees;
- have shares that are fully transferable;
- have a minimum of 100 shareholders after its first year as a REIT;
- have no more than 50% of its shares held by five or fewer individuals during the last half of the taxable year;
- invest at least 75% of its total assets in real estate assets and cash;
- distribute at least 90% of its taxable income to shareholders annually in the form of dividends;
- derive at least 75% of its gross income from real estate related sources, including rents and interest on mortgages financing real property;
- derive at least 95% of its gross income from such real estate sources and dividends or interest from any source; and
- have no more than 25% of its assets consisting of non-qualifying securities or stock in taxable REIT subsidiaries.

While a REIT has been a widely accepted vehicle for funding a property via listing on the stock market, there are limitations that should be acknowledged. For example, expenses associated with listing on the stock market are substantial, including the marketing and statutory charges, in addition to the risk that the initial purchase offering (IPO) will not be fully subscribed by investors. Also investment in direct real estate in a buoyant market may offer a higher yield at times and therefore a REIT could struggle to offer a competitive yield regardless of tax advantages. Over time many property developers have grown from relative small developers and are now large enough to be listed as a global REIT (e.g. Multiplex, Westfield).

4.2.6 Overseas investors

The property market now operates within a global economy and in today's real estate market, overseas investors have become significant participants in the property investment market. No longer is demand for property limited to a prospective purchaser's geographic location as these barriers have disappeared. For example it is commonplace for an overseas investor to be just as fluent with a property market on the other side of the world as with their own local market. At times there are other reasons, such as tax implications or a lack of perceived local market demand, as to why an investor may be interested in a market in another country. A property development may be located in a particular region; however, some of the relevant stakeholders (e.g. lender, architect) may be located anywhere in the world.

The source of overseas investment has varied since 1988. The Japanese and the Scandinavians led the way during the boom period between 1987 and 1990. The UK government's favourable treatment of foreign investment together with the lifting of the Japanese government's restrictions provided the impetus for Japanese investors, developers and contractors. The lifting of restrictions by the Swedish government on overseas investment by their property companies and life funds led to the Scandinavian interest. They were both attracted by the performance of the UK's economy and its relative stability at that time. Their development companies became involved in direct development, either in partnership or on their own account. However, many have since gone into receivership following the collapse in the market in 1990 which followed the high profile Asian economic crisis.

European investors, particularly the Germans, became significant after 1992 due to the relative performance of the UK economy against their own economies and the continuing deregulation of cross-border investment by the European Union. They were joined by American, Middle Eastern and Far Eastern investors in 1993 and 1994. All have tended, in contrast to the early Japanese and Scandinavians investors, to invest in standing investments rather than developments and are interested in 'prime' properties let on institutional leases to good tenants.

More recently global investors have not been restricted from any one particular country and this trend in many ways mirrors the explosion of information transfer and availability, primarily due to the internet. In a relatively short period of time internet marketing has become a prime advertising medium, which in turn has allowed an investor on the other side of the world to access detailed information about proposed and existing real estate developments. Importantly this will include digital photographs, three-dimensional videos and virtual demonstrations that assist overseas investors to commit substantial funds, even though they possibly have never personally laid eyes on the property in question.

Expanding the market to overseas investors also has other benefits. This includes increasing the number of prospective purchasers by enlarging the marketplace, increasing the potential borrowing capacity of the purchasers (as opposed to the lenders in the local market), and ensuring the development is perceived as truly international. Furthermore, many overseas investors are more likely to invest in larger properties at the higher end of the market rather than restricting the pool of potential purchasers to the local market only.

4.2.7 Private individuals

In reality the majority of private investors purchase property investments at the lower end of the market, with a large proportion being 'mum' and 'dad' investors who are borrowing against the equity in their principal place of residence in the form of the family home. In many instances they tend to concentrate their purchases on secondary and tertiary commercial property with accompanying high yields but often located in an area with the perceived potential for capital growth. These purchasers tend to be precluded from the 'prime' market due to the significant sums of money involved and the existence of the 'reverse yield' gap. Private individuals are attracted to high yielding properties as income will very often be in excess of interest rates. However, participation in the lower end of the property market is very risky, involving intensive management and regular voids.

Unfortunately many private investors often place too much emphasis on the relationship between return and capital outlay, therefore making a direct but false comparison with the return from a standard bank deposit. This is partly due to the lack of understanding about the fundamentals of property investment including the depreciation of the building component over time and associated risk (for example, building age and require regular maintenance). Many investors are not fully conversant with the risk reflected in the yield rate where a higher yield equates to a higher risk, not lower as per a standard cash deposit in a bank. In this scenario where there is a narrow focus on the yield only (and ignoring the long-term maintenance and upkeep costs), some private investors have a smaller initial outlay but a substantially larger cost of maintenance. This is another reason why smaller investors are predominantly at the bottom end of the perceived 'bargain' market for investment properties.

4.2.8 Joint venture partners

A development company may raise finance or secure the acquisition of land by forming a partnership or a joint venture (JV) company with a third party to carry out a specific development or a whole series of development projects. The basic principle behind forming a partnership from the developer's point of view is to secure either finance or land, in return for a share in the profits of the development scheme or the joint venture company. The joint share given to the third party will depend largely on the value of their contribution combined with the extent they wish to participate in the risk of the scheme. There are many forms and methods of forming partnerships or joint ventures for the purpose of successfully funding and completing a property development. Nevertheless it is beyond the scope of this book to examine them in detail and it is recommended that the reader consult a specialised text. However, we will briefly examine the reasons behind forming partnerships and joint ventures.

A partnership may involve any combination of sharing the risks and rewards of a scheme via many different contractual and company arrangements. In addition the partners to a scheme may take an active or passive role in the scheme. Tax and financial considerations may determine to a large extent the formal structure of the partnership arrangement. A joint venture may also take many forms, but in its 'purest' form the parties participate in the development and distribute the profits in equal shares usually by forming a joint company.

Developers are typically reluctant to share profits with third parties, unless it is the only way of securing a particular site or finance for a development scheme. Partnerships with landowners may be required if the landowner wishes to participate in the profits of the development scheme or wishes to retain a long-term legal interest in the property, preferring income to a one-off capital receipt. Local authorities and other public bodies, for example Network Rail with reference to railway line infrastructure, may only grant long-term leasehold interests to developers due to their need to retain an underlying continuing interest in the property for financial or operational reasons. It should be noted there are restrictions placed on local authorities in relation to capital receipts and the forming of joint venture companies. Most often the ground rent and profit sharing arrangements will be determined by the amount of risk the landowner wishes to bear.

If a developer is involved with a particularly large or complex development, then a prudent way of undertaking such a scheme is to spread the risk through a partnership arrangement. However, this arrangement will often be beyond the financial capability of all but the very largest companies. With such schemes one or more partners may be involved and may include the landowner, contractor or another development company. Developers may also form joint venture companies with other development companies who may have the expertise or experience required for a particular type of development which is seen as vital to the success of the scheme. There are also examples of developers forming joint venture companies with retailers to combine their respective experience and market knowledge. Previously some joint venture companies were formed on larger property developments to enable the partners to arrange 'limited' or 'non-recourse' finance off-balance sheet so the borrowings did not appear on either of the partner's respective balance sheets. The rules were tightened up on these arrangements and the opportunities to benefit from such schemes have reduced substantially or been eliminated.

Regardless of the reasons for forming a partnership to finance a scheme, it is essential for the developer to ensure the definition of the profit is clearly detailed and understood. In the case of a joint venture company the profit will be distributed through the company accounts. A developer must rehearse every possible outcome of the scheme to ensure that any partnership arrangement will work and the true intentions of the parties have been agreed and carefully documented by all involved.

4.2.9 Government assistance

At any given time there are different government grants available, largely dependent on the level of available funding and the perceived ability of government assistance to be a catalyst for change. Many of these projects are designed to encourage private developers to proceed with a proposal, such as renewal or gentrification of an older building or geographical area. In many cases the financial incentives would not be sufficient for the developer to undertake a viable project, although the government acknowledges there would be wider community benefits if a developer would proceed with the project. Examples include the construction of low cost rental housing or a new shopping centre that would provide local employment. The property developer is strongly encouraged to contact the local government body which is usually willing to provide information about current and planned future schemes in the immediate region of the property development.

Discussion points

- What are the main sources of finance which may be available for real estate development?
- What attributes of REITs appeal to property investors?

4.3 Methods of development finance

A number of different sources for development finance have already been discussed. However, there are many methods of obtaining development finance from the above sources and recent years have seen the emergence of an increasing number of innovative techniques. It is important to examine the various well-established methods and briefly look at the different finance options available.

The choice of both source and method of development finance will depend on how much equity (the developer's own capital) a developer is both able and willing to commit to a scheme. If the developer has insufficient capital then the aim is to arrange as much external finance as possible in order to meet all costs associated with the property. At the same time the priority is to retain as much of the equity as possible without giving away bank or personal guarantees. A decision has to be made as to how much risk the developer wishes to pass on to the financier in return for a share in the financial success of the scheme. The availability and choice of finance will depend largely on variables such as the company's size, financial strength, track record, characteristics of the development scheme to be funded and the duration of the scheme. Whichever method is chosen, the developer will always need to be fully conversant with all aspects of taxation.

4.3.1 Forward-funding with an institution

Forward-funding is the term given to the method of development finance which involves a pension fund or insurance company agreeing to provide short-term development finance and to purchase the completed property as an investment. This happens at the start or at least in an early stage of the development process. This method of finance reduces the developer's exposure to risk where the terms usually agreed with the institution reflect this. From the institution's point of view this method of acquiring a property investment has several advantages over purchasing a ready-made investment. From the outset it provides the institution with a slightly higher yield than a ready-made investment, reflecting the slightly greater risk. By being involved in the development process, the institution influences the design of the scheme and the choice of tenant. In addition, if there is a rise in rents during the development period then the institution benefits from the rental growth.

The proposed development must fall into the 'prime' category if the developer is to be successful in securing forward-funding, although if the proposed scheme is very large in terms of its lot size, then the number of funds in the market is reduced. Therefore, it is essential for developers to fully consider this fact when purchasing a site as it will affect the way in which they evaluate the development opportunity.

On the assumption that the developer is purchasing, or has just purchased, a site in a prime location with the benefit of a planning consent, then the next step is to approach the institutions directly or via their agents. Agents have an important role to play in the forward-funding of a scheme as they have a good knowledge of the institutional investment market. In addition, because so many of them are retained by institutions themselves they should have established good contacts over time. The developer may also have established a good working relationship with particular institutions they may have worked with in the past. Institutions themselves will tend to adopt a proactive approach and directly (or through agents) seek out the development opportunities themselves in accordance with their individual investment criteria. They will have identified through their own research the property type and location they are interested in.

The developer will usually prepare a full colour presentation brochure for those institutions which express initial interest in the scheme. The brochure will typically describe the nature and location of the development, with supporting illustrative material – in addition it outlines planning consents, site investigation reports and specifications. In this process the institutions will seek to identify the level of risk associated with the potential development, including the appeal to purchasers and the underlying level of demand that the completed development could generate.

Most importantly, developers will need to sell their track record and experience on similar schemes. For example there will be an additional risk component attached to a relatively new developer who does not have the perceived ability through their development history to undertake and successfully complete a property development from start to finish. It is essential that the developer provides an analysis of the market in terms of the balance between supply and demand for similar schemes. An initial appraisal of the scheme will also be included as a starting point for negotiations on the value and cost of the proposed scheme, with any supporting evidence such as cost plans. However, the institution will not rely on either the market analysis or the appraisal by the developer and will carry out their own evaluation to assess the risk using a variety of sensitivity techniques.

The institution, having decided the proposition fits within its investment criteria, will need to satisfy itself that the proposed development is viable, that there is a demand for the development, that the specification and design of the building is of the highest quality, and that the developer has a satisfactory track record and expertise. Once again the institution will be constantly reassessing its level of exposure to risk and seeking to reduce it, if possible. The developer may need to be able to guarantee the investor's return at the end of the development period depending on the terms negotiated, so the institution will need to examine the developer's financial standing. The institution will ask itself: is the developer able to produce the scheme both within budget and on time?

After a particular fund has agreed in principle to the forward-funding of the developer's scheme, negotiations can then commence about the financial aspects of the agreement. There are various types of arrangements that can be entered into, largely depending on variables including the current state of the property market, the nature of the scheme and the financial standing of the developer. Funds will tend to tailor the arrangements to suit the particular development and their view of the market. There are many characteristics which are typical of most deals and the variation between arrangements will be reflected in the balance of risk and reward between the parties.

We shall now examine each of the typical elements of a funding agreement.

Yield

The fund and the developer will agree at the outset the appropriate yield (and therefore capitalisation rate) applicable to the scheme. This is commonly referred to as the relationship between the total capital value and the net operating income. The yield will usually be determined after a thorough assessment of recent market evidence and the fund's perception of the current and future risk. In valuation circles, the yield or capitalisation rate is also referred to as the 'all risk yield' since it is supposed to reflect all risk associated with the property. The yield is a measure of the institution's perception of risk weighed up against the rental growth prospects. In forward-funding arrangements the yields will normally be discounted by around 1–2% (i.e. 1–2% higher) from the market yield for standing investments to reflect the additional risk the institution is taking by participating in the development process. The yield will be fixed at the agreed level. It is critical that the property developer has a clear understanding of how yields are calculated and importantly the relationship between the yield rate and risk.

Rent

What is commonly referred to as a 'base rent' will be agreed upon after a careful analysis of current market evidence, being a comparison of the newly constructed floorspace to recent lettings of similar properties in the area. At times a pre-letting will be in place with the rental level agreed with the tenant in the 'agreement to lease', although there may be provisions for a review on final completion of the property development. This scenario would be more likely to occur when there is an extended time period, e.g. a large-scale development. If the rent achieved on the scheme exceeds the base rate then there is normally a provision to share the benefit of this. This element of the rental income is known as the 'overage' and is usually shared between the fund and the developer. However, the fund may cap the 'overage' rent at a certain level because the developer will be motivated to achieve the highest rent and the fund will wish to safeguard against 'over-renting' the property, which would be detrimental to rental growth prospects in the future. In other words there is a trade-off between achieving the highest possible level of rent and retaining a long-term tenant. Usually such a tenant will quickly relocate to less expensive accommodation at the first available opportunity, thereby creating an increased void.

Costs

It is essential that the developer presents to the fund a detailed estimate of current and reliable development costs. These will be analysed by the fund's in-house building surveyors or externally appointed consultants. Possibly the institution will wish to cap the total development costs at a certain level, although allowing for interest. The developer will be under an obligation not to exceed these costs. If the maximum agreed limit is exceeded then the developer will be responsible for funding the balance. There may be a provision within the agreed development cost for the developer's own internal costs, such as project management fees, overheads and a contingency allowance. The maximum agreed development cost will typically relate to previously agreed plans and specifications. The developer will be under an obligation not to vary either the plans or specifications without the prior approval of the fund. There may be a provision which enables a variation of the agreed costs due to agreed variations.

The fund will provide the short-term finance at an interest rate to reflect their opportunity cost of money and not the cost of borrowing. This is because institutions do not need to borrow money, but regard the provision of short-term development finance as part of their investment. Money will be advanced to the developer on a progress basis, commencing with the production of architect's certificates in respect of the building costs and invoices in respect of all other costs. Interest will accrue and be rolled up until practical completion or until the scheme is fully let, depending on whether the developer is responsible for any shortfall in rent. At the same time the developer will receive any profit, calculated as the development value less the development costs advanced in accordance with the terms of the funding agreement, although the fund will keep a retention that is equivalent to or greater than the amount agreed with the building contractor, depending on the existence of any defects or work outstanding.

Depending on market conditions, the developer may be able to secure a profit on the value of the land if the value of the land at the time of the funding agreement is greater than the initial cost of acquisition.

Developer's profit

The calculation of the developer's profit can vary and ultimately will depend on the type of funding arrangement entered into. The developer and fund will agree either (a) a base rent or (b) a priority yield method of funding.

(A) 'BASE RENT' ARRANGEMENT

On completion of a scheme based on a 'base rent' arrangement the total development cost (including interest) up to any maximum agreed limit will be deducted from the agreed net development value for the scheme. The net development value is the total rent achieved up to the agreed base rent multiplied by the agreed year's purchase (i.e. a reciprocal of the yield) less the institution's costs of purchase. The balance of the calculation will represent the developer's profit, often referred to as a 'balancing payment'.

As an example, we will look at the evaluation in Chapter 3 on the assumption of a forward-funded deal with an agreed base rent. Assume that the developer has agreed with the institution a base rental value of £$1,261,971 per annum, based on a net lettable area of 3,908 m² (42,065 ft²) at a base rent of £$322.92 per m² (£$30 per ft²) and a yield of 8.25%. It is also agreed that any rent achieved above the base rent will be split evenly with 50% to the developer and 50% to the institution. If the rent achieved is £1,350,000 per annum and the development cost is £12,765,006 then the developer's profit is calculated as shown in Example 4.1.

Example 4.1 Developer's profit analysis

Base rent plus (£$ per annum)	1,261,971
50% overage (i.e. £$1,350,000 p.a. – £$1,261,971 p.a. divided by 2)	44,015
	1,305,986
Capitalised at 8.25%	12.12
Gross development value	15,828,544
Less purchaser's costs @2.75% equals	15,393,259
Less development cost	12,765,006
Balancing payment to developer (i.e. developer's profit)	2,628,253
Developer's profit as a percentage of cost	20.59%

The fund's profit is represented by the movement in the initial yield from 8.25 to 8.29%, calculated by dividing the rent achieved by the development value and multiplying by 100, i.e. (£$1,350,000/£$15,393,259 = 8.77%).

Example 4.2 Developer's profit analysis

Development cost (£$)	12,765,006
Achieved rental income (£$ per annum)	1,350,000
Fund receives first slice of rental income	
@ 7.75% of development cost	989,288
Developer receives next slice of rental income	
@ 0.5% of development cost	63,825
Developer/fund share balance of rental income 50%–50%	
(i.e. £$296,888 divided by 2)	148,444
Developer's profit is the share of rental achieved	212,269
Capitalised at 8.25%	12.12
Developer's profit	2,572,695
Developer's profit as % of cost	20.15%

The fund's profit is represented by the movement in the initial yield from 8.25% to 8.77% as in Example 4.1.

(B) 'PRIORITY YIELD' ARRANGEMENT

A 'priority yield' arrangement provides the fund with the first or 'priority' slice of the rental income before the developer takes a profit, providing a guaranteed return yield on the institution's investment. The 'priority yield' is particularly used where the costs or rents are perceived to be subject to a greater uncertainty (e.g. with lengthy schemes) and therefore present a greater risk.

Alternatively, the fund will agree with the developer the priority yield (see Example 4.2) and will receive as a priority slice 7.75% of the development costs. Then the developer will receive as the next slice an agreed percentage of the development cost – in this example 0.5% – which is then capitalised at the agreed base yield. This equates to the developer's required profit as agreed with the fund. The remaining rental income is split 50/50 or as otherwise agreed, then capitalised at the agreed base yield. Again using the evaluation in Chapter 3 we now rework Example 4.1 on the basis of a 'priority yield' arrangement. The developer's profit is calculated as shown in Example 4.2.

Developer's guarantees and performance obligations

In addition to controlling their maximum funding commitment by capping costs, the fund may require certain guarantees from the developer. Unless the scheme is entirely pre-let, the fund may require the developer to guarantee any shortfall in rental income until the scheme is fully let or three to five years after completion, whichever occurs first. Alternatively, the fund may require the developer to enter into a short-term lease for three to five years after completion. Bank or parent company guarantees may also be required to support the potential rent liability to meet any costs exceeding the agreed limit.

If the developer wishes not to provide such guarantees or enter into a lease then a 'profit erosion' arrangement may be entered into. Under this arrangement interest and costs will continue to accrue until the scheme is income producing or three years after completion, whichever is the earlier. At this point any profit due to the developer will be calculated and may have been entirely eroded through an increase in the development costs above any agreed limit or a decrease in the rent actually achieved. However, the developer will then be able to completely walk away from the development without any further commitments.

Typical of most funding arrangements is an obligation by the developer to perform, i.e. to build the scheme in accordance with the agreed specification and plans within the time and budget agreed. In addition, the developer has to ensure that the professional team performs and that collateral warranties are procured for the benefit of the fund. Throughout the development the fund's interest will be protected by their surveyor who will oversee the project, attending site meetings as an observer, to ensure the developer is performing in accordance with the agreed specification and plans.

Lettings

The developer will need to obtain the approval of the fund for all lettings. The funding agreement will usually specify on what terms the fund will be prepared to grant leases – a standard form is usually attached to the funding agreement. There may be a provision which allows a particular type of lease, for example over a particular period of time with agreed rent reviews. However, increasingly more flexible lease lengths are allowed with or without breaks. From a lessor's perspective the longer and more rigid the lease, the less risk of a void. On the other hand, from a lessee's perspective it is important that the lease is as flexible as possible to allow for changing circumstances (e.g. more staff) in the future, which would decrease the lessee's risk. The agreement may also specify an acceptable tenant. A typical arrangement may specify that the tenant's profits for the last three years must exceed a sum three times the rent or total liability (including service charges). The fund may specify that only single lettings are acceptable, although floor-by-floor lettings might be acceptable after a certain length of time from practical completion.

Sale and leaseback

As an alternative to the above-described arrangements, a sale and leaseback arrangement may be entered into whereby the developer retains an interest in the investment created by the development. This type of arrangement varies depending on variables such as the availability of land, the amount of money available and so forth. However, the arrangement is not suitable for all stakeholders, for example where institutions prefer to retain total control and chose the leasing arrangements. A sale and leaseback involves the freehold of the scheme passing to the fund on completion with the fund simultaneously granting a long lease to the developer, who in turn grants a sublease to an occupational tenant. There are many variations to this arrangement, depending on the method of sharing the rental income. Sale and leaseback arrangements may be either 'top sliced' or 'vertically sliced'. With the 'top slice' arrangement the fund receives rent from the developer in accordance with the required yield. The developer is then able to retain any profit from letting the property at a higher rent than that payable to the fund. However, with upward-only rent reviews in the developer's lease with the fund the developer's profit rent may be rapidly eroded over time. This means the developer's interest is only saleable to the fund. Therefore, the 'vertical slice' arrangement is better from the developer's viewpoint as the fund and the developer share the rental income from the property in relation to an initially agreed percentage throughout the length of the lease. Institutions are more likely to enter into sale and leaseback arrangements directly with the occupiers to create attractive property investments with tenants of good financial standing.

When undertaking a property development, the cyclical nature of the market is a major consideration and the state of the market at any one particular point in time must be considered. In a property market where there is an oversupply of space (mainly secondary), many institutions would enter into forward-funding deals on the basis of a pre-letting. Alternatively they would be prepared to enter into arrangements with a developer of good financial standing where the rent is guaranteed for three to five years by the developer. However, forward-funding deals are currently being achieved on the basis of speculative schemes provided there is a lack of 'prime' space on the market and there is proven demand, where the risk is quantifiable by the institution.

4.3.2 Bank loans

In recent years the banks have received increased competition due to the globalisation of the banking sector, as well as other types of lenders entering the market with hybrid products. The clearing and merchant banks provide short-term development finance, either on a 'rolling' or project-by-project basis, by means of overdraft facilities or short-term corporate loans secured against the assets of the development company or project loans secured against a particular development. With the dramatic increase in bank lending in recent years and the wider acceptance of debt, various different methods of bank lending have been introduced, such as development companies seeking bank finance beyond the construction period up to the first rent review. Another example is the popularity of mezzanine finance which has increasingly filled part of the gap between equity and a first mortgage.

For many developers, especially the smaller ones, forward-funding is difficult to obtain as they are unable to provide the requested guarantees. Also 'prime' properties which are acceptable to institutions often represent a very small part of the market and to some extent tend to be geographically restricted (for example in the UK, prime real estate lends to be located in or close to London and towns in the South East). Large development projects with extended planning and construction phases are beyond the capacity of all but a few of the larger funds. From the developer's point of view, borrowing from a bank allows greater flexibility and enables the developer to benefit from all of the growth, unless some of the equity is given away. The developer can repay or refinance the debt when the time is right and sell on the completed investment at a higher price. In addition, the developer will not be subject to the same degree of supervision through the development process. In a rising market where rents and capital values rise rapidly, it is more profitable for developers to arrange debt finance as opposed to equity finance for the reasons mentioned above.

Corporate loans

Development companies can arrange overdraft facilities or loan facilities with clearing banks secured on their assets. However, with corporate lending the bank is concerned with the strength of the company, its assets, profits and cashflow. Accordingly, obtaining bank loans in this way is more appropriate to investor-developers and large developers rather than trader-developers and smaller developers as they have large asset bases which can provide the necessary security for bank borrowings. Usually corporate loans can be obtained at lower interest rates than project loans.

Project loans

Alternatively, development companies can arrange project loans which are secured against a specific development project. Banks normally provide loans which represent 65–70% of the development value or 70–80% of the development cost. Developers have to provide the balance of moneys required from their own resources. The banks limit their total loan to allow for the risk of a reduction in value of the scheme during the period of the loan. In addition, insisting on an equity injection by the developer confirms their high level of commitment to the overall development itself. This equity provision is normally required at the outset of the development to motivate the developer to complete the scheme. Also the developer is totally responsible for any cost overruns. The loan to value ratio (LVR) depends on the risk perceived by the bank and can vary substantially depending on the risk profile of the borrower, the perceived risk in the project and prevailing market at the time. Clearly a pre-let or pre-sold development represents less risk than a totally speculative one. At times it has been possible to secure between 85 and 100% of development costs through various layers of bank finance using a combination of insurance, 'mezzanine' finance and profit-sharing arrangements. However, in periods of sustained downturn this high LVR would be considered as too risky by the financier. Lending conditions will vary according to the banking sector's knowledge and overall confidence in the property market (systematic risk) and in the actual property itself (unsystematic risk).

Project loans are attractive to the smaller trading companies as they are not worth enough to fund their full development programme through corporate loans. For many larger companies, these loans can be carried off the parent company's balance sheet by forming joint ventures with the bank in a subsidiary company. The borrowing associated with the property would not appear on the parent company's balance sheet. Previously this has enabled property development companies to have development programmes which would have otherwise been impossible as 'gearing' would have increased to an unacceptable level. In this example 'gearing' is commonly referred to as the relationship between borrowed money (liabilities) and the company's own money (equity). At times some investments are negatively geared, such as where outgoings exceed income. In some countries this loss can be offset against other income to reduce tax.

When the banks lend against a particular development project then this development will form all or part of the security for the loan. The developer needs to provide similar information to the bank as to a funding institution. The banks will wish to ensure that the property is well located, the developer has the ability to complete the project and the scheme is viable. This requires the bank to have knowledge of the property market either through in-house staff or external advice from firms of chartered surveyors in order to assess the risk involved. The developer will need to present the proposal to the bank in the form of a package very similar to that required by a funding institution.

However, as the bank is viewing the scheme as a form of security and not investment, it is more concerned with the underlying value of the scheme rather than the details of the specification. The appraisal, therefore, forms the most important part of the presentation together with all the supporting information and market analysis. It must reflect all of the risk in the proposal with supporting market research to reduce the unknowns. Equally important is the track record of the developer in carrying out similar schemes. The banks will also examine the financial strength of the development company, although this may not form part of the security of the loan.

The bank will employ either its in-house team of experts or external surveyors/valuers to report on the proposal and provide a valuation of the proposed scheme. Part of the process will be an analysis of the risks involved and this should be reflected in the terms offered to the developer. The bank will be concerned to ensure there is sufficient contingency and profit/risk allocation built into the appraisal to provide a sufficient margin for cost increases during the period of the loan.

Another element of the risk involved in bank loans is represented by fluctuations in interest rates which the bank may be concerned about and recommend limiting (see below). Interest rates on bank loans can be at a fixed percentage, a variable percentage or a combination of both. If the rate is fixed it is only in relation to the base rate which is often centrally controlled by the government. Generally interest rates will be higher on project finance loans than corporate loans due to the higher uncertainly and therefore the increased risk that the banks face. The interest rate margin may be less if the developer pre-lets or pre-sells the property before completion as this substantially reduces the bank's risk. Also the interest-rate margin on an investment loan would be lower than on a speculative development loan. Interest rates on short-term loans are likely to be floating whereas on long-term loans they are more likely to be fixed.

The bank should also consider how the loan will be repaid, being either by the sale of the completed scheme or by refinancing from another bank. It must be remembered that on completion of the scheme the initial rental income will usually be insufficient to cover the interest costs on the loan, due to the 'reverse yield' gap problem where yields on property investment tend to be lower than medium- to long-term interest rates.

The bank will also need to protect its 'security' by obtaining a first legal charge on the site and development. It is important for this to be recorded on the title or deed to the property and therefore any prospective purchasers undertaking a search will be aware of this liability before buying the property. In the event of default on the loan the bank will be able to obtain ownership of the development if required. It may also require a floating charge over the assets of the development company. The bank will need to be able to be legally capable of stepping into the 'developer's shoes' in the event of any default which may require the legal assignment of the building contract and any pre-sale or pre-letting agreements. Like a funding institution, the bank will also require collateral warranties (i.e. secure guarantees) from the professional team.

Guarantees may be required from the parent company or a third party if the financial strength of the development company is considered inadequate due to the perceived higher risk. A full recourse loan will involve the parent company providing a full guarantee on the developer's capital and interest payments together with a guarantee that the project will be completed.

Previously limited and non-recourse loans became an attractive proposition for developers wishing to finance their development projects while providing limited or no guarantees. With a 'limited recourse' loan the parent company may only have to guarantee cost and interest overruns. Limited recourse loans were normally granted for the construction period of a project and up until first rent review. A 'non-recourse' loan involves no guarantee with the only security for the bank being the development project itself. However, in practice the parent company is still responsible and it would be very difficult for a developer to simply walk away from the scheme without damaging their long-term reputation.

The developer needs to take account of the considerable costs involved in bank finance which are usually up-front. The developer will need to pay for the cost of carrying out their own appraisal and presentation to the bank. In addition, several fees are payable to the bank although some fees may be negotiable depending on the size and type of the loan. For example, if multiple banks are competing to fund a large development project, one bank may lower or remove the fees altogether to increase their competitiveness. In addition an arrangement fee will normally be charged by the bank to cover the cost of carrying out a valuation and assessment of the project. This fee may include an element of profit depending on the risks involved. There will also be a management fee to cover the bank's costs in monitoring the project, consisting of mainly surveyor's fees. Such charges can represent up to 3–10% of the value of the loan. In some instances there may be a 'non-utilisation' or 'commitment' fee on the part of the loan not drawn down initially as the bank will have to retain the full loan facility and cannot commit the funds elsewhere.

There are some variations on the basic project loan described above.

INVESTMENT LOANS

Development companies, wishing to retain a development, can secure the option to convert the project loan into an investment loan on the completion of the project once it is fully let, usually up until the first rent review. Alternatively, the developer may agree a combined project and investment loan from the outset. Also, the developer may be able to refinance a loan on completion on better terms than a previous short-term development loan. On investment loans, banks will normally lend up to three quarters of the value but there is always the problem of the rent not covering the interest payments. Banks wish to see the interest covered by the rental income and, therefore, can limit the loan. This may be relaxed where the property is reversionary (let below market value) or the parent company guarantees the shortfall interest. Otherwise the banks may require the developer to cap the interest rate or re-arrange the payments on the loan (see below). The interest rates on investment loans are usually lower than on substantially riskier speculative project loans and the risk to the bank will depend on the financial standing of the tenant.

'MEZZANINE' FINANCE

A project loan may be split into different layers known as 'senior' debt and 'mezzanine' debt if the developer is unable to provide the normal equity requirement or wishes to increase the amount of the loan above the normal loan to cost ratios. The senior debt usually represents the first 70% of the cost of the development scheme like a straightforward project loan. Senior debt is usually provided by the major banks and is commonly referred to as the first mortgage, since it takes priority over other forms of debt if the property is sold to reclaim funds. This debt may represent more than 70% if the development is pre-let or pre-sold. When a developer wants to borrow more of the cost of the project than 70%, additional money may be raised in the form of mezzanine finance. Also the bank may increase the senior debt exposure to 85% of cost with a commercial mortgage indemnity scheme. Mezzanine finance may involve the developer losing some of the equity, or alternatively, taking out an insurance policy. This mezzanine element is normally provided by merchant banks and specialist property lenders. As this mezzanine level of finance is more risky and often referred to as a second mortgage, the bank will charge a higher interest rate (usually 1–2% higher than the rate on the senior debt) or require a share in the profits of the development. The banks will also require full guarantees from the parent company.

If a mortgage indemnity insurance policy is taken out, it will reimburse the lender if the loan is not repaid in full and will involve the developer paying a substantial one-off premium to a specialist insurance company. The policy may cover the mezzanine layer of the loan or the entire loan. Equity sharing with the bank may involve a profit share or an option on a legal interest in the scheme. Banks willing to participate in the equity of the scheme are limited to those with sufficient property expertise. Very often, due to the tax complications, the profit share will be expressed as a fee. In this instance the bank will become part of the development team and become involved in the decision-making process.

SYNDICATED PROJECT LOANS

The necessary development finance may need to be borrowed from more than one bank. In particular, larger loans are more likely to be syndicated among financiers or a group of banks by a 'lead' or agent bank. Each bank shares a proportion of the risk of the development project depending on their initial contribution – also, their profit is commensurate with the proportion of their initial contribution. As a further complication, the loan may be just 'senior' debt or it might include a layer of 'mezzanine' finance. The lead bank, usually an established property lending bank with the necessary expertise employed in-house will arrange the syndication of banks. The lead bank may underwrite the entire facility or agree to use its 'best endeavours' to secure the syndication. It is usual for the lead bank, who may participate in the syndicate, to have the final responsibility for making decisions on behalf of the syndicate during the period of the loan.

INTEREST RATE OPTIONS

The development company may wish to protect themselves from the risk of changes in the borrowing rate over the period of the development project, particularly when the market is uncertain or there are indications that future interest rates may rise. In this instance, the developer may seek interest loans at a fixed rate, but it must be remembered that the development company may be tied to a high rate of interest for a long period, so that they may be unable to gain from subsequent reductions. On the other hand, a fixed interest rate can remove uncertainty for the lender regardless of external unknown forces affecting the interest rates. Alternatively, developers may compromise and try to 'hedge' the risk that interest rates will rise during a development, but this will always be at a cost. The usual form of interest rate hedging is the 'cap', which limits the amount of interest the developer will have to pay and is similar to an insurance policy. It is an interest rate 'option' which has to be paid for at the outset, either to the bank providing the loan or another bank altogether. For instance, the bank will reimburse the developer the cost of interest over and above the cap rate. Hedging is more difficult on a speculative development loan than on loans for income-producing investment properties due to the uncertainty of the amount of loan outstanding at any one time.

4.3.3 Mortgages

Mortgages originally provided the most common form of long-term development finance. A mortgage is a loan secured on a property whereby the borrower has to repay the capital loan plus interest by a certain date. However, not many lenders are interested in long-term non-equity participating loans such as mortgages. From the lender's point of view, a mortgage is a fixed income investment and very illiquid. Some banks, the larger building societies and many life or insurance companies provide mortgages on commercial properties. However, the availability of mortgages is limited due to the problem of the 'reverse yield' gap. It is rare to get fixed rate interest mortgages, although some life funds provide long-term fixed-term interest mortgages depending on prevailing interest rates. Demand for mortgages has been mixed in the past due to changing economic circumstances.

Mortgages may normally be granted on a loan-to-value ratio (LVR) basis of between 60 and 80% depending on the risk involved. The amount of mortgage secured will depend on the security being offered by the borrower in relation to the quality of property, the financial standing of the tenant and the borrower. Mortgage loans are normally for twenty to twenty-five years in line with the length of occupational leases, although this time period is open to negotiation.

Various methods have been developed to overcome the initial 'deficit' problem caused by the difference between rental income and interest repayments over the first five or ten years. Interest payments may be fixed for a certain period and then converted into a variable rate. Some borrowers do not want to be exposed to variable interest rates and may negotiate what are termed 'drop lock loans' which allow the borrower to switch from a variable rate of interest once the rate reaches a certain level.

4.3.4 Corporate finance

As already discussed, there are various methods available of raising equity and debt finance from institutional investors via the stock exchange, which we will now examine briefly.

Equity finance

NEW SHARES

Companies may raise money by selling shares to investors in a floatation on the stock market or the unlisted securities market. The majority of new share issues are underwritten by financial institutions for a fee, who will buy any shares not bought.

Generally speaking there are two types of shares: ordinary and preference shares. An ordinary share is a share in the equity, or in other words part-ownership of the company. Ordinary shareholders have voting rights and share in the risks and profits of the company. Profits after tax are distributed via dividends, usually half yearly. Companies may also issue convertible preference shares at a fixed dividend which, within a specific period, may be converted into ordinary shares. Preference shareholders rank above ordinary shareholders in entitlement to dividend payments. However, preference shareholders do not participate in any growth in the company profits and normally have no voting rights.

RIGHTS ISSUES

A company can raise additional capital by offering existing shareholders the right to purchase a number of additional shares in proportion to their existing shareholding at a lower price. As with new issues a rights issue is normally underwritten. The net asset value (NAV) per share will be diluted. The ability of a company to raise capital via a rights issue will depend on stock market conditions, the state of the property market as measured via property share performance and the NAV per share. In the past some companies have been able to successfully raise capital on the stock market via rights issues as property share prices performed well, while only marginally diluting the NAV of their shares.

RETAINED EARNINGS

One source of finance is the company's own resources generated by profits. However, some profit will need to be distributed to shareholders as dividends. How much of the profits is paid out as dividends or retained is up to the company to decide, but they must be aware of the interest of their shareholders in maintaining a reasonable dividend.

Debt finance

Debt finance instruments may be secured on specific property assets or the property assets of the company as a whole. Alternatively, they may be unsecured where investors have to rely on the financial strength and track record of the company.

BONDS

A bond is considered a relatively low-risk investment and often the return is also relatively low in comparison to other investment options. Effectively it is an 'I owe you' note, secured on a specific investment property or a completed and let development owned by the company. Investors in a bond receive interest on a regular basis (e.g. each year) and their initial investment is repaid at a specific date in the future. Bonds are securities which can be traded on the stock market. The interest payments (known as the coupon) can be structured to avoid the usual problem of rental income shortfall. With 'stepped interest' bonds the investor receives a low interest rate initially which rises at each rent review. An alternative is a 'zero coupon' bond where no annual interest is repaid but the investors are repaid on the redemption date at a premium. However, both these types of bonds rely on rising property values which does not always occur due to the cyclical nature of the real estate market.

DEBENTURES

Debentures are securities which can be traded on the stock market. Debentures are issued by companies to institutional investors whereby the institution effectively lends money at a rate of interest below market levels in return for a share in the company's potential growth. The money is typically lent long term, usually up to thirty years, at a fixed rate of interest and is secured upon the company's property assets. Normally, the security is specifically related to named properties, but sometimes provision is made to allow the company to substitute one property for another subject to agreement on valuation.

UNSECURED LOAN STOCK

Property companies may issue unsecured loan stock (not secured on the assets of the company) to institutions at fixed rates of interest which, within a specific period, can be converted at the option of the institution into the ordinary shares of the company. However, to reflect the higher level of risk attached to the absence of a high level of security, the interest rate is also higher. The higher interest rate also covers instances when the lender is unable to recover all or part of the loan, mainly due to the lack of security.

4.3.5 Unitisation and securitisation

The property markets and financial markets have recently developed equity financing techniques to reduce the problem of the illiquidity of property investment. They are attempting to make property investment more comparable to other investments and overcome some of the inherent obstacles including lack of a central marketplace, indivisibility (i.e. either purchase the entire property or not at all), transparency (i.e. lack of information about the product) and illiquidity (i.e. to access the money tied up in the property would normally take months including advertising, negotiation and the contract stage). Another obstacle is that to finance large individual developments with funding institutions is difficult for developers unless two or more funds become involved. At times this may mean that larger single properties are valued at a discount compared with smaller investments due to the limited number of potential purchasers.

Many global markets have accepted that securitisation and unitisation of property investment is a means of broadening the demand for property beyond the existing financial institutions who are large enough to participate – for example, as a REIT listed on the stock market that smaller investors can buy part thereof. Simply explained, unitisation means the splitting up of ownership of a property or a portfolio of properties among several investors. Securitisation is a general term used to describe the creation of securities which can be traded on the stock market, e.g. shares, bonds, debentures and unit trusts. Therefore, the creation of securities is one way of achieving unitisation. Importantly, with each method the investor receives a return in exact proportion to their original investment.

It is worthwhile to discuss the background to various securitisation and unitisation techniques that have been introduced so far.

There have been various attempts at 'unitisation' of large properties, i.e. splitting the ownership of the property into small manageable chunks, allowing several investors to invest in the property. However, due to legal complications it has proved very difficult in practice to actually divide ownership of an individual property. Earlier attempts at the unitisation of individual properties have included SPOTs (Single Property Ownership Trusts) and PINCs (Property Income Certificates). SPOTs involve a trust owning a property and spreading the ownership among investors in the form of units similar to unit trusts; these were not widely adopted and faltered in the UK due to tax problems. A PINC is a security consisting of an income certificate, a contract to receive a share of income of the property after management costs and tax, and an ordinary share in the management company that manages the property. As a security, it was capable of being traded on the stock exchange based on the concept that investors receive the benefits of ownership, in the form of a share in the income and capital growth, without owning the property direct.

Previously there have been other ways of unitising a portfolio of properties via unit trusts aimed at small pension funds and private investors. Predominantly there are two main types of unit trust: authorised property unit trusts (PUTs) and unauthorised unit trusts. PUTs invest directly in property on behalf of their investors which may include private investors. They are strictly regulated to ensure that they invest in a diversified portfolio of low risk prime income-producing property as part of an overall balanced portfolio including property securities. A PUT is treated as a company for the purposes of corporation and income tax, but is exempt from capital gains tax. On the other hand, unauthorised unit trusts are unregulated trusts investing in a mixed or specialised portfolio of properties. They are attractive to tax-exempt financial institutions such as small pension funds and charities which lack the funds to invest in property directly. Where all the investors in the trust are tax exempt then the trust is exempt from capital gains tax. Unit trusts are managed by a committee of trustees elected by the unit-holders (investors) under the terms of the trust deed.

4.4 Future trends

The property development industry is part of the larger real estate market, which is subject to changing supply and demand levels results in often clearly defined property cycles. History has shown that poor timing by a property developer can result in completion at the bottom of the cycle where rents are low and demand is scarce. After each property downturn the lenders, primarily in the form of banks, have taken a vested interest in the projects they are lending 'their money' on. Accordingly most lenders now take an active role in understanding the dynamics of the property market and the likelihood that the development will reach its full potential. Thus a borrower must provide detailed market evidence and projections in order to convince the lender that the profit levels will actually be achieved.

It is critical for a borrower to understand the role of a lender in order to borrow money on suitable terms. Simply explained, the lender is seeking at all times to decrease their exposure to risk, primarily in the form of either property-specific (unsystematic) risk or market (systematic) risk. In return for accepting a perceived higher exposure to risk, the lender will charge a higher interest rate commensurate to the level of risk. While many forms of risk are unavoidable, such as risk due to the time needed for development, often a proportion of the risk can be reduced by the borrower. For example, pre-letting or pre-selling a development will remove risk associated with both the final rent/sale price as well as the likelihood that the property will remain void after completion.

At any given time there will be a myriad of financially strong companies, new players embarking upon development schemes on the basis of pre-lets and, in some cases, speculative schemes. There are a large number of established and emerging lending products that are ideally suited for each project and are actively seeking to lend money. At the same time there are property developments that may not succeed and many lenders will avoid these projects, forcing the property developers to rethink their proposal or even the whole project in the current property market climate. Many vacant sites are testament to property developers waiting for the optimal time to initiate a proposal or re-approach a lender when the market is on the rise.

Property developers have a wide variety of lenders and associated products available to them, although the market is constantly changing and adjusting to its own supply and demand forces. In the future the property developer must be constantly seeking to keep abreast of changes in the lending market, especially when considering changes in taxation and legislation. Only then will the property developer be able to develop a competitive product to realistically compete with other property developments, especially after considering that the interest and borrowing costs are such a large proportion of the overall costs associated with any property development. Overall it can be argued that banks will remain as short- to medium-term debt financiers, although securitisation and unitisation are becoming more readily accepted in both the property and equity markets at the same time. The rapidly growing acceptance of REITs throughout the world should continue, especially since this investment medium has the ability to overcome many of the negative benefits associated with direct property investment and is ideal for larger pension funds.

4.5 Reflective summary

There are a variety of sources and methods of financing property development both in the short and long term. The choice and availability of funding will depend on the nature of the scheme, how much risk the developer wishes to share, and the confidence of both financial institutions and the banks in relation to the underlying economic conditions at any particular time. The most secure route from the developer's point of view, provided the development is 'prime', is forward-funding with a financial institution, a method which combines both short- and long-term funding. However, if the developer wishes to retain flexibility, either wishing to retain the investment or sell it when market conditions are favourable, then debt finance is more appropriate in the short term. The terms and method of debt financing will depend on the financial strength of the developer and the value of the security being offered. A pre-let scheme being carried out by a financially strong developer represents the best proposition. The greater the risk, the less likely the developer will be able to obtain debt finance on favourable terms, unless the developer either contributes its own capital or shares the eventual profits. If debt finance is used then both the developer and the financier must have regard to the availability of long-term finance and the requirements of property investors. Property has to compete with other forms of investment which offer more liquidity to the investor, so funding and valuation techniques have to be developed to improve the attractiveness of property as an investment.

Chapter 5

Property cycles

5.1 Introduction

> The actual, private object of most skilled investment is to beat the gun, as the Americans so well express it, to outwit the crowd, and to pass the bad, or depreciating, half-crown to the other fellows
>
> John Maynard Keynes 1936

There are no disagreements that property cycles exist; however, they receive relatively little attention in the property profession. Due to the nature of property development and the critical element of 'time' within the development framework, an understanding of property cycles is essential where these two primary reasons are both related to risk. First, it will allow an opportunity for property developers to maximise their return (i.e. lower risk), either to end purchasers or renters, by completing the project at the best possible time for release onto the market. Second, at the other end of the spectrum it will help to avoid the worst possible completion date (i.e. higher risk). A developer who understands the nature of the property cycle is able to plan a project ensuring (as much as possible) that it is completed and released into the market at the best possible time, further reducing the risk and maximising the return on the development. Even though property cycles are an accepted component of property markets throughout the world, it can be said that there is often a relatively poor understanding of the status of a property cycle within submarkets where a property development is occurring.

This chapter discusses the concept of property cycles and informs the property developer about the theory behind property cycle behaviour. Most often, a property developer will be operating in a market which has little available information about prevailing property cycles. For example, What is the length of time between each peak in each property cycle? What is the amplitude of each property cycle (the difference between the highest point and the lowest point in each cycle)? Where is the market positioned within the current cycle? Company reports and forecasts may provide answers to these questions but the smart developer will also carry out his own research to establish the impacts of the cycle on their business and adjust their development strategy accordingly. Only then will a developer minimise their exposure to risk and maximise the likelihood of success. The focus of this chapter is on how to identify cycles in a particular marketplace and to minimise the risk to a property developer due to cyclical market behaviour.

5.2 Existence of property cycles

As a starting point, property cycles can be defined as:

Processes which repeat themselves in a regular fashion.

However, the question that then arises is:

If these processes are so regular, why can't property developers predict them?

This question is more applicable to some property markets than others. For example, the commercial property market has always been particularly prone to the boom–bust cycle which causes very destabilising economic effects (Scott and Judge 2000). Even so, predicting the timing and magnitude of these rises and falls in the marketplace has never been achieved to absolute perfection in any property market throughout the world.

Surprisingly 'cycles' are traditionally much more complex than at first may be anticipated. It has been demonstrated that there is substantial heterogeneity in terms of duration, amplitude and co-movements (Stanca 1998). Furthermore, the variations can differ substantially across cyclical episodes and recessions are by no means mirror images of expansions (Stanca 1998). The perfectly symmetrical example of a cycle in Figure 5.1 would therefore be totally unrealistic as it does not promote any irregularities or traditionally sharper 'bust' periods.

The nature of the property market, the behaviour of the stakeholders and the process of property development, collectively form an ideal environment for property cycles. The starting point is to consider both the limited supply of land and also the level and type of demand, where both aspects are considered in the following sections. Following on, there are other property market characteristics which contribute to property cycles' behaviour as listed here.

Limited availability of reliable property data

In comparison to markets for other investment goods or assets, both the amount and quality of information available about transactions in the property market are substantially below average. For example if you were starting a new company which was to be listed on the equity market, there is a wealth of information easily accessible to everyone about existing companies, including financial data. The same applies for financial information about all companies listed, including their exact share price and

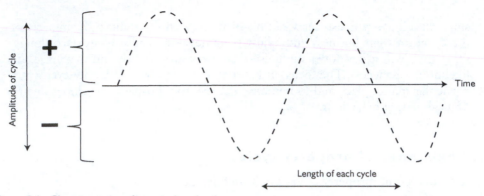

Figure 5.1   Characteristics of a typical cycle phase

the number of shares traded. The same statement applies to other investment mediums including the monetary exchange rate which is updated by the minute. Other examples include the gold price, the oil price and so forth.

Also affected by some other factors, it is relatively difficulty to find out information about the operation of a particular property market with a high degree of reliability. This is partly due to the confidential nature of property transactions and difficulties in gaining access to data on individual transactions between two other parties. Probably the largest barrier is the lack of a central marketplace for the property and real estate market, which makes it possible for property transactions to occur between two parties anywhere and not be openly disclosed. In most instances the transaction will also be subject to some form of tax; however, there is no requirement for government bodies to disclose the details. With the increasing importance of privacy legislation, improving access to this data becomes even less likely.

It is possible for the successful property developer to partly overcome this barrier by (a) purchasing property transaction data from a third party or from the government (where legal) or (b) relying on networking skills and established trust relationships to access this information. The advantages associated with networking cannot be under-estimated from a property and real estate perspective.

A time lag between the purchaser–seller transaction occurring and the release of this information into the public domain

Unlike other financial assets which transfer between a buyer and a seller, the actual period of time between (a) the initial agreement to transfer a property between two parties and (b) the completion of the transaction, will be weeks, often months. However, only at (a) is the information available (if at all) to third parties. At times this period may be extended to multiple months or any agreed time period. Due to the period required to conduct due diligence checks of the subject property (e.g. confirm ownership, encumbrances, checks with local authorities on future planning proposals which may affect the completed development, finance approval, etc.), arguably there will always be a time lag between (a) and (b). Note the use of online searches can reduce the period between agreement (a) and transfer (b) and release onto the market (c); however, the period required for due diligence searches can never been removed. The main effect on property cycles is that the purchasers are making buying decisions today (a) however, this information is only available at (c), a considerable period later. If there is a scenario involving a proposed property development (i.e. a purchaser pays a deposit and then pays the balance upon completion) where the time between (a) and (b) can be up to a year or more, the market can have changed substantially over this period. In turn this will cause an imbalance between supply and demand as decisions are based on out-of-date information.

Unique nature of each parcel of land and also each building

Unlike other assets (gold, money, etc.), each parcel of land is unique. This is due to the combined factors of location, land and the buildings both on and surrounding each plot. This difference provides challenges for analysing supply and demand interaction. In theory, land parcels sited close together should have a similar value. However, there are many attributes which can increase or reduce demand for a specific plot of land, such as the area of the lot, shape, topography, distance to transport and services (e.g. retail shops), view, access and planning/land use restrictions.

A relevant example is a property's actual view of the ocean or a river, where a premium exists for a property with a 'view' but a 'close view' cannot be compared. On the other hand a location near a high voltage overhead transmission line (HVOTL) will often decrease the value but that reduction will depend on how close the HVOTL is to the plot in question and whether the pylons are visible (Sims et al. 2009). These factors can influence the degree to which a parcel of land is affected by market trends which complicate (and increase the error rate of) an analysis of the market.

A building, especially a larger building such as a multi-level office or retail shopping centre, is relatively unique and difficult to compare with market benchmarks. While certain residential products, such as units/flats or detached housing, may be practically identical in design and construction it remains true that larger buildings have 'points of difference' between other buildings. The result is that property developers face a challenge in determining how demand for a proposed new building (development) can be measured. If a direct comparison cannot be made with an identical product then an amount of error will need to be factored into the analysis.

Property and real estate is a large 'lumpy' asset

In comparison to property and real estate, other competing assets in the investment market are divisible into smaller components. This applies to cash, equities and other forms (e.g. gold, silver). However, land and buildings are 'lumpy' assets where a multi-level office building cannot be developed in stages in a similar manner to a residential land development. The problem therefore arises with a new proposed property development, say for example a high-rise building or a large residential apartment block. During the development phase it is not possible to release part of the development onto the market, mainly due to logistical challenges relating to the construction phase. As a building is constructed from ground level upwards, the building can only be occupied when all construction has ceased. This creates a problem for the developer as the entire asset is released onto the market on a particular date. The downside is that the new supply of accommodation is released onto the market in large amounts at once, rather than in smaller components, and if several buildings are released onto the market at the same time then this can easily cause an oversupply situation, irrespective of demand.

The highest and best use of land is constantly changing

The property and real estate market is in a constant state of change, both in terms of supply and demand. For example when a town was first settled, it was accepted that housing was established in the centre. As the town expanded to city status and there was a need for industry and other associated uses, the housing was usually relocated to the outer areas of the city. The dynamic changing nature of land use in a city is accepted; however, this has direct implications for supply and demand fundamentals. So demand for a certain land use today (e.g. light industrial) may be gradually decreasing at the same time as another competing land use (e.g. retail) is increasing in the same location. It would be an ineffective property development if it was designed and constructed for the old land use, i.e. the focus was placed on the existing, albeit declining, land use rather than the future land use. The 'visionary' skills of the successful property developer come into play here.

The property developer must closely monitor the changing nature of the real estate market and make their own judgement as to what will be well received into the marketplace. Often this will require an application for planning permission for a change of land use; however, it must be supported by a well-argued submission and be within the existing planning framework.

Land is physically only in one location and can't be moved

A land parcel is fixed in its location and regardless of how many modifications are made to any structures built on or in it (such as converting an office building to a residential land use), the locational characteristics of land cannot be changed. From a property cycle perspective this means the land component (and associated structure) will be subject to the prevailing market conditions within which the land is located. An example is a proposed property development located on an inner-city allotment. If the inner-city or downtown area is going through an urban transition phase and there is a high crime rate and associated stigma attached to a precinct, the perception of the new development is likely to be affected. The impact of such factors cannot be altered by the developer and such a downturn in market sentiment is unavoidable. Regardless of how large a hole is dug or how tall a building is, the location of a parcel of land cannot be relocated to a different market.

Most investment in property is by individuals who trade infrequently

Most landowners trade infrequently and therefore they have limited knowledge about property and real estate at any particular point in time. While the largest land use in urban society is housing and most landowners only own a single parcel of land, in general the property and real estate market has undertaken relatively little research about supply and demand fundamentals. This is in direct contrast to other investment mediums (e.g. term deposits, exchange rates, equity markets) where there is a large amount of information available about past, current and predicted future trends. Therefore, individuals are at times making property and real estate decisions with very limited information and knowledge about the surrounding market conditions at a particular point in time. From a property cycle perspective this can equate to demand for a new property development in the construction phase; however, upon completion the same purchasers may seek to sell their investment and then 'flood' the market with resales. This may result in a sharp price decline due to the lack of knowledge by investors about economic market fundamentals.

5.3 Types of property cycles

There are as many different property cycles as there are different property markets; however, this discussion is focused on the theoretical framework of property cycles. Therefore, each property developer needs to conduct their own research into each individual market to identify the length and amplitude of the cycles in their prevailing market area. A generic property cycle does not exist.

The underlying structure and operation of the property and real estate market in most regions causes some form of cyclical behaviour to be observed. Even though cycles are an accepted part of the on-going operation of the property market, they are often poorly understood. This is partly due to the challenges in identifying and examining reliable information about the property market in which the project is located. In isolation this aspect is often the main difference between a good developer and a poor developer, being also linked to the ability to survive a market downturn until the next upswing occurs.

A starting point for property cycle analysis is to consider this statement: 'everything on the planet and including the planet will return to dust', especially in light of the built environment (i.e. buildings and structures) which form a major component of a property development. In this context it can be argued that every item experiences a period of rising (i.e. growing, increasing) and then falling (i.e. shrinking, decreasing). Some goods (e.g. equity markets) repeat this cycle many times where others just undertake one cycle (e.g. a disposable item which cannot be recycled). The former also applies to most property markets where an on-going and circular boom–bust cycle occurs. Understanding why this occurs in a particular market sector is critical.

Historical evidence confirms that property markets throughout the world experience cycles based on the changing relationship between supply and demand. Many of these boom–bust phases can be somewhat predicted and (especially with the benefit of hindsight) are inevitable. A market downturn as part of a property cycle is of no surprise to a good property developer who will be ready with a survival strategy. An extreme example of a property cycle is based on the effect of a new mining company which has just commenced mining in a township. Initially the discovery of mineral deposits (e.g. gold) will cause a sharp upward swing in the value of property in the town. After a period of stabilisation where supply is increased to meet demand, prices will then even out. In some instances the mine itself closes and the property becomes practically worthless (or even of negative value) and the whole area becomes a 'ghost town'.

Originally the notion of a 'seven year' cycle was encouraged as a representation of the difference between good and bad harvests (Niehans 1992). This period of seven years is still accepted in some markets as the common length of a cycle although it appears to be a rudimentary means of allocating such a defined time period. According to Burns and Mitchell (1946) there have always been at least four distinct characteristics of business cycles:

1   can only be found under a capitalist society and not under other systems;
2   not limited to a single firm or industry, but is economy-wide;
3   one cycle follows another, marked by a similar sequence of events; and
4   cycles differ in length and amplitude.

A large amount of cycle research has identified four prominent cycles which require further discussion.

(a) Short-term cycles

Commonly accredited to Clement Juglar (1819–1905), these short-term cycles refer to a period of between seven to eleven years and were based on historical financial information. In many circles the Juglar cycle was also known as the fixed investment cycle (Tylecote 1994). Juglar studied business conditions and confirmed that their actions were not completely random, but generally operated within a number of regular cycles. Importantly Juglar convinced economic science that advanced economies inherently develop in cycles, and persuaded the business world of the existence, persistence and pervasiveness of these cycles (Niehans 1992). Juglar clearly described three phases in the cycles, namely prosperity, crisis and liquidation. It was also shown that 'every cycle in every country is seen to be, to some extent, historically unique, and in its length has to be explained in light of its particular features' (Niehans 1992, p. 554). Juglar was the first person to use a combined time series analysis of factors such as interest rates, prices and central bank balances to analyse a well-defined economic problem (Tvede 1997). Furthermore, Juglar was the first person to acknowledge that depressions were adaptations to situations created by the proceeding prosperity.

(b) Medium-term cycles

Brought to prominence by Kuznet was the medium-term cycle, typically over a period of fifteen to twenty-five years. It has strong econometric support (Solomou 1987) and downswings can be triggered by sharemarket crashes. The Kuznet cycle has been explained by the dynamics of investment in buildings and land, and in other long-term assets such as stocks and shares (Tylecote 1994). For example, the Wall Street crash in 1857 was a clear Kuznet downturn with upswings influenced by changes in immigration and rainfall variations.

(c) Long-term cycles

Early research by Kondratieff promoted the existence of long-term cycles, with a complete cycle stretching over a period of approximately forty-five to sixty years. This theory was based on a number of 'long waves' where there was a serious of oscillations with an initial over-investment in capital, then a recession, until the invention of new technology eventually led to a new spurt of investment (Tvede 1997).

An overview of the long-term cycle would be as follows:

1   first cycle commenced with the Industrial Revolution;
2   second cycle followed the railroad boom;
3   third cycle started by electrical power plants and reinforced by industries such as automobile, steel and textile industries.

(d) Extremely long-term cycles

Although rarely discussed, this type of cycle is based on international relations and refers to four generations over a period of 100–120 years (Modelski 1987). Significant events include the timing of global wars and variations in the balance of global power. According to Modelski (1987) these four generations can be broken down as follows:

- 1st generation – a global war takes place with the emergence of a new world power controlling world trade and economic relations;
- 2nd generation – a period of uncontested leadership;
- 3rd generation – delegitimation occurs; and
- 4th generation – a period of de-concentration occurs with a serious challenger emerging. The 1st generation then follows soon after.

The property cycles described are typical examples of cycles that may operate in a particular market and many cycles will be loosely relatedly to the shape in Figure 5.2. At the same time it is commonplace for multiple cycles to be operating simultaneously in the same market and this may further complicate the property cycle analysis as undertaken by the property developer. Reference to Figure 5.2 highlights a very simplified observation in a market over time (measured in years) where there three distinct cycles. The shortest cycle (a) is repeated approximately every four years although it has the smallest level of volatility (amplitude). The next longest cycle (b) is repeated every twelve years although it has a much larger level of volatility. The third longest cycle (c) is repeated approximately every fifty-six years; however, it has an extremely large level of volatility.

Property developers should give further consideration to some additional observations on Figure 5.2, as discussed here:

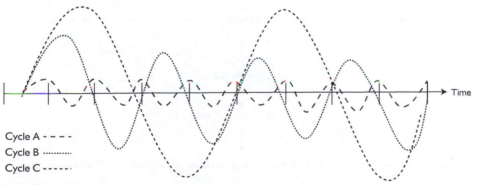

Cycle A - - - -
Cycle B ··········
Cycle C - - - - - ·

Figure 5.2 Multiple cycles in a single market

- The theoretical aspects of a property market will not convert directly into practice at all times. Many successful developers have no formal university education, but they are able to make reliable judgements on a daily basis. While theoretical models such as Figures 5.1 and 5.2 are useful as a decision-support tool, they lack the 'human factor' and 'gut instinct' input which forms such a large part of property development.
- The cycles in Figure 5.2 (and in many models) and been smoothed to remove some of the volatility. For example the largest cycle in Figure 5.2 appears to commence an upward cycle; however, it actually decreases and then increases again before reaching its peak. The challenge for any type of analysis is to determine when it is the optimal time to adjust the data. This can only be resolved with skill and experience since all data can be manipulated to some extent.
- The shape of each cycle in itself will vary. Often reference is made to a 'boom–bust' cycle and this typically refers to the 'bust' or severe downturn that occurs. The downside shape of the cycle will not necessarily mirror the upside shape of the cycle – it would be unreasonable to expect this simplistic formation.

5.4 Business cycles and structural change

As many property developments and investments are closely linked to the economic and business frameworks, there is usually a relationship between property cycles and business cycles. This relationship is highlighted in Figure 5.3 where there is a clear relationship between (a) the property market, (b) the overarching 'real economy' and (c) financial markets or the 'model economy'. Consideration should be given to the

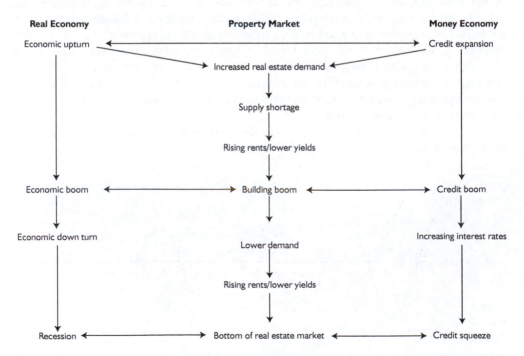

Figure 5.3 Property market, financial and economic framework (Source: adapted from Barras 1994)

direction of the arrows in Figure 5.3 where the property market has either a passive (one-way) relationship or alternatively there is a two-way relationship. An example is an economic upturn or downturn which will affect the property market; however, it is a one-way process. On the other hand the property market has a two-way relationship with both a recession and a credit squeeze.

An example is between a high level of interest rates in the business cycle and the negative effect on the property developer which limits their capacity to borrow funds for the development. The interaction of the short-run business cycle with property cycles creates great variability in a developer's plans and the ability to progress schemes at different times. The developer also needs to be responsive to the more evolutionary changes which occur in occupier preferences as a result of long-term changes in the structure of the economy.

(a) The business cycle

Research in the last decade or so has established the nature of the link between the economic, or business, cycle and the property market. Useful references on this complex topic include various papers on building cycles by Richard Barras (for example see Barras 1994).

Three important cycles have been identified, all of which exhibit different periodicity: the business cycle (which drives the occupier market), the credit cycle (which influences bank and institutional funding) and the property development cycle itself. A simplified version of Barras' analysis follows, beginning with the business upturn:

1   Strengthening demand, rising rents and capital values trigger the start of the new development cycle upswing.
2   If credit expansion accompanies the business cycle upswing, it can lead to a full blown economic boom. The banks may also fund a second wave of speculative development activity.
3   However, because of the long lead times in bringing forward new development, supply remains fairly tight and values continue to rise.
4   By the time the development cycle reaches its peak, the business cycle has already moved into a down-swing, accompanied by a tightening of monetary policy to combat the inflationary effects of the economic boom.
5   As the economy subsides, the demand for property declines; rents and values fall as a result and the vacancy stock increases in supply.
6   As the economy moves into recession, the fall in rents and values continues, property companies are hit by the credit squeeze, bankruptcies increase and the development cycle is choked off.

As Barras also pointed out, the experience in Britain has been that a property boom of the scale of the late 1980s is typically followed by a more muted development cycle. Banks and investors are struggling with debts incurred during the most recent recession and are disinclined to fund speculative development. At the same time, oversupply of property built in the previous boom is sufficient to meet demand during the whole of the following business cycle. It is only when supply is exhausted, at the start of the next business cycle upswing that the speculative development cycle takes off once more.

Clearly, the likely success of development projects will be influenced by where in the cycle they are started and completed.

(b) Structural change

Underlying the short-term business cycle are longer-term shifts in occupier requirements which result from structural changes in the economy. For example, the recent expansion of demand for very large warehouses has resulted from strategic reorganisation within the retail and logistics industries, helped by the increasing availability of sophisticated information technology. Similarly, the changes in working practices among office occupiers, again encouraged by developments in information technology, will most likely generate demand for new kinds of office building in the future. Developers (and investors) who monitor these long-term changes can begin to create new types of product ahead of the rest of the market; equally they can avoid being left with buildings which have a diminishing 'shelf life'. This is one area in which property market research can be of great use.

5.5 Surviving a market downturn such as a GFC

In most developed countries the media provides regular commentary that the real estate market is either (a) experiencing a 'boom' in real estate values or (b) has entered a downturn in the cycle. Since the market constantly fluctuates within this cycle the central question to be asked is not 'if' a market downturn will occur, but 'when' will this downturn occur? (See for example long-term cycles in Figure 5.2.) However, despite evidence that severe market downturns, such as the Global Financial Crisis (GFC) in 2007–8, and the Asian Financial Crisis (AFC) in 1997, are also a regular occurrence, many real estate developers ignore the likelihood this will occur in the lifetime of their business. This natural cycle results largely from over-exuberance and also the misguided belief that markets and associated values will always increase in value. The inevitable downturn tests the survival skills of real estate developers who must deal with minimal demand for new projects and little or no cash inflow into the business while somehow managing to keep on top of their existing repayment liabilities (i.e. cash outflows). The lack of cashflow, at a time when financiers are especially risk adverse with tightened lending criteria and associated higher interest rates, usually forces a property developer to downsize or cease operations with a loss of goodwill. A careful understanding of supply–demand fundamentals coupled with retention of access to cash will assist developers to survive until the next inevitable upswing in the market. The key to survival is experience and a reliable business or organisational strategy.

The cyclical nature of real estate markets can have a positive or negative effect on the operation of a real estate developer. Bad property development decisions (e.g. initially paying too much for vacant land, over-extended selling period for completed units, etc.) can be hidden in a rising market which may cover such inefficiencies for a period of years. On the other hand a market downturn (i.e. 'bust') will test the survival skills of every developer who will seek to wait out the decline until the next upturn. Even though history has consistently proved that there will always be a level of uncertainty about the future, developers often miss the opportunity to protect themselves by resisting overexposure to debt or gaining an understanding of the interaction between

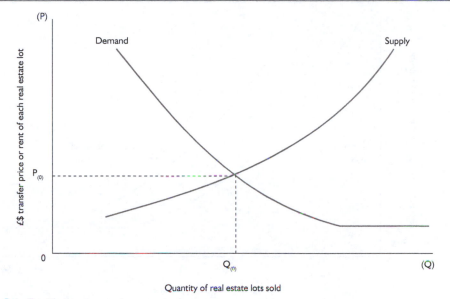

Figure 5.4 Equilibrium in a real estate market

supply, demand and the current position within the cycle to enable them to build this into their calculations.

It is widely accepted that upturns and downturns in property values are regular and unavoidable events in real estate markets. It is relevant here to refer to standard economic theory which allows a property developer to model varying scenarios based on movement along supply and demand curves, although the most relevant to understanding a market downturn is the shift or movement in demand. Understanding such models based on the interaction between supply and demand is often second nature to a successful property developer who will be able to identify (i.e. probably already knows) a certain price point when, for example, a real estate product can be priced out of the market resulting in no demand. The emphasis should be placed on the point at which there is a decreased level of demand where the following question needs to be asked: how will the agreed sale/rental value affect demand? As observed in Figure 5.4 the optimal scenario for a property developer is when supply equals demand, i.e. avoiding an under-supply or over-supply relationship. Therefore, the quantity demanded $Q_{(0)}$ equals the cost of each real estate unit $P_{(0)}$. Unfortunately this optimal scenario, commonly known as 'equilibrium', is rarely achieved in reality as there is constant movement (supply, value and demand) of real estate which upsets this balance. A property developer should be familiar with the shape of the demand and supply curve for their particular target market such as the office market. For example in Figure 5.4 there is a minimum ceiling for real estate units to be sold where it is not possible to purchase or rent a real estate product below a certain price.

The model in Figure 5.5 highlights the effect that an external a structural change (e.g. higher or lower lending interest rate as set by the government) can affect the real estate market and also then affect the level of market demand. Often this is an external shock and is not as predictable as a market downturn in a smaller cycle (see Figure 5.2). For example, a decrease in demand may be due to a catalyst such as an

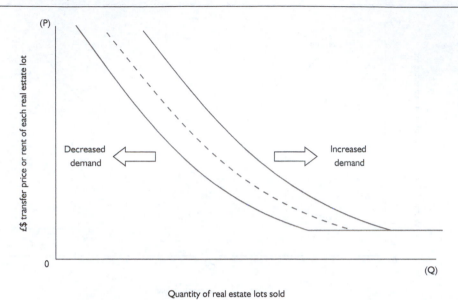

Figure 5.5 External shift in demand for real estate

oil shortage crisis, a stock market crash, a country at war or pressure on a particular currency (e.g. Eurodollar) (see Figure 5.3). This catalyst starts a domino effect which snowballs through many other markets including real estate markets. Usually this also adversely affects unemployment levels which then filters into the real estate market, e.g. lower demand for office space, ability to repay a mortgage.

After there is a downward shift in the demand curve (Figure 5.5), this creates the over-supply scenario in Figure 5.6 where this is a clear gap between the original $Q_{(0)}$ and the new $Q_{(1)}$. At the same time there is a decrease in the agreed sale price of each lot, down from the original $P_{(0)}$ to a lower $P_{(1)}$. In order to return towards equilibrium there are only two realistic options, these being either to (a) increase demand or (b) decrease supply. Since the real estate market is lumpy and not able to quickly reduce supply (i.e. it may take years), the property developer must be able to survive a sustained long-term period of depressed prices until the market recovers to a supply equals (or is less than) demand scenario.

In real estate markets there is a commonly used adage that 'knowledge is power' and therefore a successful developer will not freely share their survival skills. An analogy can be drawn here with a magician who is typically very reluctant to share their secrets with third parties. Accordingly it is not possible to list here a successful remedy in the form of an action plan. A large market downturn arguably strengthens the operation of the market in the post-crash phase where property developers who recover are stronger and better prepared. The poorly prepared property developers do not manage to survive, often due to cashflow problems or shortages.

Every successful property developer has a formal plan for surviving a major market downturn and prospering on the other side of the cycle during the upturn. The key is timing. In mainstream society one of the most popular sayings when referring to property is 'location, location, location' with seemingly less regard for other market characteristics. However, a successful real estate developer would argue that the most

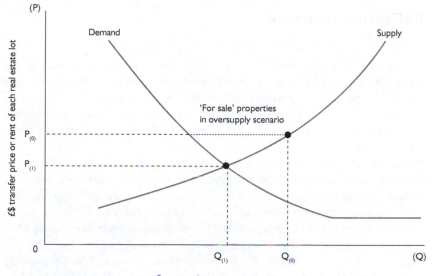

Figure 5.6 Over-supply scenario due to lower demand

important three words are actually 'timing, timing, timing'. This statement emphasises there is usually long-term demand for most locations where there is already an established property market; however, this demand will only occur when the timing is optimal. But the question remaining is: timing of what part of the market? There will be a different timing for each of the following submarkets as based on (for example):

1   different land uses, e.g. residential, industrial, retail;
2   different locations, e.g. country, state, city, suburb, block, street;
3   state of the local economy, e.g. employment, GDP; and
4   how the political parties are operating and other factors, e.g. impending election.

As the property market is clearly cyclical, the successful approach for property developers is relatively straightforward: resist selling in a market downturn when supply exceeds demand and the profit margin is reduced. This is not always easy to avoid, especially when many companies face cashflow problems at some stage. As property developments are typically funded with the help of a financier and a large exposure to debt, this creates a situation which then revolves around servicing the debt. In a large market downturn a financier is also under pressure due to other bad debts (i.e. other failed companies unable to repay debt) and therefore the financier increases the pressure on the property developer to repay debt. The only viable solution is for the property developer to have reliable access to cash which will be available until the market recovers. At times this may take years, hence a substantial cash supply, or alternatively another source of finance, is essential for survival. A seasoned property developer will reduce their exposure to finance loans during a market downturn; however, this strategy relies on knowledge about timing of the market downturn.

5.6 Reflective summary

It is essential to have a good understanding of property cycle behaviour in the market where a specific property development is being proposed. Unfortunately there isn't a generic property cycle but this research area, although relatively complex, should not be ignored if a developer is seeking to maximise returns and minimise exposure to risk. Both objectives should remain the primary aims of a successful property developer.

Even though property cycles are an accepted part of market behaviour they are poorly understood. A property developer, if correctly informed by accurate research into prevailing property cycle behaviour, may at times delay their commencement date to ensure the development is released onto the market at the highest point in the cycle (Figure 5.1). Nevertheless there are usually multiple cycles in operation in a property market (see Figure 5.2) and selecting the optimal timing can often be the single largest challenge in the area. The property cycles are unpredictable when looking to the future and are adversely affected by the decisions of others, such as a competing property developer's decision to delay (or not to delay) their development.

Property cycle behaviour also has direct links to other cycles including the economic and financial cycles. The timing of other cycles can adversely affect the property cycle and must be factored into the long-term planning of the property developer, as far as it is possible to do so. Historical market downturns, such as the global financial crisis in the last decade, highlighted the interlinked nature of the property market from an international perspective. Property developers must consider the available information and make judgement calls based on their data; however, their final decision will be largely based on their experience and judgement. The availability of cash to meet loan repayments during a market downturn is often the difference between survival until the next upturn. While a financier may offer to lend a large financial sum at a comparatively low interest rate in prosperous times, it is critical to model the ability to service this loan over an extended downturn period with associated higher interest rates. As much as there is a large base of information on cycle behaviour, the property market itself operates largely on the decisions of humans with their associated perceptions. Therefore, an understanding of property cycles will assist as a decision-support tool; however, it cannot incorporate all future changes in the marketplace and should be assessed on a development-by-development basis.

5.7 Case study – establishing a track record as a successful property developer

Wilbow is an established and highly respected property developer currently involved in a wide variety of residential lot developments in the Dallas/Fort Worth and Houston areas of Texas. Wilbow has been the recipient of numerous accolades and won business and industry awards for excellence in commercial property development including the

Plate 5.1 Wilbow Corporation logo

Urban Design for Tribute at Mills Branch and the McSam award for Marketing of the Heritage residential community. Established in 1988, Wilbow Corporation, the US Operations, is a wholly owned subsidiary of Wilbow Group Pty. Ltd.

Wilbow Corporation (Plate 5.1) is owned by the Bowness family whose founder, Bill Bowness, established as a leading property developer in Melbourne in 1976. Mr Bowness has some forty years of experience in the real estate industry, having served as president of an Australian investment firm from 1971 to 1976 prior to forming Wilbow. The firm grew into one of the leading land developers in Australia, and through a review of international markets, expanded its operations to the United States. Through investments made in Hawaii and the west coast during the early 1970s, Mr Bowness travelled frequently to the USA, which led to Texas visits and the eventual decision to establish a corporate presence in the Dallas/Fort Worth market.

Company's property development philosophy

The mission statement of Wilbow Corporation is as follows:

> We develop properties into the best presented estates within the respective submarkets, through the formation of mutually beneficial relationships with municipalities, engineers, contractors, lenders and home builders.

Insightful quotes from the Chairman and CEO William Bowness

This section refers to comments from William Bowness about surviving as a property developer through property cycles.

> We played the late-80s boom as hard as we could as we knew it would come to an end and we prepared ourselves accordingly.

> I learned during my banking experience that if you look after your bank, keep them informed, let them know what is going on, keep your word, they will take care of you.

> What breaks developers is not a bad buy but the holding costs.

> Understand the economic and political scene, understand that the world changes and interest rates go up.

But one of the great things of a bust for a professional developer is it puts the part-timers – the would-be, could-bes – out of the marketplace.

I have rolled the dice a great deal but have always had a plan A, plan B or plan C.

Achieving best practice in property development

The material growth envisaged for the US Operations is now being realised after the global financial crisis. Wilbow strategically acquires new properties to supply the acute shortage of new home sites in the most desirable locations, including Flower Mound, Colleyville, SH 114 corridor. In addition, Wilbow maintains its current portfolio of communities and successfully attracts new additional homebuilders to existing communities.

Property development involves a myriad of disciplines that include finance, engineering, market research, building sciences, landscape architecture and much more. A successful approach combines these skills into a cohesive product. Wilbow's success is determined by their people. A staff of highly qualified and flexible personnel is assembled in Dallas. A network of award-winning consultants who are leaders in their field are employed to complement the in-house team. Expertise, ideas and information are exchanged between Wilbow staff, external consultants, builders, lenders and market strategists and also city leaders to introduce new solutions and innovations to the market. This provides Wilbow with the rare opportunity to exchange global best practice initiatives. The company's ethic of detailed planning and design helps eliminate the unexpected and results in properties of the quality and style for which Wilbow is known (see Plates 5.2 and 5.3).

Qualifications span surveying, civil engineering, architecture, landscape architecture, valuations, accounting, building, banking and real estate. They are backed up by technology and internal systems that help identify projects with the most potential and accurately track their progress to assist in financial control. Wilbow's staff actively participate in a wide variety of industry associations. This participation provides Wilbow with direct involvement in industry issues, causes and objectives.

(This case study is based on information sourced from www.wilbow.com (2013) with the consent and input from the Chairman and CEO William Bowness.)

Plate 5.2 Completed dwelling in the Wilbow Corporation development

Plate 5.3 Well-planned Wilbow Corporation development

Chapter 6
Planning

6.1 Introduction

The role of planning in property and real estate markets is designed to co-ordinate the efficient use of resources, being the limited supply of land either vacant or improved. Another critical objective is to restrict conflicting land uses from being located directly next to each other, such as building a new oil refinery next to an existing residential area. The process of planning involves many stakeholders who are both directly and indirectly associated with the proposed development. While some cities are criticised for planning legislation which is too prescriptive and generally considered too inflexible, other cities have a complete absence of planning where the market determines land use and development takes place on an ad hoc basis. A successful property developer will understand the concept of planning. While planning legislation is commonly at the local level a developer should understand all aspects of planning law which have the potential to affect land use and the development potential of a particular site.

Planning policy is constantly evolving in response to both environmental and political factors such as the increasing importance of sustainability in the built environment and the broader society. In many countries it is necessary for developers to obtain an individual planning consent for virtually every property development project. It is beyond the scope of this book to cover in detail all aspects of the planning system and readers are referred to other texts for further reading. The aim of this chapter is to provide an appreciation of the principals involved in the planning process which are a vital part of development practice. The planning process and the associated legislation will ultimately determine the type of development allowed on any site and thus influence the value of the completed development. Criticisms are often that the planning system and process adds uncertainty to development in the form of time delays, potential cost increases and at times, unpredictable decisions which may even be overturned on appeal. These are issues that concern developers who have in the past misunderstood the time and resources required to progress through the planning process before any development can actually begin.

6.2 Planning and the environment

The higher profile of environmental concerns has led to sustainable development goals being increasingly demanded and hence incorporated in development proposals, the wider involvement of third party consultees but also frequent delays in the consideration of planning applications. As a consequence of the debates held at the 1992 Earth Summit in Rio de Janeiro and United Nations Bruntland Commission, many international governments have been committed to the concept of 'sustainable development', i.e. 'meeting the needs of the present without compromising the ability of future generations to meet their own needs' (WCED 1987). Over time there have been increased linkages between town planning and environmental law and regulations and these close ties have continued in many aspects of land use planning and property development.

The details of the planning legislation and its interpretation vary between regions, even adjoining towns, and are heavily influenced by the political party in governance. For example a 'green' party would have an above-average sustainability agenda which may affect the requirements imposed on a new property development in that area. The advent of social media has repeatedly demonstrated the groundswell which can occur when a property development may affect environmentally sensitive land, although sections of the public feel the planning laws have not adequately considered the full consequences of their decision on the environment. The developer must be fully aware of the needs of the community and corporate social responsibility (CSR) is often undertaken by the developer as well.

Other aspects regarding planning and the environment are covered in more specialist books and websites listed in the references at the end of the book. This chapter comprises a broad summary of how the planning system operates and its implications for those involved in the property development process. A large number of government agencies and departments are involved in the planning system. This includes dedicated planning departments depending on the size of the town, region, state/county/borough and country. Their aim is to control activities classified as development.

6.3 What is development?

The concept of 'development' has varying perceptions for different stakeholders in the real estate market and also in wider society. Some sectors of society are anti-development at one extreme (e.g. based on the 'sustainability' argument), regardless of the need to redevelop and reinvent to increase the level of efficiency in the urban market, e.g. less traffic congestion, fewer vacant disused properties. Other arguments are based on the desire to retain and preserve all older buildings in their original state, regardless of advances in design and construction approaches. Cycles exist at all levels (see Chapter 5) and all improvements will eventually become physically obsolete, regardless of other factors. As a result undertaking development is an essential part of our society. Loosely explained, development can also be described as 'renewal', 'renovation', 'reusing' or 'updating'; however, there are as many different types of development as there are shades of grey and it is difficult to encapsulate development with a single definition.

In relation to planning legislation the following activities are generally classified as development:

- the carrying out of building, engineering, mining or other operations in, on, over or under land, or
- undertaking any material change of use in any buildings or other land.

The concept of development can be divided into two categories, one being the physical operations such as building or engineering works and the other being the making of a 'material change of use'. The discussion in this book about property development tends to focus on the physical activities, but the question of use is important. Indeed it is possible to argue that planning control is basically one of land use, because once the use of land has been determined, the question of precisely what is built on it is arguably a matter of detail.

Not all activities will require planning permission, for example in the UK some minor building works come under the heading of 'permitted development' and do not require planning permission, such as building a boundary wall or certain changes of land/building use (e.g. if the changes are within the same *use class*). However, most types of *development* do need permission.

6.4 The planning application

Making an application

If a property developer is carrying out a development for which planning consent is required then usually they must apply to the local planning authority for consent. Since not all development activities require consent, it is prudent to check local planning legislation to establish whether or not permission for a proposed development is required. In some regions efforts have been made to minimise planning constraints and barriers by removing certain types of development from planning control and the need for planning consent. Where it is not clear whether planning consent is required, the planning authority can be approached for informal advice. It is also possible to formally establish whether a proposed development, use of land or building is lawful by applying to the local planning authority for a formal 'certificate of lawfulness'. This will state whether the proposals would constitute development for which a planning consent is required. The local planning authority must give a formal decision within a reasonable timeframe (e.g. eight weeks) and there is normally the option for a right of appeal against the decision.

In most cases anyone may apply for planning consent in respect of any property. Furthermore it is not necessary for a particular applicant to have any legal or financial interest. However, if the applicant is not the owner of the property then it is normally a requirement for a notice advising of the application for planning consent to be served on:

a. the freehold landowner, or
b. any lessee with an unexpired term (e.g. at least seven years still to run), or
c. any occupier of agricultural property.

Most planning authorities stipulate a particular form which must be used to serve this notice, where the applicant must advise the planning authority of the names and addresses of the people on whom they have been served. In certain instances of property development (e.g. buildings for various types of public entertainment or buildings exceeding certain heights), these applications are usually publicised by the local planning authority who will place a notice in the local press stating where any member of the public might inspect the plans of the proposed development. Where the development might have a significant effect on neighbouring property, the authority is usually obliged to draw the attention of neighbours to the application. Any owner or lessee or occupier of agricultural land upon whom notice is served, or any other member of the general public, has the right to make representations to the local authority. There are usually different types of application which can be submitted to a local planning authority dependent upon the type of development which is proposed and the location of the site concerned. Examples of the main types of property development applications are as follows:

1   *Outline planning permission application* An outline application is generally used early on in the development stage to establish whether or not the proposed development is likely to be approved by the planning authority before any substantial costs are incurred. The application is made to establish the principle of a particular form of development outside conservation areas, without the need to deal with the matters of the siting of the buildings, their design, external appearance, landscaping or the means of access into the site. These 'reserved matters' can either in whole or in part be left for the future submission and determination of the planning authority in the event of outline permission being granted. A typical example of such an application would be for the residential development of a greenfield site on the edge of a settlement. The amount of detail the applicant is required to include in an outline planning application may vary between planning authorities; therefore, the applicant would be wise to check the requirement with the local authority first.

2   *Full or detailed application* A detailed application will seek not only to establish a land use principle but also approval for all the 'reserved matters' listed above. The application is comprehensive and in the case of a residential development would include not only the information about the location of the site but also the layout of houses and roads, the design of the dwellings themselves, the principle landscaping proposals and all the relevant technical information. As the design of new buildings and other reserved matters are particularly important considerations in conservation areas (see below) it is mandatory to submit full applications in such areas.

3   *Changes of use* Legally an application for the change of use of land or, more commonly, buildings is regarded as a full application rather than an outline. This requires an applicant to submit full details of their proposals to the planning authority. Changes of use applications are normally, although not always, related to where such permission is required to change from one category of land use to another. A typical example would be the change of use from industrial land use to residential land use such as a block of flats or units.

4   *Applications for heritage listed building consent or conservation area consent to demolish* All proposals to demolish buildings in conservation areas requires specific permission and thus proposals for development in a conservation area must also, if demolition is envisaged, be accompanied by a separate application for conservation area consent to demolish. Similarly any changes proposed to heritage listed buildings, i.e. those specifically identified by the planning authority and relevant government bodies as being worthy of protection, must be the subject of specific applications.

A developer wishing to carry out building or engineering works should consider applying for an 'outline' planning consent before the detailed application. However, the decision on which application to submit first will depend upon the following:

- the location of the site
- the issues involved in the proposed development, and
- the nature of the developer's legal interest in the site.

For example a developer may only have an option to acquire the freehold of the site or a conditional contract which is triggered by the granting of a satisfactory planning consent. Preparation of design drawings for a large building project can be time consuming and very costly in comparison to an outline planning application. Although an outline planning application is a relatively quick and inexpensive method of determining whether the proposed development is likely to be granted full planning permission, it must still contain sufficient information to describe adequately the type, size and form of the proposed development. The local planning authority will reserve for subsequent approval the reserved matters and will attach conditions to an outline planning consent to cover such issues. A typical condition (which would be one of many on an outline permission) is as follows:

> Before any development is commenced, detailed plans, drawings and particulars of the layout, siting, design and external appearance of the proposed development and means of access thereto together with landscaping and screen walls and fences shall be submitted to and approved by the local planning authority and the development shall be carried out in accordance therewith.

Costs

Fees are normally payable to the local planning authority in respect of applications and the appropriate fee calculated in accordance with scales prescribed (usually reviewed annually) must be paid at the time the planning application is submitted. Most often the fees are charged on a sliding scale depending on the size and nature of the development. As noted, the planning authority is required to determine each planning application within a reasonable timeframe (e.g. an 8-week statutory period) which runs from the date the application is registered by the determining authority together with the correct fee. If no fee or an incorrect fee is deposited with the planning application, then the reasonable timeframe period will not start to run until the correct fee is paid.

Processing the application

Planning applications must be made on forms provided by the local planning authority and are often available electronically online. The forms are self-explanatory but many applications are delayed because applicants either do not complete the forms accurately or fail to provide all the information required. When the application form is submitted, the planning officer (a qualified person employed by the local planning authority who will handle all matters arising from the application) should be asked to confirm that the form gives all the information required and that no additional information is needed from the applicant and also that no supporting documents are necessary.

The application will take its turn in a queue to be processed and checked; the proposed development site will be inspected by planning authority staff and there is normally consultation with other organisations, such as the highway authority or the water authority on site restrictions, or with amenity societies in areas of special environmental interest. The planning authority usually consults other appropriate authorities before giving planning permission. For example, the authority in charge of the main roads and freeways will be consulted on the design of any new roads which are ultimately intended for adoption as public highways, or the local Environmental Health Officer might be consulted on potential noise or pollution problems, or the Health and Safety Executive might be consulted on 'hazardous' or potentially explosive processes that may be proposed in the development.

The process of making a planning application will vary depending on the planning legislation within a country or region. For example a typical process in England is summarised in Figure 6.1.

There are various matters about which it is important that developers should satisfy themselves by means of direct discussions with the appropriate authorities. The more important of these are the adoption of roads and footpaths by the Highway Authority where it is appropriate for them to be adopted; the acceptance by the local authority of public open space provided within any development; and the necessary approvals under the building regulations, fire regulations or any other statutory provisions which relate to the development in question, e.g. the respective Factories Acts, the Shop Acts and Public Health Acts. Developers will agree directly with the relevant Highway Authority the design and specification for all new roads and footpaths and enter into the necessary adoption agreement. They will reach a similar agreement with the appropriate water company in respect of sewers. A planning permission in no way constitutes an agreement for adoption.

During this process the staff of the planning authority may discuss with the applicant points that need clarification or problems that are revealed. At this consultation stage, it is often possible to make modest corrections or alterations in the application to avoid such problems, but more radical alterations might require another round of consultations, or a new application. At the time of submitting the application, it is useful to ascertain the date of the council's planning committee meeting at which the matter will be considered and to check at the appropriate time that it has been placed on the committee agenda.

The local planning authority pays particular regard to the provisions of any approved development plan and any other material considerations in reaching its decision.

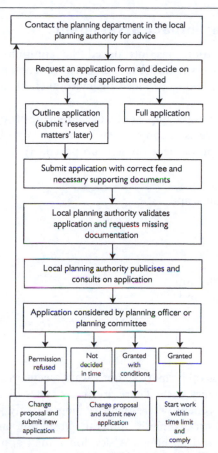

Figure 6.1 Adaption of the planning process in England (Source: adapted from www.planningportal.gov.uk)

However, if it is proposed to grant a planning permission which will constitute a substantial departure from the development plan, the local planning authority must advertise the planning application and give opportunity for objections to be made. The relevant government body overseeing the operation of the planning system will be advised of the proposals and may decide to determine the application (i.e. to arrange a public inquiry to hear the arguments for and against a particular proposal). In the absence of any intervention by a government body, the local planning authority may grant permission.

Discussion points

- Under what circumstances might it be prudent for a developer to make an application for outline planning permission?
- What other local authority departments consider the impact of the planning application, and why is this?

6.5 Environmental impact assessment

In many regions legislation has introduced the compulsory requirement for an environmental impact assessment (EIA) or an environmental impact study (EIS), where a developer is often required to produce an environmental statement (ES) as part of a planning application.

This statement examines the impact of a development proposal on the environment in its widest sense. There is a comprehensive list of factors which can be assessed dependent upon the type of development proposed. Typically this can include the following aspects:

1   Impact on the landscape
2   Visual impact
3   Ecology
4   Geology
5   Archaeology
6   Air and water pollution
7   Contamination of land
8   Noise pollution
9   Wildlife conservation
10   Agriculture.

Some projects, such as oil refineries, motorways/freeways and major power stations require assessment in every case. There are many types of developments which are typically subject to assessment where they are judged likely to give rise to significant environmental effects, for example, mineral extraction, and major infrastructure projects such as roads and harbours. A developer may apply to the local planning authority for an opinion on the need for environmental impact assessment prior to a formal planning application. A developer has a right of appeal if they disagree with a local authority's request for an environmental statement. The relevant government body may make a direction that an environmental impact assessment is required or in certain cases the developer may voluntarily prepare an assessment to allay any anticipated fears on environmental matters. The main difference to the normal planning process where an environmental impact assessment is required is that there are additional publicity requirements and an extended period (e.g. sixteen weeks in the UK) for the relevant planning authority to determine the application.

The developer can save time and a great deal of energy by consulting the relevant planning officer of the district council or appropriate authority at the outset, before any formal planning application for a reasonably sized development is submitted. Through early consultations it might be possible to agree the development proposals in principle with the planning officer. Planning authorities generally have the power to delegate various powers of decision to their planning officer (although they might not always do so), thus applicants should always ascertain whether the planning officer is authorised to grant the necessary planning consent or whether they are only able to make a recommendation to the planning committee. Planning authorities usually aim to give a formal decision within a reasonable period (e.g. eight weeks) of receipt of the application, although in practice any substantial application is likely to take longer than the recommended 'reasonable' period to reach a decision. If consent is then refused or granted subject to conditions to which the applicant has objection, an appeal may be made to the overseeing government body. If no decision is given within the prescribed reasonable time period, the applicant can assume that planning consent has been refused and may appeal to the higher government body. The right of appeal is not universally applied by developers if they consider that further negotiations beyond the reasonable period (e.g. eight weeks) is likely to result in planning permission being obtained.

Even though consultation with the planning officer does not guarantee that a proposal is acceptable, applicants should at least thoroughly understand the views of the planning authority by ensuring that they have a working knowledge of all the relevant development plans and other policies so that they are in a better position to consider how their application may be approached. Some planning authorities appoint advisory committees to comment on certain types of development. Most commonly, architectural advisory panels are set up to give advice on the architectural merits of proposed developments which are to take place in sensitive areas. Planning authorities must also consult local residents' organisations or amenity committees of various kinds. Amenity societies are normally consulted when development is proposed within conservation areas and the local planning authority has a duty to advertise the receipt of planning applications for developments in those areas. A local conservation area committee might exist and its comments will carry weight. The planning officer should be able to indicate whether it is possible or advisable to consult such committees or voluntary organisations at the informal pre-application stage.

With environmental issues now very high on the political agenda, it is becoming increasingly important for developers to consult with the relevant pressure groups at an early stage. Although planning authorities are advised by the government not to exercise detailed control over design (except in conservation areas), such issues are commonly disputed. This is because matters of design are subjective and capable of being interpreted in several different ways. The particular sensitivity which is associated with applications for development within defined conservation areas or affecting listed buildings arises because of the special protection afforded to such areas and buildings by the planning legislation. Planning authorities have powers to define conservation areas, i.e. areas of special architectural or historic interest which it is desirable to preserve or enhance, and this process can take place in conjunction with the preparation of a local plan or in its own right. Once defined, all planning applications for development within the area must be detailed and, as set out above, separate applications must be made for any demolition. For example it may be the prevailing government policy that all proposals for development within the conservation area must preserve or enhance its character. Accordingly there is commonly much debate between planning authorities, conservationists and developers over the design merits of schemes in conservation areas in the context of government policy.

If the development proposals entail the demolition or alteration of the character of buildings which are on a list of buildings of special architectural or historic interest prepared by central government (e.g. listed buildings), then special procedures are necessary. A specific consent for the demolition or alteration of the listed building must be obtained before any work can be carried out. Applications for the necessary consent must be made to the local planning authority and advertised, so that any member of the public may make representations. There are a number of reputable groups with an interest in the protection of listed buildings which in turn are consulted by planning authorities with proposals for alterations or demolitions of listed buildings. Developers may have to spend a considerable time discussing their proposals with interest groups, in an attempt to reconcile their commercial requirements and the sensitive refurbishment of a listed building.

If buildings appear under threat from development proposals, a local government authority may be able to issue a temporary listing also known as a 'building preservation notice' which has immediate effect and lasts for an extended time period (e.g. six months) to enable the respective government body to decide whether to formally list the building concerned. If the building is not subsequently listed, compensation may be payable to the developer for consequential loss.

Local planning authorities also consult archaeological experts if required. If the site is of archaeological interest, then the developer must allow an archaeological dig or a watching brief by archaeologists to take place before development commences. The period of the dig and any compensation or contribution from the developer is often a matter of negotiation between the developer and the interest group. For example under certain legislation specific areas of archaeological importance can be entered and excavated for an extended time period (e.g. up to six months) but often developers offer contributions to speed up this process. Here again, the development process may be delayed and, therefore, it is prudent for a developer to consult with archaeological groups. The general policy guidelines on archaeological matters should be considered prior to purchasing a site. Many archaeologists now act as private consultants to developers. Trees on the development site may be protected (for example by a Tree Preservation Order or TPO), it would therefore be necessary to obtain consent with the relative authority before felling or cutting them down. Normally the necessary application is made to the local planning authority but in certain cases the consent of a higher level of government (e.g. Forestry Commission) is necessary.

Advertising the development

In most regions there are strict regulations enabling planning authorities to exercise a very tight control over all external advertising, including site boards, on the property itself which promotes any development. This is frequently a matter in respect of which the planning authority will have a policy document. Normally the greatest weighting is given to considerations of visual amenity and public safety in those cases where advertisements might possibly distract the attention of road users.

6.6 Planning application approval

In certain regions some types of planning applications, although often relatively minor in nature, might be decided by the planning officers of a planning authority using powers delegated to them by the planning committee. Lesser items, such as house extensions or advertisements, might be dealt with quickly in this way. More important or substantial applications will be decided by a group (e.g. a planning committee) of elected representatives of the local government authority, who will meet regularly and normally have comprehensive powers to make a decision on the application on behalf of the council. The committee will be advised by planning officers whose advice is based upon their knowledge of the area and its problems and policies, and upon consultations they have had with such bodies as the highway and water authorities. Their professional advice which takes the form of a report will also be influenced by their knowledge of planning case law and how the council's policies operate.

Normally a planning committee will seek and listen to the advice of their professional officers and will also take note of any representations made by members of the public. However, it is important to note that the committee is not obliged to accept the officers' recommendation and may make a decision against the recommendation. Development is often highly contentious and public comment might come to the committee from individuals or groups or in the form of petitions. The committee must consider the variety of advice and representation it receives and make a decision to approve or refuse the application or perhaps, in some cases, to refer it back to the applicant to seek a modification.

Public interest in planning is often strong when the existing status quo might be disturbed. This statement is particularly true for a high profile site which may not be protected by historical legislation but has broader perceived value to the local residents. The planning committee is a committee of elected councillors who, as part of the democratic political machinery, recognise the impact that the decisions they may make has on their political standing. It is important for developers to appreciate, particularly where large controversial schemes are being prepared, that the political make-up of a specific planning authority, can, in practice, have a major influence on the eventual decision that is made.

6.7 Planning consent

Planning permission may be qualified in various ways and we briefly consider here the more important of these. Permission may be granted subject to a variety of conditions, for a limited duration of time or for the personal benefit of certain people or organisations. With regard to time limits, there are two aspects to consider. Every detailed planning permission will lapse unless the development is commenced within a set time period (e.g. five years) from the date the original permission is granted or such other time as the planning authority stipulate. The relevant legislation (e.g. a planning Act) includes provisions which change the duration of the planning permissions and consents. With outline permission, the necessary application for approval to the various reserved matters must be made within a time period (e.g. three years) of the outline consent being granted. The development must commence within say five years of the date of outline approval or within say two years of the date of detailed planning approval, whichever is the later, subject to the imposition of any other specific time limit by the planning authority. Only a very limited amount of work on site (known as 'material operations') is necessary to prove that development has commenced to meet the time-limit requirements. Decisions in the court systems at times have decided that in order to show that a planning permission has been implemented, there must be a genuine intention to carry out the development concerned when a commencement is made on the site. This 'test' has caused developers problems, particularly if planning permissions are obtained at the height of a property boom and are not implemented and are then due to lapse during an economic slump. Developers must choose either to renew permissions (which require renegotiations with the planning authorities) or to let existing consents lapse. To deal with the evasion of the time-lapse provisions attached to planning permissions, local authorities usually have powers to serve a completion notice, the effect of which is that planning permission will lapse unless the development is completed within a reasonable time. In addition there are usually rights of objection for the people who are affected by them which are considered by the relevant government body depending on the location of the property development.

The second aspect of time limits is when the planning permission remains in force for a limited period and at the end of that period any buildings or works which have been erected must be removed and any use authorised by the permission must cease. At the end of the limited period of planning permission things will revert back to the state which existed before the permission was granted. On the expiry of limited period planning permission, it is open to the applicant to make a new application for the retention of the buildings or the continuance of the planning use. Note that planning permissions for limited periods are of limited value.

Conditions may be imposed limiting occupation to a particular type of occupier. These fall into two broad classes. There is the condition which limits occupation of a building to someone engaged in a particular trade or vocation. Perhaps the best-known condition of this type is limiting the occupation of agricultural cottages to those engaged in agriculture. The other and more restrictive type of condition, although rare, is one which limits occupation to a particular occupier personally. The ability to impose limitations on the type of occupiers sometimes enables a planning authority to grant a permission which otherwise it would not be prepared to consider.

The powers of planning authorities to impose conditions on a planning permission is an accepted part of property development and planning authorities will provide guidance on the use of conditions in respect of transport, retail development, contaminated land, noise and affordable housing. Other matters covered include use of conditions for design and landscape, truck access, 'granny' annexes, staff accommodation, access for disabled people, holiday occupancy and nature conservation. Planning conditions are normally only imposed where they are necessary, relevant to planning, relevant to the development to be permitted, enforceable, precise, and reasonable in all other respects. A typical example might be a condition requiring landscaping, proposed as part of a scheme, to be carried out in the next planting season available and maintained thereafter to the satisfaction of the local planning authority. There is the same right of appeal against the imposition of a condition on a planning permission as there is against the refusal of planning consent.

Discussion points

- What are the various qualifications that may be attached to planning permissions?
- How can these qualifications affect a developer's proposals?

6.8 Planning agreements or obligations

While the planning authorities prepare the policies and plans to establish the planning framework for a particular area, this in itself does not bring about the implementation of the development plan. Implementation depends upon landowners and developers who initiate development proposals within the planning area. Having prepared the planning framework, the local planning authority normally waits for developers/landowners to produce specific planning proposals and to make planning applications. In addition to the power to, where appropriate, attach conditions to the grant of permission it is possible for planning authorities to enter into legally binding agreements with developers which enable development proposals to come forward in circumstances where the authorities could not rely solely on their statutory powers of control, i.e. planning conditions. Planning agreements (referred to as planning obligations) between a developer and a planning authority may be made under the provisions of the current legislation in the region. The concept of planning obligations refers to both planning agreements and unilateral undertakings. Agreements might be made to phase the development of land to accord with the dates when various public services will become available or improved road access will be provided. Similarly, agreements might be made with regard to the provision of land within a comprehensive development area for public open space or amenity purposes. Where there is inadequate infrastructure, e.g. a lack of main sewers or adequate road access, it might be possible for an agreement to be entered into requiring the developer to make financial contributions towards the cost of making available the infrastructure such that development may proceed. With regard to the provision of infrastructure, it should be remembered that agreements may be entered into with other authorities (relating to the adoption of roads and sewers, etc.) as well as the local planning authority.

If developers feel that a local planning authority is attempting to exert undue pressure on them to enter into an agreement which will impose unduly onerous burdens, their remedy is to make a formal planning application and take the matter to appeal if planning consent is not granted. However, in many cases it will be to the advantage of developers to offer a planning obligation themselves if they consider that an appropriate contribution of a facility or infrastructure will enable the development to take place at a very much earlier date than would otherwise be the case and they may offer a 'unilateral undertaking' to be bound by such an obligation. This can be considered as material by the planning authority in determining a planning application and at an appeal to a higher government body following a refusal of consent due to failure to agree terms on an agreement with the planning authority. Agreements bind landowners and successors in title and are registered on the legal title of the land affected.

There has been an increasing use of planning agreements, particularly involving local authorities with infrastructure burdens. Contributions towards infrastructure, whether they are given in cash or by way of sites for various public authority purposes, will obviously be reflected in the amount which a developer will be prepared to pay for the land for the development. However, often the developer has to assess the likelihood of an agreement and its cost prior to site purchase which adds uncertainty.

The extent to which a planning authority should, when considering a planning application, negotiate with a developer in order to obtain some material benefit for the community (referred to as 'planning obligation') has been a matter of controversy, especially where the burden can be so onerous as to prevent or restrict development due to unacceptable profit margins. Communities can at times have an input into the circumstances in which planning agreements are considered to be reasonable. As a general rule, planning agreements should only be sought where they are necessary to the granting of permission, relevant to planning and relevant to the development to be permitted. There are five basic tests of reasonableness:

1   relevant to planning;
2   necessary to make the proposed development acceptable in planning terms;
3   directly related to the proposed development;
4   fairly and reasonably related in scale and kind to the proposed development; and
5   reasonable in all other aspects.

Some feel that it is wrong for a planning authority to bargain to obtain a material benefit in return for a planning permission and that such a practice brings the planning system into disrepute. Many feel that developers should not be subjected to ad hoc demands from planning authorities which are seen as potentially dangerous precedents in the nature of local taxes on development. However, others argue that, in moderation and with common sense, planning gains can facilitate development in circumstances where the authority is not able to provide needed facilities and the planning gain makes a contribution to the welfare of the community in which the development takes place. Remember that the applicant has a right of appeal if planning consent is refused for any reason, and where a developer has submitted a unilateral obligation to enter into an agreement on specific terms, the nature of the those terms and their appropriateness is examined in detail.

6.9 Breaches of planning control

Local planning authorities often have wide powers to ensure that no development which requires planning permission takes place without permission being granted. It can also ensure that no unauthorised uses are allowed to continue unless the planning position is regulated and that all development permitted is carried out in accordance with the conditions which the authority has attached to the permission. These powers are included in the relevant legislation for the region in which the property development is located.

Where development has allegedly occurred without permission the authorities are empowered to first require information concerning the development, the identity of the owner of the land involved and other pertinent matters to be divulged to them. They can serve an enforcement notice on the landowner/developer that identifies the breach of planning control, the action required to remedy the breach and a time limit to carry out the necessary action. There is a right of appeal against the notice on one of seven grounds. The most commonly used ground is that planning permission should be granted for the development concerned and the lodging of an appeal on that ground is regarded by the higher government body, which consider the appeal, as often being a deemed planning application. Other grounds include stating that the alleged breach has not taken place and that the means required by the planning authority to rectify the situation are unreasonable.

Additional powers are available in the form of breach of condition notices, where authorities can require that conditions on a planning permission are complied with. There is no right of appeal against such a notice, as the notice is then pursued to the court system via a criminal prosecution which can be defended. More immediate powers of action are contained in a 'stop notice' which literally requires a specific activity to cease, normally as a forerunner to enforcement action being taken. If an appeal is successful against enforcement action where a stop notice has been served, there is a compensation procedure.

6.10 Planning appeals

In the event of the local authority refusing planning permission for a proposed development, or in the event (subject to the discretion of the applicant) of the authority either taking a longer period (e.g. more than eight weeks) to determine the planning application or granting permission subject to a condition which aggrieves the applicant, there is usually a right of appeal. To be valid, an appeal against a refusal of planning permission must be lodged within a designated time period (e.g. six months) from the date of refusal. In deciding whether to lodge an appeal against a refusal of permission, an appellant must be satisfied that there is sufficient evidence to suggest that the local planning authority's reasons for refusal is inappropriate. Planning authorities are required by central government to ensure that reasons for refusal are sound and clear-cut, and the onus at the appeal is placed on local authorities to demonstrate that the refused proposals would cause demonstrable harm to interests of acknowledged importance. In assessing whether reasons for refusal are capable of being set aside on appeal, the appellant (and their advisers) would examine the relevant planning policies and any other material considerations.

Once a planning appeal has been lodged, jurisdiction of the case passes to the relevant higher government body. Often there are three alternative methods of appeal which can be pursued by appellants:

1 *Written representations* This is an exchange of written statements with the appellant's statements putting forward the case in support of the appeal being allowed and the local authority's statement in reply seeking to justify the reasons for refusal. Both statements are considered by an inspector who visits the site and determines the appeal on the basis of the written evidence and the site visit.

2 *Informal inquiry* This usually takes the form of a meeting between the appellants and the local planning authority with the inspector acting as chair. Written statements of case must be submitted in advance of the hearing. Having read the statements, the central issues are identified which require discussion at the start of the hearing and these issues are aired between the parties. Following the hearing, a site visit is held and the appeal is determined on the basis of the written statements, the hearing discussion and the site visit.

3 *Formal public inquiry* This is a quasi-judicial inquiry where the appellant is represented by a barrister, solicitor or qualified person who presents the case for the appellant, calling expert witnesses to give evidence in support of the appeal. The planning authority is similarly represented, usually by the council's solicitor, who calls expert evidence supporting the refusal, often from the government's planning officer. The witnesses are cross-examined by the opposing party and the inspector, having listened to the evidence presented at the inquiry and following the site visit, then makes his decision or recommendations.

The three methods of appeal vary according to the prevailing legislation in the region. Often there is a provision for costs to be awarded against either party at an inquiry if there has been unreasonable behaviour. This definition normally applies to local planning authorities who have unreasonably withheld planning permission for proposals that were clearly acceptable or where grounds for refusal have been withdrawn at a late stage. In certain cases where such proposals have been brought forward by developers to appeal, such as which are clearly contrary to an adopted development plan and where there are no material considerations suggesting the plan should not be followed, costs can to be awarded against the appellants.

With every increase in economic and property development activity in each property boom a higher number of planning applications are subsequently refused, therefore the number of appeals rises. Conversely, with a downturn or recession the level of activity declines and there are fewer appeals. As an increased emphasis has been placed on developers to forward a detailed development plan, developers have been advised to pursue schemes in the context of their involvement in the development plan process outlined earlier as a means of facilitating a favourable policy framework in which to submit an application. The time taken for appeals by written representations and informal inquiries are usually substantially shorter, but the delays which occur in the decision-making process are clearly a factor to consider when making a planning application which is likely to go to appeal. If the planning system is unable to respond quickly enough some development opportunities and any associated economic growth potential will be lost.

6.11 Future perspective

In that the planning system is a slave to the political and legislative system, the future is difficult to predict. The development industry is generally dissatisfied with the delays and uncertainty which it experiences and sometimes the ad hoc nature of the decisions made, particularly where local politics holds sway over professional logic. Various proposals have been put forward to improve the planning system and developers must ensure they understand the process thoroughly.

With a strong political involvement there has been an acceptance that planning plays a key role in delivering sustainability and economic growth. The current trend is for increased population growth, increased globalisation, economic migration of peoples and increased need to adapt to and to mitigate the impacts of climate change. The planning system is required to have increased flexibility and responsiveness in the system, increased efficiency in the planning process and a more efficient use of land. As far as developers are concerned it appears that some of the historic problems remain in the system, those of delay for example, while newer and complex issues, like sustainability, have arisen and need to be addressed in property development.

6.11.1 Recent changes to a planning system – the UK as an example

In an effort to reduce delays and bring forward development, especially in the recent economic downturn, there have been significant changes to the planning system in England with the introduction of the Localism Act 2011 and the National Planning Policy Framework (NPPF) in 2012. These changes have been designed to simplify the planning system to make development easier to bring forward and reduce the amount of control central government has on planning decisions in local areas.

Localism Act 2011

The introduction of the Localism Act in November 2011 effectively handed back decision-making powers to local councils, communities and individuals in relation to neighbourhood planning, housing and a wide range of issues related to local public services. One of the key aims of the Act was to reform the planning system to make it more democratic and more effective.

National Planning Policy Framework

To support the Localism Act the government introduced the NPPF in March 2012 which replaced all Planning Policy Statements and Planning Policy Guidance notes ('PPS' and 'PPG' respectively) except PPG 10 (Sustainable Waste Management) and PPG 21 (Tourism) in addition to other guidance notes.[1] It also replaces Circular 05/2005 relating to Planning Obligations. The NPPF has reduced planning guidance down to a 50-page document which establishes a focus on sustainable development, by introducing the Presumption in Favour of Sustainable Development. The aim of the document is to provide a framework within which local communities and their governing bodies (local councils) can produce their own local and neighbourhood plans (Local Development Framework (LDF)), which take into account the needs and priorities of their communities. The LDF is a collection of documents outlining how planning and development within that area will be managed.[2]

In determining planning applications in England, local planning authorities must have regard to their LDF. The local planning authority (LPA) is responsible for deciding whether a proposed development should be allowed to go ahead. LPAs can also grant planning permission retrospectively for planning applications submitted **after** development work has been carried out.

6.12 Reflective summary

The planning system is based upon decision-making in various levels of government with particular emphasis placed on the government authority at the local level. For most forms of development involving householder applications and minor changes of use, the system generally works in a straightforward manner. However, for more major forms of development, the political nature of the system commonly means that the simple exercise of completing an application form correctly is insufficient to obtain planning permission. Consultation between developers and council officers, local interest groups, highway authorities and others on anything but the smallest project is now regarded as critical. Combined with the increasing role of the development plan and its adoption process, this has changed the way in which developers regard this system. However, the use of computer technology and internet access to planning schemes and online forms has assisted the developer's access to information. This discussion has overviewed the broad principles that provide a good basis upon which to approach the planning system. It should be noted, however, that like valuation, tax or building surveying, planning matters are a specialism in which developers often employ expert advice. The heightened concern of government and others in relation to environmental matters is greatly influencing the planning system as well. This concern has placed town planning in its widest sense in a high place on the political agenda. The planning system should be perceived as a stakeholder in property development rather than a hurdle or barrier that must be overcome.

Notes

1 The Minerals Planning Guidance notes and many of the Secretary of State's letters to the Chief Planning Officers since 2002.
2 Such plans must have regard to the Localism Act 2011, the Town and Country Planning Act 1990, sections 19(2)(a) and 38(6) of the Planning and Compulsory Purchase Act 2004 and section 70(2) of the Town and Country Planning Act 1990. In relation to neighbourhood plans, under section 38B and C and paragraph 8(2) of new Schedule 4B to the 2004 Act (inserted by the Localism Act 2011 section 116 and Schedules 9 and 10) the independent examiner will consider whether, having regard to national policy, it is appropriate to make the plan.

Chapter 7

Construction

7.1 Introduction

In the development process, a property developer's second major financial commitment is to place a contract to construct the development. From this point forward, some of the earlier flexibility will go, although this does, to some extent, depend of the procurement route selected.

Construction is a crucial stage in the development process and the key aim is to construct a good quality building that performs on time and on budget. 'Time, performance and cost' is the mantra. The selection of the procurement route needs to be decided early on as it has an effect on the composition and the size of the professional team. After the initial brief is decided, a schedule of accommodation is prepared and the broad design constraints are decided, the choice of building contract can be made.

The first part of this chapter provides an overview of the different types of building contracts and procurement routes, with the second part explaining the management process from pre-contract to the post-contract defects liability period. The roles of key personnel such as the project manager are explained. In addition, the risks associated with building contracts are identified along with strategies to either reduce risk or to transfer it to the contractor or other parties. The procurement options of public-private partnerships and partnering are also covered.

7.2 Procurement

The decision on the form of contract will depend on the developer's requirements and the size and complexity of the development, and, again, time, cost and performance are key influencing factors. The developer must determine whether cost management, time or building performance is the highest priority, as each will favour different procurement options. Different types of developer and stakeholder attach varying degrees of importance to time, cost and performance (quality) and these factors will influence the procurement strategy to be selected (see Table 7.1).

Table 7.1 Examples of prioritised criteria by client type (Source: adapted from Kelly et al. 2002)

	Owner occupier (%)	Speculative developer (%)	Investor (%)
Performance (function / quality)	45	20	50
Time (certainty or speed)	25	50	30
Cost (certainty or price)	30	30	20
Total	100	100	100

The selection of an appropriate procurement strategy has two elements: analysis and choice. The developer, usually with their consultants, must assess and set the project priorities and perhaps more significantly, their attitude to risk. Second, the developer must consider all possible procurement options, evaluate them and choose the most suitable. The factors which have to be taken into account at this stage include:

- factors outside the control of the professional/project team such as interest rates, inflation and legislation;
- client resources;
- project characteristics;
- ability to be flexible and make changes;
- risk management;
- cost of construction;
- time allocation; and
- performance and quality.

Clearly there may be conflicts between these factors and priorities need to be set to ensure that the procurement strategy selected gives the client/developer most control over the factors that are of greatest importance. There are a range of checklists that developers can use to determine which procurement route is most suitable and readers are directed to the RICS website[1] for further guidance.

Developers must decide whether the building design is to be carried out by (a) an architect and the professional team or (b) the contractor. For example, it may be necessary to shorten the pre-contract time by overlapping the design and construction elements of the scheme. Alternatively, early completion may be achieved by using fast-track methods of construction. Of crucial importance in deciding on the form of contract is the likelihood of changes to the design during the contract and the need for flexibility. An additional important factor is the extent to which the developer wishes to pass risk onto the contractor. A public sector developer will be also concerned to achieve value for money to meet the requirement for public accountability.

There are no definitive rules about choosing any particular form of building contract. A developer may use any of the available types of contract and adapt it to suit their own particular requirements, provided that it is generally acceptable to building contractors. However, there are advantages in using the forms of contract typically used in the building industry, where there is tested and practical experience, so that the strengths and weaknesses of that particular type of contract are known.

From the developer's perspective, building contract arrangements (often referred to as 'procurement methods', i.e. methods used to both design and construct a scheme) may be broadly divided into three main categories although with many variations in each.

The first category is based on the use of the traditional standard form of contract (in the UK this type of contract was developed by the Joint Contractors Tribunal (JCT)) which provides for a main contractor to carry out the construction in accordance with the designs and specifications prepared by the developer's own professional team – upon whom they must rely for the quality of design, adequate supervision of construction and suitability of the building for the purpose for which it is designed. This procurement option is also referred to as a 'design-bid-build' contract.

The second category is the 'design and build' contract which is frequently used in preference to the traditional contract. Here the contractor is responsible for the construction, and also for the design and specification. The contractor takes full responsibility for ensuring that the building meets the requirements of the developer and is fit for the purpose for which it is designed.

Third there is management contracting, arguably based on some methods of construction in the USA, which was used by some of the larger development companies on complex developments. Like the JCT contract, the professional team is responsible for the design and specification. However, the building work is split into specialised trade contract packages and the management contractor (for a management fee) co-ordinates and supervises the various subcontractors on behalf of the developer.

7.3 The traditional approach: design-bid-build

This type of contract tends to be used on relatively straightforward small to medium-sized schemes. The developer is able to use a number of variations and amendments to tailor the contract to the needs of the project. In the UK this procurement route is traditionally based on the contract as drafted by the JCT, which comprises a number of bodies: the Royal Institution of Chartered Surveyors, the Royal Institution of British Architects and the British Property Federation (representing developers and property owners). This contract has been amended several times and despite much criticism continues to be used in the UK mainly for small to medium-sized schemes. Many countries will have produced their own standard form contract for use in property development projects.

The following discussion describes a typical design-bid-build contract. Developers appoint their own professional team who are responsible for designing the building to meet the specified requirements, for supervising the construction phase and for administering the contract. The architect, together with the project manager if appointed, lead the professional team and call in whatever other team members they need to deal with such matters as structural design problems and the provision of mechanical and electrical services. Quantity surveyors or construction economists should be appointed at the outset, not only to provide a cost estimate but also to provide cost-planning services. Frequently project managers have a quantity surveying background, which enables them to maintain a keen eye on the financial management of the development. The role of the professional team in designing and administering the scheme under this type of contract is described in the next section.

7.3.1 Role of the professional team members

Unless a planning consultant is appointed, the architect is responsible for obtaining planning permission and all other statutory approvals such as building regulation approvals and fire certificates. The architect is responsible for the design of the buildings in terms of aesthetics and functions, all in accordance with the developer's brief and budget. The architect also is principally responsible for the management of the contract, although it is supervised by the project manager if one is appointed. The architect does not supervise the building contract on site on a full-time basis. Accordingly, a developer may also appoint a clerk of works or resident engineer at an extra cost to carry out a full-time 'on-site' supervisory role.

The quantity surveyor is responsible for preparing estimates of building cost, preparing the Bill of Quantities (i.e. measured specification of materials and work to enable the contractor to submit a price) and, during construction, for preparing valuations of work (usually monthly) upon which the architect issues the 'interim' and 'final' certificates. The quantity surveyor should be appointed as early as possible within the development process to advise on cost management and the merits of alternative forms of construction (e.g. steel frame or concrete frame) and should also provide estimated cashflows of the building contract expenditure. The quantity surveyor reports on the cost of construction and measures actual payments against the estimated cashflow. Their role is to explain why the actual cashflow differs from the original estimate and to prepare revised estimates for the remainder of the project. The quantity surveyor is also responsible for estimating the cost of possible variations in design so that the development team can decide whether or not they are merited.

Other members of the professional team may include structural engineers and mechanical and electrical engineers. The structural engineer works closely with the architect and quantity surveyor in assisting with the design of the structural elements of the building, calculating loads and stresses, and advising upon how the design of the building should be modified to accommodate them. A mechanical and electrical engineer advises on all the services required such as electricity, gas, water and the design of the heating, lighting and plant installations, and, where air-conditioning is to be provided, they will also be responsible for designing the system and liaising with the architect, so that it can be incorporated into the design. Increasingly, they have a key role in reducing the environmental impact of the in-use phase of a building's life cycle through the integration of sustainable building services.

Provided that the contractor executes the building work in a good and workman-like manner in accordance with the architect's drawings and specification in the Bill of Quantities and with any instructions subsequently given to the contractor by the architect, the contractor will not normally have responsibility if the building is not suitable for the purpose for which it was designed. This is irrespective of whether the unsuitability is attributable to faulty design or to some physical inadequacy in the structure. Developers must turn to their architect and other professional consultants for a remedy. On occasions, the respective responsibilities of the professional consultants for an inadequacy or defect in the building are not clear-cut. In such circumstances, developers find themselves in a complex situation, dependent upon the outcome of the arguments between professional consultants. Such situations may be avoided by the exercise of care in the selection of the team of professional consultants and, where a highly complicated or specialised or sophisticated building is involved, it is helpful if the professional consultants have had previous experience of dealing with that particular building type. The quality of the team has a significant influence upon the success of the development project.

Usually, all the professional team including any project manager, are appointed on a percentage fee basis, either by negotiation or in accordance with their professional body's suggested scale fee. The percentage fee typically relates to the final building contract sum and, from the developer's perspective, this does not provide a financial incentive for the professional team to ensure that the building is constructed on time and within budget, although the professional's reputation is at stake along with their chances of future work with the same developer. A developer may negotiate a fixed fee basis for appointment. However, the developer must be aware that the professional concerned will invariably include an element in their fee proposal to cover the risk for extra work, and therefore the developer may not necessarily gain any advantage.

Developers will require the design team (i.e. architect, structural engineer, mechanical and electrical engineer, and sometimes the quantity surveyor) to enter into deeds of collateral warranty for the benefit of investors/purchasers, financiers and tenants. Collateral warranties extend the benefit of the developer's contract with the professionals involved. Typically, they require the professional practice or company to warrant that all reasonable skill, care and attention has been exercised in their professional responsibilities and that they owe a duty of care. In addition, the professionals are required to warrant that they have not specified deleterious materials. They will also need to provide evidence of sufficient professional indemnity (PI) insurance cover from their insurance company. It can often be difficult and time-consuming to procure these warranties in a form acceptable to all parties, as financial institutions and banks require almost total responsibility from the professional design team. The professionals, in turn, are increasingly resisting deeds of collateral warranty in the form required by financial institutions and banks due to the restrictions placed on them by their insurers in relation to their professional indemnity policy. It can become very complex for the developer acting as agent in the middle. For example, professionals may refuse to sign these agreements under seal, because they wish their responsibilities to last for six and not twelve years. Additionally, they may wish to limit the assignability of their deeds of collateral warranty to the first purchaser and first tenant of the completed development. They may also insist their liability is restricted to remedial costs of any defects, not consequential or economic loss. It is important, at the very least, for the developer to ensure that every member of the design team signs a warranty before they are appointed.

Some developers have turned to insurance in the form of latent defects insurance known as decennial insurance, especially on larger schemes. Latent defects insurance is relatively expensive, typically 1–2% of the building contract sum, but it has the advantage that the insurer is responsible for pursuing remedies with the professional team. The insurer assumes responsibility for repairing the property should an inherent structural defect be discovered which renders the building unstable or threatens imminent collapse. The insurer typically covers a project for an extended period after completion (e.g. ten years), provided that an audit has been carried out before construction begins and an independent engineer reports on the design and construction of the building. The developer has to pay for the fees of the independent engineer. The insurance policy is totally assignable to tenants and purchasers, who seem to be increasingly demanding that such an insurance policy is in place. Residential property development in many developed countries is automatically covered by a mandatory scheme enforced by the government. This protects individual residential purchasers after they take possession (and all associated future risk and responsibilities) of an individual property or when part of a larger property development has been transferred from the developer to an individual purchaser. In a normal instance the seller, following the transfer of ownership, has no additional obligations associated with the property.

The majority of the design work on this type of contract is carried out prior to the appointment of a contractor and has a considerable impact on the final cost. Once the contractor has started work, variations requiring revised instructions to the contractor frequently result in increased costs or delays to the contract programme. Revised instructions are often caused by the developer making a variation, or the architect issuing late instructions as the design is inadequate or incomplete. Thus, the relationship between the developer and the architect is crucial at the design stage. Developers should establish positive and realistic cost limits, which should be relayed to the architect and the professional team. They should ensure that the architect thoroughly understands, with the aid of a written brief, their requirements in respect of all the aspects of the building and its usage, the standard and type of finishes required, the services needed and the date for completion. Above all, if possible they should avoid changing their mind, unless this is necessary to secure a particular tenant or improve the value of the scheme. It is vital that the architect is chosen with great care as much depends on the architect producing the working drawings for the contractor on time and the developer must determine whether an architect has the capability and resources to deliver the design element effectively.

Discussion points

• Who are the key professionals that a developer needs to bring into the development team?
• What roles does each professional play in the team?

7.3.2 Choosing the contractor

Once the detailed design is completed, the quantity surveyor prepares the Bill of Quantities, which specifies and quantifies the materials and the work to be carried out in great detail. For example this will include the number of door hinges and other architectural hardware. All Bills of Quantities must be checked as any errors can involve additional cost due to later variations to the contract. In addition, provisional sums and contingency sums must be included. After this, contractors are invited to submit tenders for carrying out the work based on the drawings and the Bill of Quantities. However, there is nothing to prevent a contractor being asked to price the Bill of Quantities. This may happen when the developer has employed a particular contractor (e.g. an in-house contractor) over time, is pleased with the quality of their work and therefore prefers to re-employ them. It may be that the contract contains a great deal of specialised work for which one contractor has an outstanding reputation and they might be chosen on that ground to carry out the work, subject to an acceptable pricing of the Bill of Quantities. If this route is taken it may be advantageous to appoint the contractor earlier in the development process to advise the design team on the practical aspects of the design.

When choosing contractors to be invited to submit competitive tenders for carrying out the work, there are several issues to be considered. It is necessary to limit the total number of contractors invited to submit competitive tenders. Usually approximately six contractors are adequate for even the largest contract; this is because the pricing of a Bill of Quantities for a large job is time consuming and expensive for contractors who are not keen to submit competitive tenders when there is, in their opinion, an unreasonably large number of tenderers.

If the work is specialised, contractors skilled in that type of work are selected. Sometimes it will be preferable to use a large national contractor, while at other times local or regional contractors are favoured. Some contractors have a reputation for producing work of high quality, others for producing work quickly and on time, and this can be a vital consideration for the developer's cashflow. Still other contractors have a reputation for submitting keen tender prices, and there are those who have a reputation for expertise in formulating claims for extra payments on any and every occasion during the contract. The architect or developer may feel that some contractors are entirely dependable, while others may have let them down on a job in the past. Unfortunately, high-quality work, speed and low cost are a very rare combination. Developers are often guided by their architect, project manager and/or quantity surveyor on the selection of the contractors for the tender. Before a contractor is included, they should be asked whether they are willing to tender for the job. There are times when contractors are fully extended and are unwilling to tender for work. For a variety of reasons some contractors may not be interested in tendering for work in a particular locality.

The prices submitted by each of the contractors are examined by the quantity surveyor and the project manager to ascertain what is offered. The quantity surveyor prices the Bill of Quantities independently and this is used for comparison purposes with the contractor's tenders. Each contractor prices each element of the work, setting out the applicable rates for each element of work or unit of material. The contractor's priced Bill of Quantities forms part of the contract documentation and is used by the quantity surveyor to value the work carried out by the contractor.

The reliability and financial stability of the contractor are vital considerations. Therefore, when contractors are chosen, it might be advisable to take out a 'performance bond'. The contractor takes out a performance bond with an insurance company which guarantees to reimburse the employer for any loss incurred up to an agreed amount as a result of the contractor failing to complete the contract. Financiers may also insist on a performance bond for their benefit. The failure of a contractor is a major disaster from the developer's perspective as long delays occur while the legal position is resolved and another contractor is found to complete the work. The new contractor might ask a considerably higher price for completing the job than was contained in the original contract. Furthermore, if defects later appear in the completed building, it can be very hard to apportion responsibility between the original contractor and the contractor who takes over the job. Thus it is easy to understand why employers ask for a performance bond, even though the extent to which their losses are reimbursed is limited and the cost of the performance bond is usually added to the contract cost.

7.3.3 Paying the contractor

The method of payment for the works has a substantial impact on the developer's cashflow position. It also has significant impact on the contractor's cashflow; the contractor should consider the method of payment when preparing the price for the work. Thus, the method of payment must be made clear when the contractor is invited to tender. In many instances the architect authorises monthly payments based on the value of work certified by the quantity surveyor. Usually a certain percentage (3–5%) of the total value of the work undertaken is retained until the end of the contract. This is known as the 'retention'. This arrangement best suits the contractor who obtains payments for the work carried out irrespective of when the building is ready for occupation. The developer has to pay out very substantial sums of money over a considerable period of time before obtaining the benefit of a completed building at the end of the contract.

The ideal arrangement for developers is for the whole of the contract price to be paid when the building is handed over, so that they do not part with their money until the time when they should be receiving an income from the building or have the benefit of occupation of it. It must be remembered, however, that if it were possible to make such an arrangement (it is extremely rare), contractors would increase their tender prices by one means or another to take account of interest and the additional risk. In the case of a large contract spread over a period of time, some contractors might not be able to finance the work easily without payments from the developer. Some compromise might well be devised for payment to be made in certain set stages – the last payment on completion and handover of the building is weighted to give the contractor an incentive to get the building completed. The method of paying for the work has to be related to the circumstances of each contract. The contractor is more likely to be flexible if they are a partner in the scheme and stand to benefit in terms of a profit share.

7.3.4 Calculating the cost

An accepted approach is for the contractor to submit a bid on either a 'firm price' or a 'fluctuations' basis. The 'firm price' means that although the cost of labour and materials used in carrying out the work may fluctuate with the market, the contract sum will not be varied to take account of these fluctuations, whereas the 'fluctuations' basis means that once the contract is awarded to the contractor any increase or decrease in labour and materials is added to or subtracted from the contract sum. Under both types of contract, there is a clause allowing adjustments to be made to take account of alterations in cost due to government legislation. However, the developer may delete such clauses, particularly when the tender market is very competitive. It is vital to note that 'firm price' does not mean the contract sum once fixed will not alter. Quantity surveyor's re-measurements, architect's variation orders and instructions and extensions of time may affect the cost.

Developers and their professional advisers must decide on what basis they wish contractors to prepare their competitive bids in order that the contractors submit prices on the same basis. Developers do not always find it easy to decide which basis is likely to be to their advantage. The risk of fluctuations in the cost of labour and materials during the contract cannot be avoided; the question to be decided is whether the risk is to be borne entirely by the contractor as in a firm price contract, or whether the risk is to be borne by the developer as in a fluctuations contract. If contractors have to prepare their bids on a fixed price basis, they will add something to their prices to cover themselves against the risk of increased costs. To tackle this problem contractors are often asked to quote prices on both a fixed and fluctuations basis; leaving the developer to decide in the light of the differing prices which basis is likely to prove most advantageous during the whole of the contract period. Each contract must be judged on its own merits against the background of tender market conditions.

There are two main alternative methods of calculating the cost. First, a cost-plus contract, where the contractor is paid on the basis of the actual cost of the building work ('prime cost') plus a fee to cover their overheads and profit. The fee might be fixed or a percentage fee calculated with reference to the final building contract sum or the initial estimate. Second, a target cost contract might be negotiated or established by tender. A target cost is agreed with the contractor plus the contractor's fee. Any savings or additions to the target cost are shared by the parties.

7.3.5 Duration of the contract

The date agreed in the contract for the completion of the building is not certain as the contractor can apply for extensions of time for a number of reasons. Some of these reasons which justify an extension of time entitle contractors to recover additional loss or expense that they may have suffered as a result of that extension. Extensions of time usually result in an increase in the cost which will certainly include the contractor's 'preliminaries' (overheads such as insurance, cost of plant hire, etc.) The impact of an extension is felt twice by the developer: first, it affects cashflow and, second, it increases costs.

The main reasons for extension of time which entitle the contractor to recover additional loss and expense are:

1   Inadequacy and/or errors in the contract documents, the drawings and/ or the Bill of Quantities. This may be due to professional incompetence in the preparation of the documents or new legislation might be introduced during the contract which requires amendment to the drawings. Additionally, the relevant building control officer for the local authority (responsible for providing building regulation approval) or the local fire officer (responsible for issuing a fire certificate) might impose conditions on their approval which necessitates design changes. Unforeseen ground conditions (in the case of a cleared site), hidden structural problems (in the case of a refurbishment) or contamination issues can cause additional expense and delay. This underlies the necessity for a thorough site investigation and/or structural survey before the tender documents are prepared.
2   Delay by the architect in issuing drawings or instructions.
3   Delays caused by tradesmen directly employed by the developer.

Additional reasons which may be included in the contract terms (depending on the results of negotiation with the contractor on the standard clauses), entitling the contractor to an extension of time, but not to recover any additional loss or expense, include:

1   Failure by the nominated subcontractors: on almost every building contract some work is subcontracted to specialists. When contractors find and appoint their own subcontractors, they are responsible for delays caused by them, so that the developer does not suffer. Architects can nominate subcontractors to undertake certain elements of work. The reasons for doing this may be the high quality of previous work with the architect, or expertise in designing and constructing elements of the work, e.g. the structural steel work.
2   Bad weather: architects should ensure an accurate record is taken of weather conditions on site.
3   Strikes and lockouts.
4   Shortage of labour (tradesmen) or materials.
5   Damage by fire where it is the contractor's responsibility for insurance under the contract.
6   'Force majeure', e.g. earthquakes, floods, storms, etc.

Developers may be compensated for a delay for which the architect has not granted an extension of time under the terms of the contract. The compensation is in the form of liquidated and ascertained damages at a rate agreed in the contract to cover the developer's loss. However, the agreed rate can fall short of the developer's true loss in terms of the overall development cashflow.

A developer cannot assume that the work will be carried out within the time set out in the contract and for the exact contract sum. A key advantage of many contracts is flexibility with regard to the way the price for the job is to be fixed and its elasticity, which enables the type and quantity of work within the contract to be varied and yet leave the quantity surveyor free to negotiate the final price for the job at the end of the day. However, there is a disadvantage with regard to the flexibility of this type of contract: it does not discipline the developer into making early clear-cut decisions as there is always the scope to make late variations. Also the competence and efficiency of the professional team is vital to the successful outcome of the contract. Inadequacies of the plans and specifications are not realised until the building contract is underway. Also the professional team is not motivated sufficiently to control costs and delays, as their fees increase in proportion to the final contract sum. This type of contract may lead to a very confrontational situation with the contractor if the tender documents are inadequate or many variations are made. Also, many contractors have a tendency to use the claims procedure as a negotiating ploy to claim additional money to cover any losses they have made on the contract. It is for this reason alone that many developers use design and build contracts. Developers must ensure that they are in control and are kept closely advised as to the likely financial outcome of the contract at all times. The best arrangements for dealing with this are examined in the section on project management.

Overall the key advantages and disadvantages of the traditional approach are summarised in Table 7.2.

Table 7.2 Advantages and disadvantages of design-bid-build procurement

Advantages	Disadvantages
Competitive equity and fairness	Strategy is potentially open to abuse, resulting in less certainty
Design-led – facilitates high quality design	
Reasonable price certainty based on market forces with reduced unknowns	Overall construction timeframe may be longer than other options because there is no parallel working
Acceptable strategy in terms of public accountability and transparency	No 'buildability' input from contractor
Well-known procedure and proven approach	The strategy can lead to adversarial relationships between the parties
Flexibility – changes are easy to arrange and with limited effect on value	Approach is not fully promoted by all stakeholders

---

Discussion points

- For what reasons might a developer select a traditional (design-bid-build) contract for a development project?
- What are the two types of bids that a contractor can submit?

7.4 Design and build

This type of approach is unique. Typically the contractor actually assumes the risk and responsibility for the design and construction of the scheme in return for a fixed-price lump sum. Design and build is a fast track strategy. Its use has become more widespread due to dissatisfaction with other types of contracts and the problems encountered with splitting design and construction responsibilities. Design and build was originally used on simple and straightforward schemes but is now used on most types of building projects. It is widely used by public sector clients for hospitals and schools for example.

The contract is based upon a performance specification by or on behalf of the developer. Here 'performance' means the various requirements which the building must meet. However, the developer's requirements must be set out clearly, and in as much detail as possible, in the performance specification if this type of contract is going to work from the developer's viewpoint. The responsibility for the design and the precautions that should be taken to ensure that the finished building meets with all the various statutory requirements and suitability for the purpose for which it is designed, rests wholly with the contractor.

Performance specifications vary from being fairly simple to very detailed, depending on the nature of the scheme. For example, if a developer wished to develop a site with very simple standard-design warehouse units the specification might include a schedule of floor spaces for units of different size, with an indication as to how much office accommodation is to be provided, the total amount of toilet accommodation, the services to be put into the building, the floor loadings and the clear floor heights, together with an indication of the total yard area. On that simple performance specification, contractors would be asked to submit schemes for the erection of their own standard-design units to meet the requirements, together with their price for carrying out the contract. The complete responsibility for obtaining all the necessary statutory approvals, and for designing and erecting the buildings and ensuring that they will be suitable for the purpose for which they are required, rests with the contractor.

However, the performance specification tends to be extremely detailed and will specify the materials to be used by the contractor. It is the contractor's responsibility to comment on any materials specified by the developer if they consider them to be unsuitable for their designated purpose, before the contract documentation is entered into.

It is possible to arrange for complicated buildings to be erected under a design and build contract. However, in such cases, the performance specification is critical and has to be carefully prepared by a team of professional advisers. In the case of a complicated building, which does not conform to a standard design, developers are entirely dependent upon the adequacy of their own performance specification to ensure that they get a building which meets their requirements.

A developer using this type of contract may appoint a specific contractor with which they have successfully worked before or who has expertise in constructing buildings similar to the one proposed. Alternatively, a developer will go out to tender to appoint the most suitable contractor for the job based on their design, specification and price. Typically, the tender list will be short (two or three contractors) due to the considerable amount of work involved by the contractor. As with many contracts the track record and financial stability of the contractor is an important factor. Some contractors specialise in design and build, building up valuable experience in constructing and designing, while some larger contractors have 'design and build' divisions. The contractor may employ all the necessary skills in-house or more usually employ external architects and engineers under the supervision of their own in-house project managers.

A developer may appoint a quantity surveyor to carry out the usual pre-contract activities such as advising on the most suitable form of contract and preparing initial cost estimates. The quantity surveyor will perform similar duties during the contract as those on a traditional contract including valuations for interim payments and the final account. In some cases, the quantity surveyor may take on the role as the developer's representative (often referred to as the 'employer's agent' in the contract), and effectively project manage and administer the contract. The quantity surveyor may agree the letters of appointment and deeds of collateral warranty with the professional team, and chair and minute all project meetings (usually the role of the architect in a contract).

Under a design and build contract the contractor submits drawings and specifications to the developer for approval, who can check to see just what type of building will be built, what services will be provided, and so on. The contractor is responsible for designing and constructing the scheme in accordance with the approved drawings and specification. In a simple package deal where a standard type of building has already been erected by the contractor in a number of locations, examples of which can be inspected and the occupiers asked for their comments on the adequacy of the building, developers will know the product and any feedback on its suitability. In such circumstances, the design and build contract is advantageous where the buildings are of a simple, straightforward nature, with often repetitive design elements, and can be built to a standard design which has been used by the contractor elsewhere. The advantages are that the design time and cost can be greatly reduced: the contractor is working to their own designs with which they are familiar, and they may use various standard types of component, which they can buy advantageously and which they are used to using on site. The contractor's own designs will undoubtedly reflect the contractor's practical experience of putting up buildings. The result should be that the contractor works more efficiently and speedily, and thus more economically, so that the price of the building to the developer ought to be lower.

The advantages of this type of contract for the developer are that while it is possible to provide for fluctuations in the contract price, and there are various alternative ways of paying for the buildings as the contract proceeds, usually a lump-sum fixed price is agreed: the contractor is committed to provide the buildings for a known cost and takes the risk. Clearly, contractors allow for these risks when preparing their price but the developer is reassured to know that the price is fixed. It may be changed only by variations issued by the developer or changes in legislation. When the tender market is competitive, clauses in the contract dealing with changes in legislation may be deleted. The developer does not run the risk of becoming involved in endless professional arguments if the design or construction of the building is defective; it will be entirely the responsibility of the contractor to see that matters are remedied.

There are disadvantages – for example, the developer does not have the same detailed control over the design, and if the developer requests alterations during construction the cost might be increased out of all proportion – in short, the developer does not have the protection of flexibility in many contracts (for example the traditional JCT contract). It can be argued that the final cost of the buildings under a fixed-price, lump-sum package deal may be higher due to the risk which is carried by the contractor but, in practice, this is often offset by the advantages to the contractor of using his own design and standard components. Another advantage is the likely achievement of an overall saving in time due to the overlapping of the design and construction processes. Furthermore the developer will save money on professional fees as involvement by professionals will be less.

There may be types of development for which the design and build type of contract is not suitable and some developers have suggested that the aesthetics and quality of the finished building may be lower compared with a more traditional contract. To overcome this problem a developer can appoint an architect to prepare the initial drawings and sketch designs under what is known as a 'develop and construct' contract. The contractor is then responsible for developing the design as part of their tender submission. However, under this arrangement the developer must ensure that the design responsibility is adequately defined.

Alternatively, the developer may appoint an architect and engineer and novate their appointment contracts to the contractor. By novating the contract the contractor steps into the shoes of the developer and becomes the client of the professional concerned under the terms contained in the original contract. Thereafter, the contractor takes control of the design process and the professional is liable to the contractor. A potential conflict of interest may arise as the architect/engineer may continue to treat the developer as their 'employer' on the basis of their long-standing relationship. It is important to ensure that the professional and the contractor develop a good working relationship. The developer will still insist on deeds of collateral warranty with each professional in case the contractor goes into receivership. The advantage to the developer under this variation is that they can appoint their choice of architect

Table 7.3 Advantages and disadvantages of design and build procurement

Advantages	Disadvantages
Developer deals with only one construction company and therefore inherent buildability is achieved	Only a limited number of firms offer design and build, so there is limited competition
A higher level of price certainty is achieved before construction commences (provided the developer's requirements are adequately specified and changes are not made)	Developer has to commit before detailed design is completed
	In-house design and build forms are an entity, so compensation for weak parts of the firm is not possible
Reduced total time of project due to overlapping activities	There is no design overview unless separate consultants are appointed for the purpose by the developer
Fewer overall uncertainties in the overall process	Preparing an adequate brief can be difficult
	Difficulties comparing bids as each design, programme and cost varies
	Design liability is limited by the standard contract
	Changes to project scope can be expensive

and engineer, with whom they have a good working relationship, while retaining the advantages of the standard design and build contract. A key drawback of the design and build contract for developers (whichever variation is adopted) is the lack of flexibility. Developers must decide before the contract is signed on their exact requirements, as major variations can usually only be made at a considerable cost. In summary, the key advantages and disadvantages of the design and build approach are briefly as shown in Table 7.3.

7.5 Management contracting

With this procurement strategy, a management contractor is engaged by the developer to manage the building process and is paid a fee. Management contracting, initially developed in the USA, became more widespread in an international context because developers were impressed with the fast track methods of construction. The management contract is generally used on larger complex development projects where the developer requires speedy construction at competitive prices with the flexibility to change the design during the contract. In summary, the building contract is split into specialised contract packages, either by trade or building element, and let separately under the supervision of the management contractor.

The developer appoints the professional team to prepare the drawings and specification for the project. The quantity surveyor prepares a cost plan based on the drawings and specification. The actual cost incurred by the management contractor ('prime cost') is paid by the developer having been certified by the architect and monitored by the quantity surveyor. The developer pays a fee to the management contractors for their services in managing the various separate contracts. The developer may not always appoint the contractor with the lowest fee proposal. It is crucial that the contractor has management contracting experience and sufficient staff with the right skills. The fee may be a lump sum or a percentage of the contract cost plan. The 'prime cost' includes the amounts due to the various contractors of the various parts of the project, plus the management contractor's own on-site costs. The construction work is carried out by the various contractors, who enter into a standard contract with the management contractor based on detailed drawings, specifications and Bills of Quantities. The selection of the contractors should be carried out by the developer and the professional team in consultation with the management contractor. The architect has the power to issue variations known as 'project changes'.

The management contractor should be involved at an early stage with the professional team in advising on the practical implications of proposed drawings and specifications, and the breakdown of the project into the various separate packages. The management contractor is paid in relation to interim certificates issued by the architect, including instalments of the management fee. The disadvantages are that the developer has to pay the management contractor's fee, as well as the professional team, and the management contractor is not responsible for the actual building works. The developer has no direct contractual relationship with the various contractors carrying out the work. Accordingly, the developer must enter into design warranties with each contractor that is capable of being passed to purchasers, financiers and tenants. However, the management contractor may have to pursue the remedies of the developer in respect of any breaches by the various contractors. The developer has to reimburse the management contractor in settling or defending any claims from the contractors, unless the management contractor is in breach of the contract or of their duty of care. The liabilities and hence risk of the management contractor is limited. They are not responsible for the payment of any liquidated damages for any cost overruns if caused by reasons outside their control or by delays due to the various contractors. It is essential that the management contractor, developer and the professional team co-operate since the developer has to pay extra for the management contractor's expertise and experience.

Note that the management contract itself is not a 'lump-sum' contract. The management contract is based on the contract cost plan prepared by the quantity surveyor, which is only an indication of the price. However, the contracts the management contractor enters into with the various contractors are usually based on the standard 'lump-sum' contract. Therefore, the final cost is based on the contracts with each specialised trade and it may bear no relation to the cost plan within the developer's contract with the management contractor. Cost control is essential under this contract and it must be very tightly managed by the project manager and quantity surveyor as there is no direct incentive for the management contractor to keep within the cost plan as their fee is directly related to the final building cost. The success of this type of contract in terms of cost control depends on the ability of the management contractor to appoint the various trades within the budget of the cost plan. However, the final cost is whatever the cost is to the management contractor (including the fee) and there is no penalty for exceeding the cost plan.

The developer has to accept a very high degree of risk with this type of contract. Developers who have used this type of contract have found cost control the biggest problem as there is no tender sum. There is no control over the delays caused by the individual contractors. There are extra costs involved in duplicating site facilities for the management contractor and the various contractors. However, the main advantage of management contracting is speed as projects are usually completed more quickly than on traditional contracts where full detailed drawings have to be prepared before the contract commences. This speed is achieved by the flexibility of dividing up the contract into separate elements, overlapping the design and construction of each element. The 'packaging' of the contract allows the developer, to some degree, to control costs and delays as contracts agreed to later on in the process can be varied to suit. Pre-construction and construction times can be reduced when compared with other contracts. Developers are concerned that this type of contract does not effectively control the development's design and quality standards suffer as a result.

Due to some negative experiences with management contracting due to the disadvantages noted above, a variation known as 'construction management' was preferred. With construction management, trade contractors are placed directly by the developer and a construction manager is appointed for a fee as part of the professional team appointed at the same time, not necessarily afterwards like the management contractor. The construction manager acts as the developer's agent and the appointment of a project manager is required to co-ordinate the professional team. Their fee is usually percentage-based with an additional lump sum for the provision of site facilities. Their role is to manage and co-ordinate all the various contractors, review design proposals, control costs (i.e. a fixed budget), control the contract programme and be responsible for quality control. The administration of claims for payment by the contractors and variations are their responsibility, although the final account with all the contractors is administered by the architect and quantity surveyor.

The advantage of employing a contractor on the professional team is to bring their experience and expertise into the design stage at the beginning. A contractor may employ 'value engineering' techniques (i.e. detailed studies of the cost-effectiveness of alternative materials and methods of construction) to review the design process. However, the developer has to ensure that the contractor has the relevant design experience, otherwise the advantage of strict design control will be negligible. The developer has to have much greater involvement in this type of contract arrangement and it is the job of the construction manager to ensure that developer makes firm decisions at the appropriate time. It is a very management intensive contract resulting in higher staff and fee expenditure on behalf of the developer compared with other types of contract.

The key advantage is the saving of time, achieved by overlapping the design and construction of each package and involving the construction manager at the beginning of the design process. It is appropriate where an early completion of the scheme is crucial. However, despite the developer having direct control over the various contractors, cost control remains a problem. There is still no guarantee of what the final cost of the scheme will be, although incentives may be used to increase the contractors' share of the risk but at a cost to the developer. This method may be used by developers if they wish to maintain flexibility, take advantage of fast-track methods of construction and retain control while accepting greater risk. They could reduce this risk if they were confident of their exact requirements at an early stage and the pre-contract period was long enough to allow for detailed design. Overall the key advantages and disadvantages of the management contracting approach are summarised in Table 7.4.

A similar strategy to management contracting is the design and manage contract. Here a contractor is paid a fee to manage and assume responsibility for the works and also the design team. The advantages are: early completion because of overlapping

Table 7.4 Advantages and disadvantages of management contracting procurement

Advantages	Disadvantages
Time-saving potential for overall project time	Imperative that brief is detailed and clear
Increased buildability potential	Higher level of price uncertainty
Breaks down traditional adversarial barriers	Relies on a good quality team
Parallel working is inherent	
Work packages are let competitively	
Flexibility – changes can be made provided the packages affected have not been let and there is little impact on those already let	

activities; the developer deals with one firm only; it can be applied to complex buildings; and the contractor assumes the risk and responsibility for integration of the design and construction. On the other hand, the disadvantages of this approach are: price uncertainty (not achieved until the final work package is let); developer loses control over design quality; and the developer has no direct contractual relationships with the works contractors or the design team – thus making it difficult for the developer to recover costs if they fail to meet their obligations.

7.6 Project management

The appointment of a project manager is not necessary for every project. A project manager tends to be needed for large and complicated rather than small, simple projects. Often, developers act as their own project manager with 'in-house' staff or employ one of the professional consultants to exercise the management function. Typically, a project manager will receive a fee representing 2–3% of the final building cost, depending on the extent of the role and the complexity of the scheme. However, the developer may appoint a project manager on an incentive basis linked to whether the final cost is within budget. A development company may be asked to take on the role of project management on the basis of a fee, either fixed or related to the profit of the development, by an owner-occupier or property investment company for instance. Project management in this context has a much wider definition to include the management of the entire development process.

Project management is an occupation and project managers may be architects, quantity surveyors, real estate managers, valuers/agents or have a building/contracting background. The project manager should be appointed at an early stage to be able to advise the developer about detail relating to the type of building contract applicable to the development, and to be involved in the development brief and the design discussions. Also, the project manager should be able to advise the developer on the selection of the professional team, particularly those who have previously worked with the project manager. The professional team should complement each other and work well together. The project manager's role is to act as the client's representative when co-ordinating the professional team and liaising with the contractor. The project manager is concerned with the overall management of the project and is not involved in carrying out any part of the project. The project manager needs plenty of common sense, administrative ability and a good knowledge of construction.

The management objectives must be clearly defined in consultation with the project manager and made known to everyone in the project team. The objectives are to ensure that the finished project is suitable for its intended purpose, that it is built to satisfactory standards, that completion occurs on time and that the project is carried out within the budget. The project manager is often responsible for appointing the professional team on behalf of the developer and will agree the fees, letters of appointment and deeds of collateral warranty under the guidance of the developer. The developer should ensure that the project manager is supplied with copies of all the funding documentation entered into with any financier of the scheme. The documentation will include the plans and specifications agreed with the financier, and the project manager should ensure that these are complied with throughout. If alterations are necessary, then approval from the financier will be formally required. The project manager is responsible for ensuring that arrangements for the disposal of the building, either the letting or the sale, are carried out efficiently and satisfactorily.

It is essential to examine the role of the project manager through the pre-contract and contract stages. The following is based on a traditional JCT contract (typically used in the UK).

Discussion points

- What are the benefits of having a project manager in the development team?
- Under what circumstances might a developer decide to appoint a project manager?

7.6.1 Pre-contract preparations

The project manager should check that the developer has the necessary legal title on the site, whether it is freehold or leasehold, and that vacant possession of the whole site is available immediately. All restrictions on the site should be carefully checked (e.g. underground services, easements and rights of light or support), and compared with the proposed scheme so that the building work will in no way interfere with them. The project manager should arrange, if not previously carried out by the developer, all of the necessary ground investigations, structural surveys and site surveys, and communicate the results to the rest of the professional team. It is important that all the site boundaries are clearly defined, and that a schedule of condition of the boundary fences, adjoining roads and footpaths, etc. is prepared. It may be necessary to negotiate 'rights of light' or 'party wall' agreements with adjoining landowners/occupiers.

The architect is responsible for ensuring that all the necessary statutory approvals have been obtained, such as planning permission and building regulations. The fire officer should be consulted early in the design process and the architect should ensure that the design is in accordance with all relevant legislation. The architect is responsible for assuring the project manager that all necessary statutory consents have been obtained. It is most important for the project manager to obtain unqualified assurances on these matters because, in practice, many expensive delays are caused as a result of one or other of the statutory consents not being obtained before the contract starts. Sometimes there are circumstances which might persuade the project manager to allow a contract to start before all the statutory consents have been obtained, but in so doing the project manager and the developer must realise the risk that is being taken.

7.6.2 Preparing the contract documents

The project manager's most important job is to ensure that the contract is not allowed to commence without adequate documentation. Incomplete drawings are probably the most common cause of delays and cost increases. If a contract is started before all the drawings are completed and the architect is unable to provide all the drawings to meet the contractor's required time schedules, the consequences can be serious. The project manager needs to be absolutely satisfied with the availability of the drawings by the architect and that sufficient staff resources within the architect's firm are in place. If a contract is started before the drawings are fully complete, which is often the case, a detailed schedule must be obtained from the architect, showing exactly when the outstanding drawings will be delivered to the contractor. Before the building contract is placed, the architect must obtain from the contractor a written statement confirming that (provided the drawings are supplied in accordance with the architect's schedule) there will be no claims for delays due to lack of drawings, and getting this matter right at the outset cannot be overemphasised.

Project managers must also be satisfied that the Bill of Quantities is as complete and accurate as possible. The quantity surveyor will measure the quantities off the architect's drawings, so again this stresses the need for their accuracy. Some items in the Bill of Quantities may be described under the headings of 'prime cost' (meaning actual cost) or 'provisional sums'. 'Prime cost' items usually cover materials or goods that generally cannot be precisely defined. 'Provisional sums' items cover elements of the work which it is not possible to detail properly and evaluate at the time the contract is entered into. The contractor is required to allocate a sum of money against these items. The project manager must understand why the prime cost and provisional sums items have been included in the Bill of Quantities, and be satisfied that it is impossible to make the detailed provision at the outset. Quantity surveyors should be questioned to ensure that they have received adequate information from the architect to enable them to prepare their Bills with complete confidence in their accuracy.

If a pre-letting has been achieved then it is important to include any specific requirements of the tenant within the contract document. Furthermore, such requirement should be clearly referred to as the 'tenant's specification' so there is no doubt. There may be a situation where the tenant subsequently alters their specification which delays the main contract on the 'developer's specification'. If a claim is subsequently made by the contractor then it can be apportioned to the tenant for payment.

7.6.3 Appointing the contractor

If it is proposed to invite competitive bids from selected contractors, the project manager should agree with the architect and the developer the names of the contractors who will be invited to tender. When the competitive tenders have been received and evaluated, the job is normally awarded to the lowest tenderer. However, there has been a move away from automatically accepting the lowest tender with the advent of 'best value' approaches whereby other factors are considered and the lowest tender may not offer value for money in the long term. Some clients appreciate that paying more in the short term has much greater long-term benefits. The quantity surveyor compares each tender against their priced Bill of Quantities. The project manager decides whether a performance guarantee bond has to be obtained by the contractor. Once satisfied on all matters, the project manager then authorises the placing of the building contract. All the contract documents should be ready, so that the contract may be signed before work actually starts on site, although in practice work often starts before the documents are signed, on the basis of a letter of intent, but this should be avoided. The project manager will discuss with the architect reasons for wishing to appoint any nominated subcontractors and, if appropriate, then authorise their appointment.

7.6.4 Site supervision

The project manager should be continually satisfied about the arrangements made by the architect for site supervision during construction. The size and complexity of the scheme may merit the appointment of a full-time site supervisor, such as a clerk of works or a resident engineer or indeed a resident architect. The architect should also arrange for progress photographs to be taken periodically on site, so that a clear visual record of the state of the contract at any time is always available to supplement the architect's own reports on the progress generally.

7.6.5 Construction period

When the contractor has taken possession of the site, the project manager ensures that the works are carried out on schedule and that the overall cost is kept within the budget. To carry out his duties effectively, regular meetings of the project team are held. The frequency and composition of the meetings depend upon the size of the particular job and may vary at different stages of the job. The project management meetings are often arranged on a monthly or fortnightly basis. The project manager, the architect and the quantity surveyor form the nucleus of the project management team. If the project manager is also controlling the letting or sale of the project, then the surveyor/valuer/agent is normally a member of the team, particularly in those cases where the purchasers or tenants might wish to have special works carried out. If the scheme is being financed externally, then the fund or bank's representative or appointed advisers will also attend the project meeting to fulfil their monitoring role. The contractor may be invited to attend the part of the project management meeting at which the progress on site is discussed. Often, the project manager attends separate site meetings with the architect to be kept informed of building progress. All project management meetings must be professionally conducted and accurately minuted.

Typically, at the beginning of a project meeting the minutes of the previous meeting are considered and any matters arising dealt with. The architect presents a report on the progress of the work, indicating what parts of the work are ahead of, or behind, schedule and comments on the overall progress of the job. The architect should state any difficulties which have arisen at every meeting and whether the contractor is delayed as a result of lack of information. The architect should also report as to whether any variation orders or instructions have been given to the contractor and, if so, their likely effect on the progress of the work. The project manager will learn independently from the contractor or through attendance at site meetings whether the contractor is being delayed by lack of information or materials/labour.

The quantity surveyor then presents a report on the financial situation, indicating whether or not the work of measurement on site is well up to building progress and whether any variation order or architect's instructions have been given which affect the cost of the job. The quantity surveyor should indicate the position with regard to prime cost and provisional sums and present an overall summary as to how the cost of the job so far compares with the contract sum. The quantity surveyor should also indicate any factors which might increase or decrease the cost of the job at a future date.

When appropriate, the surveyor reports on the progress with regard to the disposal of the property and on any requests for special or extra work received from prospective purchasers or tenants. Then the practicability and advisability of carrying out those special works are discussed. Ideally, purchasers or tenants should take over the completed building in accordance with the original design and specification, carrying out required special works at their own expense once the building has been handed over to them. However, it is not always possible to insist on such an arrangement, as it may be necessary to carry out such works in order to secure the letting or sale.

Then the project manager summarises the overall financial situation, particularly with regard to payments to the contractor, compares them with the budget, checks on dates of handover and compares the estimated date for the receipt of income or capital payments with the budgetary expectation. These are matters of vital importance to the developer's cashflow. If it appears that the project is running behind schedule, then methods of speeding up the work to recover the position are considered, together with the implications for cost. Usually, there is a liquidated damages provision in the building contract and the question of its enforcement has to be considered. In practice, liquidated damages are often inadequate to compensate the developer for losses incurred as a result of the delays, because if the true cost is written into the contract documents at the time of the invitation of tenders, contractors would increase their tender prices out of all proportion in order to safeguard themselves against a the risk of a heavy liquidated damages claim, which might in fact never be made.

This summary of project management arrangements where a JCT contract is used is intended to illustrate the basic principles involved, which apply generally to other procurement methods. The project team may be much larger and a management contractor or a construction manager may be part of the team on building schemes of a complicated nature. Accordingly, the project management is that much more intricate. On the other hand, where design and build contract methods are used, the role of project management is simplified accordingly, and may be taken over by the quantity surveyor. In the case of a lump-sum, fixed-price design and build contract, project managers are essentially concerned with quality control and progress. They may inspect the buildings during construction or arrange for a professional adviser to do so. Periodic meetings with the contractor to discuss building progress and the achievement of the handover dates should enable them to fulfil their role.

7.6.6 Handover of the completed development

A short time before the date for completion and handover of the building from the contractor to the developer, the architect prepares a 'snagging' list, indicating all the minor defects that must be remedied before handover occurs. It is useful for the developer's surveyor and the intending occupier's representative (if known) to accompany the architect to ensure that all are satisfied with the snagging list. At the outset of the contract, the project manager will have confirmed that the building works are adequately protected by the contractor's own insurance arrangements. The contractor's insurance no longer protects the building once it has been handed over, so it is vital for the project manager to ensure that the developer has adequate insurance cover from handover until the insurance cover provided by the occupier takes effect.

If the development has been pre-let or pre-sold to an owner-occupier, then the occupiers and/or their contractors may wish to have access before formal handover by the main contractor working for the developer. From the developer's perspective, this situation should be avoided, unless it is necessary to secure the deal. If an occupier wishes to gain early access to attend to fitting-out works, then arrangements should be documented clearly in the contract. Ideally an occupier's special requirements should be incorporated at an early stage into the design process or if the occupier is secured after the building contract has started, then the occupier should be allowed access only after practical completion of the building. It is important to include the developer's base specification for the scheme into any documentation, so that any changes which lead to an increase in cost or delays can be attributed to one party or the other. If there is an overlap between main contractor and fitting-out contractor, the project manager should ensure that the occupier arranges adequate insurance. Problems can occur when the fit-out contractor's work affects the work of the main contractor. The project manager, with the architect, needs to attribute and resolve problems quickly.

The quantity surveyor should then be asked when any outstanding re-measurement work will be completed and be in a position to agree the final account with the contractor, so that the architect may issue a final certificate. The JCT contract will have provided for a certain percentage of the total cost to be retained by the building owner until the end of the defects liability period, often six months from the date of practical completion (twelve months in the case of any electrical and mechanical element of the contract). Special maintenance periods may be agreed for particular parts of the work (e.g. landscaping). The contractor is responsible for remedying any defects (other than design) which have occurred during the defects liability period, provided that they have not been caused by the occupier. It is vital that the buildings should be carefully inspected at the end of the liability period, because if there are any obvious defects at that time which the architect does not identify, it may well be assumed that the architect was prepared to accept the building subject to those defects.

The importance of inspecting the site and its immediate environs on the handover date should not be overlooked. If during the building contract any damage has been caused to adjoining property (for example, damage to boundary walls and fences is not unusual), then the contractor must remedy it. Inspection of the roads, footpaths, kerbs, grass verges, etc. immediately adjoining the site is carried out to see that the contractor remedies any damage, otherwise the highways authority might subsequently ask the developer to bear the cost of any remedial works.

The architect should produce as 'built drawings' a building manual and maintenance schedule to assist the occupier by giving a comprehensive schedule and description of all components (taps, locks, fastenings, sanitary ware, etc.) that might need replacing at some future date, together with recommendations for regular maintenance work to preserve the building fabric. Similarly, manuals and operating instructions for services are provided by the services engineers.

Where an occupier is not taking possession immediately, then the developer is responsible for a vacant building and a programme of regular cleaning and maintenance should be instigated. Whatever physical arrangements are reasonable and necessary to protect the property against vandalism should be made. An example is in a shopping development, where un-let shop units will have a neat hoarding put across the frontage immediately before the handover date. Consideration is often given to the issue of employing security guards or patrols. Adequate insurance cover should be in place to give protection against fire and loss due to the damage of property. Public liability insurance should also be arranged to protect against claims from injured third parties.

7.6.7 Monitoring construction progress

The project manager's objective is to produce the building on time and within budget for the developer client. Therefore, it is important to examine how the project manager reports to the developer on construction progress and cost. Any delays in completion or increase in costs will affect the profitability of the development; therefore, it is essential that a developer is kept regularly informed on progress and cost. The developer will need regularly to update the cashflow appraisal prepared at the initial evaluation stage to assess profit.

Every project manager will have their own method of reporting but it is important to agree with the developer at the outset the information required. The best method of reporting uses charts and graphs to compare actual progress and cost against the original estimates. This is typically undertaken by using one of the specialist software packages designed for project management. The starting point should be the appraisal used at the site acquisition stage so that actual costs and progress can be compared against the estimates made at the time of acquisition. This means the developer can easily clearly identify changes in costs and progress, instead of reading through pages of written information. The charts usually have written comments on them giving reasons why costs have increased/decreased or why site progress is behind schedule. Most software packages generate an interactive Gantt chart which adjusts process time and /or costs in response to changes. Once the developer has assimilated the information in the charts and graphs, further questions can be asked of the project manager as to the reasons behind identified increases in cost or delays in progress. In particular, it should be clearly understood who is able to authorise the expenditure of money. Every member of the project team must know whether they are able to spend money and, if so, what authorities they must obtain.

Typical reporting methods include the following:

(i) Bar (or Gantt) chart

The bar chart is a calendar showing the development programme in weeks or months. The programme is divided into tasks and the period during which each of these is to be carried out is shown on the chart. An example of such a chart is shown in Figure 7.1. The chart includes pre-contract activities as well as the contract programme, which are equally important to monitor as any delay will impact on the start of construction.

This shows how crucial it is for the project manager to be involved in the development process from the beginning. The chart indicates when each task is to start and finish. It shows how the tasks overlap and the work that should be in hand at any time. From time to time the programme, and the bar chart, may need to be amended but a comparison of what has been achieved against what the chart shows gives the developer and project manager a simple yet instant test of progress.

The bar chart can be used to indicate when information or decisions are needed by the project manager from the developer and by the contractor. This is vital as lack of information or instructions is one of the main causes of delay. The bar chart demonstrates that delay in one activity can affect the whole programme. Once a delay is identified it is important for the project manager to advise the developer what effect it will have on the overall programme and how time can be made up in other activities. It is vital for the project manager to issue the bar chart to the entire professional team, so that each member of the team can identify the target dates they have to work to.

Once the contractor is on site, the bar chart may be substituted with the contractor's own bar chart, which identifies the timescale for each trade involved on site. It is important that the project manager receives the contractor's bar chart regularly so that the overall bar chart can be updated and amended as necessary. The developer may not need to know the progress of each individual trade on site and it is often sufficient to break down the contractor's programme, into substructure, superstructure, finishes and external works. However, the project manager must be able to produce the contractor's bar chart at any time, as in some cases the developer will need to know the detailed programme. For instance, the developer may need to know when the area in the building identified as a show suite is ready.

Another method of monitoring progress and highlighting the importance of providing information and decisions is to prepare a chronological timetable of events. However, the bar chart is the most instant way of comparing progress against original estimates.

(ii) Cashflow table and graph

A cashflow table and graph is prepared by the project manager. The purpose of the cashflow table in Figure 7.2 is to estimate the developer's flow of cash payments throughout the development period. The developer can use this to prepare a cashflow appraisal, which can be regularly updated throughout the development. The importance of the cashflow has already been discussed in Chapter 3. The combination of the table and a graph can provide another means of checking progress by comparing actual with estimated payments. However, it is less effective at measuring progress than either the bar chart or a development timetable. Estimates of cashflow often have to be revised and there is a danger that they do not really highlight problems of delay until the last months of the contract.

(iii) Financial report

The project manager's financial report may typically look like the example shown in Figure 7.3. It is based on the quantity surveyor's cost reports and payments already

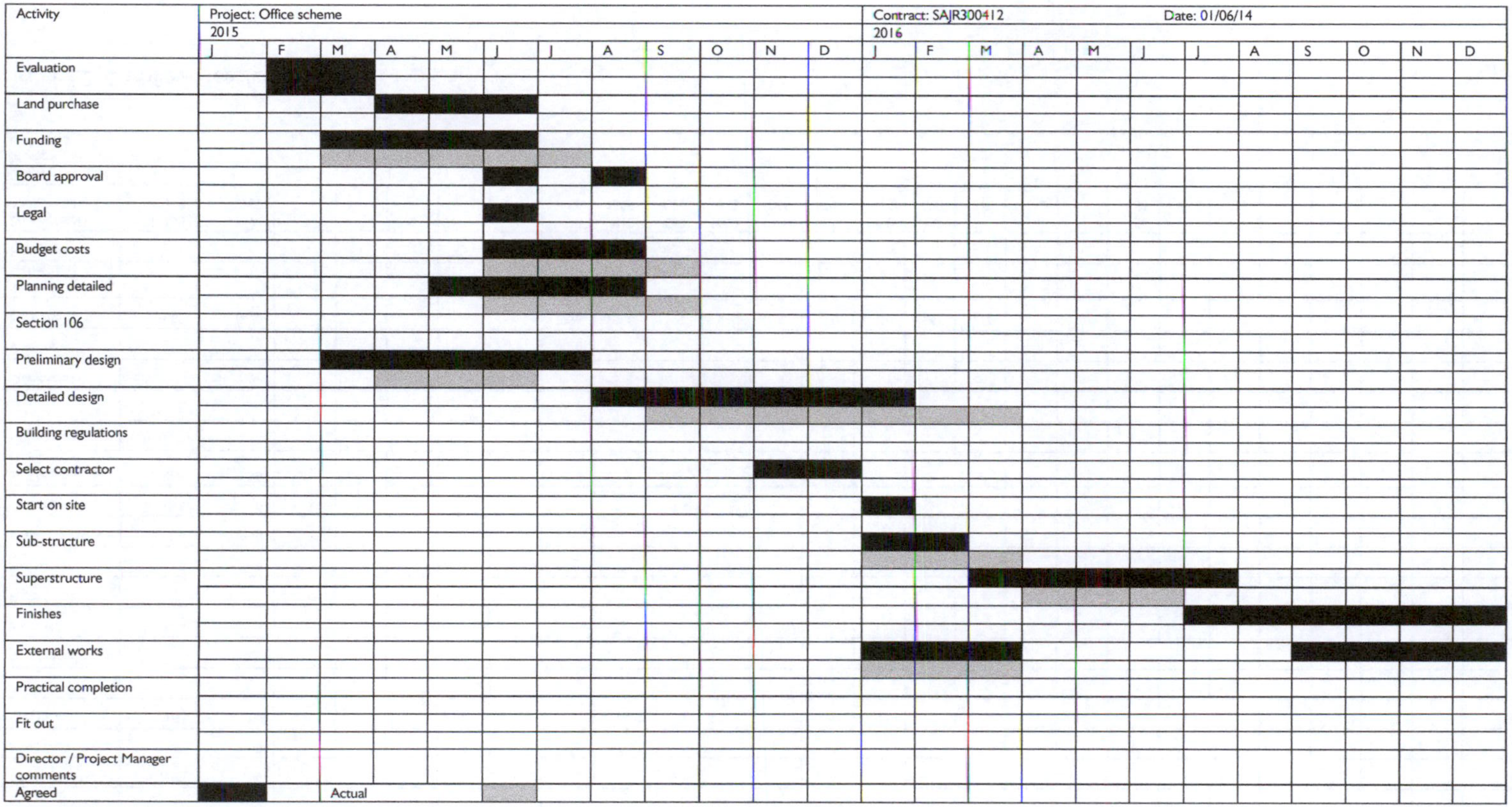

Activity	Project: Office scheme												Contract: SAJR300412						Date: 01/06/14				
	2015												2016										
	J	F	M	A	M	J	J	A	S	O	N	D	J	F	M	A	M	J	J	A	S	O	N	D
Evaluation		██	██																					
Land purchase				██	██																			
Funding			██	██	██	██																		
Board approval			▓	▓		██		██																
Legal						██																		
Budget costs						██	██	██																
Planning detailed					██	██	██	██	▓															
Section 106																								
Preliminary design			██	▓	██	██	██																	
Detailed design								██	██	██	██	▓	▓											
Building regulations																								
Select contractor										██	██													
Start on site													██											
Sub-structure													██	▓										
Superstructure															▓	██	██	██						
Finishes																			██	██	██	██	██	██
External works													██	▓						██	██	██		
Practical completion													▓											
Fit out																								
Director / Project Manager comments																								
Agreed	██		Actual			▓																		

Figure 7.1 Overall development programme – Gantt chart

Project: Office scheme | Contract: SAJR300413 | Date: 05/06/14

Fees (000's)	Budget	Total To-date	2015 J	F	M	A	M	J	J	A	S	O	N	D	2016 J	F	M	A	M	J	J	A	S	O	N	D
Architect	450	336						30		30			30	30	20					12						
Structural Eng'r	180	90						24		22			22	22	12					12						
QS	180	90						22		24			22	22	12					12						
M&E Eng'r	176	68						34																		
Proj Man	180	80								40			44							20						
Acoustic																										
Landscape																										
Party wall																										
Right of light																										
Site surveys	6	6																								
Ground surveys	16	30								18																
Planning	6	6																								
Building Regulations	44	26																								
Others																										
Demolition																										
Enabling																										
Main contract	9000	1670						1396	1598	922	948	1056	636	450	141	216					276					
Statutory authorities																										
Fitting out																										
Budget total	10198	2698						918	810	852	1080	1170	1194	720	430	216			56		270					
Actual total	10398	2262						1506	1598	1056	948	1056	754	450	220				56		276					
Building / Project manager comments																										

Figure 7.2 Cashflow: fees/construction

	Project Office Scheme	Contract SAJR300412						Date 05/06/14
	Board approval date	Revised report date		Revised report date		Revised report date		Director / Project Manager comments
	16/10/13	11/06/14	+(-)	11/06/14	+(-)		+(-)	
Site start	09/02/13	09/02/13		09/02/13				
Completion	22/01/14	29/02/14	– one month	29/02/14	– one month			Agreed extension of time, see memo.
Net lettable area	38250	38250		38250				Client to measure in August.
Building contract value	9,000,000							For cost breakdown see quantity surveyor report no. 4.
Demolitions								
Enabling works								
Substructure			+106,068		+106,068			Increased cost of piling due to ground conditions.
M&E services					–13,480			Saving due to design change.
Finishes		7,906,068		7,882,388				
External works								
Preliminaries								
Inflation								
Contingencies	200,000	200,000		200,000				30,912 of contingency not expended.
Statutory services								
Tenant works								
Claims (unsettled)								Delay caused by late receipt of drawings from structural engineers.
Instr. (not priced)								Client's variation to finishes needs to be finalised.
Pending instructions								
Others agreed ext. of time	46500	46500	+46500	46500	+46500			
Client's variation				30,000	+30,000			
Total	9,000,000	9,156,508	156,568	9,169,088	–169,088			

Figure 7.3 Financial report: building costs

made to the contractor as certified by the architect. It enables the developer to identify variations in costs throughout the contract.

The project manager should advise the developer of the reason for the cost variation. Any variation in cost from the original contract value may be due to a claim from the contractor, architect's instructions to the contractor or variations required by the developer. We have already examined the circumstances under which a contractor can make a claim for additional costs. Claims may be based on the inadequacy of the drawings and/or the Bill of Quantities. In addition, they may be based on delays in the architect issuing drawings or instructions. The project manager must ensure that claims and variations are kept to a minimum if costs are to be kept in budget and should monitor the activities of the architect and ensure they keep to their drawing schedule. The project manager together with the quantity surveyor must advise the developer of the cost of any variation proposed to ensure the developer is aware of the implications. No revision should be made without justification. The developer must know why cost estimates have to be revised. The project manager must maintain a scrutiny of costs and question any decision which has a cost implication.

(iv) Checklist

Most project managers prepare a checklist of the main activities throughout the development. An example of such a checklist is shown in Figure 7.4. The checklist defines the main activities applicable to the particular development, some of which will require approval by the developer. It should highlight information required by the project manager from the developer. It is essential for the project manager to identify the decisions needed by the developer and by what date. Developers should know if the progress of a development is being held up because a decision is required of them, and the implications of any delay in the decision. The developer/project manager relationship is a two-way one and both should ensure the other is kept fully informed at all times.

The above-mentioned methods of reporting are typical but every project manager will have their own preferred method. While regular reporting on progress and cost is a way of keeping the developer informed, it also provides the project manager with an essential tool especially if the developer insists on regular reporting in the manner shown above. The project manager will know whether the aims are being achieved and it will bring into sharp focus the targets that need to be achieved and the problems that need to be tackled.

The project management of a development through the construction process, whether carried out by the developer or through the appointment of a project manager (or other professional), is about teamwork and motivating the team to work together. Problems must be sorted out before team members resort to a blame culture and become entrenched. The project manager must anticipate delays by ensuring constant communications with the professional team and the contractors. Contracts and paper communications should not be relied on; there is no substitute for personal contact. Overall, project managers should fulfil their role efficiently, constantly considering cost and time. They need the ability to lead and motivate the professional team and the contractor. This, again, shows that the property development process is all about the interaction between people.

Activity	Approved by	June 2014 Project Manager	June 2014 Client	July 2014 Project Manager	July 2014 Client	Director / project manager comments
		Projects Office Scheme		Contract no. EBC 2408		Date 05/06/14
Clients brief		✓	✓	✓	✓	
Select / appoint architect		✓	✓	✓	✓	
Select / appoint QS		✓	✓	✓	✓	
Select / appoint structural engineer		✓	✓	✓	✓	
Select / appoint M&E		✗	✗	✗	✗	All agreed except one item.
PM appointment		✓	✓			All agreed except arch.
Deeds of collateral warranty		✗	✗	✗	✗	
Letters of intent						
Development / feasibility appraisal			✓		✓	
Site boundary / ownership agreed			✓		✓	
Appoint rights of light						
Appoint party wall						
English Heritage agreement						
Summary of funding documentation			✓		✓	
Funds/banks surveyors approval						
Tenant requirements						
Agreed contract programme		✓	✓	✓	✓	
Planning drawings / application		✓	✓	✓	✓	
Section 106 agreement						
Freeze design / stage report		✓	✓	✓	✓	
Building Regulations application		✓	✓	✓	✓	
Certificate of readiness		✓	✓	✓	✓	
Summary of insurance requirements		✓	✓	✓	✓	
Summary of building contract conditions		✓		✓		
Appoint building contractor		✓	✓	✓	✓	
Agreement with statutory undertakers						
Signed receipt for maintenance manuals		✗	✗	✗	✗	
Client decisions (major items)		✗	✗	✗	✗	Meeting arranged 17/07/14 to finalise.
Finishes board – urgent		✗	✗	✗	✗	

Figure 7.4 Checklist to monitor primary activities

Discussion points

- What are the main issues that developers need to consider during the construction phase of a project?
- What are the major risks the developer faces during this stage?

7.7 Public-private partnerships

Collaboration between public bodies, such as local authorities or central government and private companies is known as a public-private partnership (PPP). Therefore, public developments are now able to consider PPP as a method of procuring buildings and infrastructure. In the public sector, there are three main procurement approaches: PPP, design and build, and prime contracting. The rationale for PPP is that private companies are more efficient and better managed than public bodies. In bringing the public and private sector together, the aim is that the business community's management and financial skills will lead to better value for money for taxpayers. The Private Finance Initiative (PFI) was created the early 1990s. Governments and local authorities traditionally paid private contractors to build roads, schools, prisons and hospitals out of tax money. Under PFI, contractors pay for the construction costs and then rent the finished project back to the public sector. This enables government to get new hospitals, schools and prisons without raising taxes. The contractor is allowed to keep any cash left over from the design and construction process, in addition to the 'rent' money. Critics say that governments are mortgaging the future and that the long-term cost of paying the private sector to run these schemes is more than it would cost the public sector to build them itself.

PFI is in its infancy for hospitals and schools but it is a well-established way of paying for new roads and prisons. The complex nature of PFI contracts and the political obstacles can result in controversial schemes, which in turn means that progress in some areas can be slow. It has been argued that trade in public services could ultimately benefit the private sector substantially.

PFI has broadened the concept of public-private co-operation. If privatisation is a take-over of a publicly owned entity, PPP is more like a merger, with both sides sharing the risks and seeing the benefits. With health and education accounting for a large proportion of the GDP of many countries, the rewards for industry of opening up the public sector to private finance are huge. However, governments are not always clear how far they want to go in these areas in the face of opposition, and critics argue that taxpayers will end up paying for PPP developments. There are claims that some PFI projects have been sub-standard with private companies taking short-cuts to maximise profits. Another PFI criticism is that firms make their profits by reducing employees' wages and benefits, whereas PPP supporters maintain that some hospitals and schools would not be built if it was not for private finance and assert that PFI will lead to increased quality in public services. Performance-related penalties, now part of most PFI contracts, will ensure improvement in standards. PFI is a fast, effective, and in the short term, an economic way of getting new projects built.

7.8 Partnering

Clients, designers and contractors have evolved their relationships with developments and developed different types of business relationships; among these is partnering. Under this type of arrangement, parties to a contract work towards agreed goals which will benefit all concerned. Partnering thrives in an atmosphere of trust and openness and fails when co-operation is absent. In summary, partnering is a business relationship for the benefit of all parties – built on trust, openness and respect. While the contract establishes the legal relationship between parties, partnering establishes the working relationship. Some refer to it as the traditional way of doing business where a person's word was their bond. Partnering can be used on small, large or complex projects and is promoted as a win-win way of doing business for all parties. Partnering is based on an ethos of teamwork and in the case of construction projects goals are achieved through a teamwork approach to:

- design control and efficiency
- minimising pre-construction budgeting and approvals periods
- maximising efficiency of the construction period (and completion dates)
- problem solving co-operation
- cost control reporting and reconciliation, and
- agreed conflict or dispute resolution procedures.

For partnering to succeed, a number of key principles have to be adopted by the parties – for example, a commitment to, and value placed on, a long-term business relationship, thereby ensuring a willingness to work towards longer-term goals, such as a reduction in project times and improvement in building performance (quality). In the partnering organisations there is a requirement to develop an environment for long-term profitability and to encourage innovation. In addition, partners commit to improved project buildability and a lowering of project costs through the process of value management. The establishment of project organisational structures and clear lines of communication is required to reduce conflict and disputes and ensure successful outcomes. Advocates claim the successful outcome is a project constructed in less time, costing less and of higher quality than would otherwise be realised through traditional procurement routes. With a partnering approach, it is argued, developers and contractors are able to define the project better and identify risks prior to commencement to avoid time delays and cost overruns and the subsequent poor relations between parties. In summary, the key elements of partnering in construction are as follows:

- Commitment from top management – a jointly developed partnership charter is not a contract but a commitment.
- Equity – all stakeholders' interests are considered in establishing mutual goals and there is a commitment to win-win thinking.
- Trust – teamwork without trust is not possible, and personal relationships are developed to build trust and understanding about each stakeholder's risks and goals.
- Development of mutual goals/objectives – at partnering workshops, mutual goals and objectives are established and the means by which to meet them are identified.

- Continuous evaluation – to ensure implementation, stakeholders agree to a plan for periodic joint evaluation based on mutually agreed goals.
- Timely responsiveness – saves money and can stop a problem growing into a dispute. Methods of discussing issues are developed prior to project commencement to reduce conflict risks.

Partnering workshops are set up at the commencement of the project and involve all team members. The workshop, a planning session to establish the partnering tools and to problem solve critical design/construction issues, typically takes two days to complete. The project charter is agreed to be a most visible tool which sets out the team performance goals and mission for the project. It is periodically reviewed to ensure that it has continued appropriateness. It is vital to agree criteria to measure whether the partnering is successful. The second stage is, therefore, to agree a team report card and the benchmarks for measuring success. The third and most important tool developed at the initial workshop is the issue resolution process. The process should encourage communication and creative problem-solving. Roles and responsibilities are defined and communication points are agreed. Without a commitment to principle-centred leadership from management, partnering will flounder. Follow-up meetings are held every three or six months during the project and it is in the follow-up meetings that partnering is actually delivered and results produced. Similarly, at the end of the project, an evaluation should be undertaken so that lessons can be learned by all stakeholders for future projects.

7.9 Reflective summary

The developer's aim during the construction process is to produce a good quality building on time and within budget. The choice of procurement is critical and is dependent on project size and complexity as well as the developer's attitudes to risks. Three main types of building contract are available: (1) traditional contract format, (2) design and build, and (3) management contracting, although there are many variations on each depending on the exact contractual arrangements and the role of the professional team. Each contract has its main advantages: the traditional contract for its flexibility; the design and build for cost control; and management contracting for speed. Disillusionment by developers with the traditional contract has led to an increased use of the design and build contract where the contractor is responsible for design and construction, avoiding the problem of cost increases due to the late production of drawings by the architect. To overcome the problem of quality control with the design and build contract the appointed contractor has taken over the appointment of the developer's chosen architect through a novated contractual arrangement. Whichever method is used the success of the building contract, in terms of achieving the aims above, relies on good control, leadership, and firm early decision making by the developer and the project manager. The professional team and the contractor should be motivated towards the same goal, resolving any conflicts and problems before they arise. One factor impacting considerably on the construction phase in the UK is the shortage of labour, particularly in plumbing, electrical and plastering trades. Other locations will have their own skills shortages. In the public sector PPPs and PFI are affecting procurement. The advent of partnering is also affecting some developments with distinct advantages for long-term working relationships.

7.10 Case study – modular construction

This case study overviews an entrepreneurial approach to providing accommodation. It is accepted that property development includes a proportion of entrepreneurism and this can also affect the selected construction approach. An example is the use of container modular home designs which are designed to reduce construction costs and, in certain circumstances, provide a viable and affordable alternative to traditional on-site built conventional homes. This example is based on approaches to building two container-modular homes being either (a) fully modularised (Plate 7.1) (b) or a hybrid modular construction (Plate 7.2). A fully modularised container home requires minimal on-site construction work whereas hybrid modular construction has greater flexibility in the arrangement of its floor plan. Note that both approaches are promoted as economical in a property development scenario; however, they do not receive automatic planning permission or approval nor are they fully accepted by the broader market. Substantial market research would be undertaken to ensure adequate demand for this innovative construction approach.

With this design the containers are used to create single-room modules that can form a flexible multi-room system in many different combinations. This system must adhere to the spatial limitations of the container modules and it is essential for

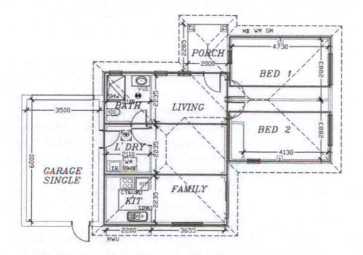

Plate 7.1 Example of modular home floor plan highlighting the individual modular units

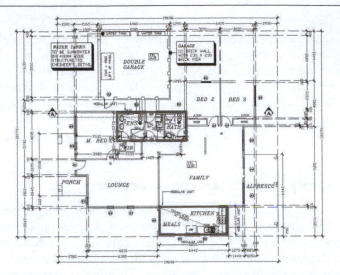

Plate 7.2 Example of hybrid modular home floor plan highlighting the two modular units

vertical loads to be distributed exclusively through each corner. Furthermore, when modules are combined they act holistically as a single structure to resist horizontal loads and then transfer these loads to foundations through unit-joints. The design of a container's structure dictates that it is most stable when positioned horizontally and supported at four points. Depending on the construction system used, the load-bearing capacity of shipping containers limits the number of containers that can be stacked to a range between 3 to 8 containers, therefore being 3–8 levels high. To create spatial and functional linkages like doors, windows and hallways, different sized openings are necessary. In theory the containers can be configured to create a self-supporting building where containers create the interior space while also providing the structural support. However, it is also possible to use containers solely as supporting structures by stacking them to create external wall structures. These wall structures can then be roofed using a secondary supporting element which is supported by the external wall structure. This then creates habitable positive space between the external container walls.

Container buildings will always require a foundation in addition to their own sub-floor structure. The type of foundation used is dictated by the site conditions and also by planned geometry and the service life of the building. In the case of mobile buildings, demountable foundations are preferred as they may be uninstalled and relocated with minimal disturbance to the site. For a permanent structure, design engineers will recommend a conventional slab or footing design as per the design loads and site conditions. Containers are generally classified as light building models which in turn affects the thermal performance of the container structure during summer and winter. The absence of solid high thermal mass material results in container buildings having a low heat capacity. As a result small climatic changes lead to excessive temperature fluctuations of the interior.

To compensate for the lack of thermal mass, container modules are fitted with high quality insulation before lining the interior with plasterboard, although the downside of this modification is a marginal loss in interior area and volume. To further enhance the thermal performance of the container modules, thermal breaks are installed to prevent thermal bridging at connecting corners. Fire protection is another important factor that affects the safety of the structure. Building regulations provide stringent guidelines that designers must abide by to ensure the safety of the occupants and the building itself. Even though steel does not burn, when compared with concrete a steel construction generally has a very low resistance to fire. This is because steel is a good thermal conductor and suffers a loss of strength at high temperatures which can happen at temperatures as low as 500 °C. In contrast, concrete is stable up to 1000 °C. Therefore, additional measures must be taken to meet fire protection requirements and increase the fire resistance of components; one suggested treatment is to use gypsum plasterboard or plaster coatings that foam up in the event of a fire.

Economic benefits

Similar to most modular constructions, container buildings are erected over a substantially shorter construction period compared with conventional buildings. This direct time saving can reduce investment costs, therefore making container constructions more cost-effective. Depending on the standard of fittings required, construction costs are often lower in comparison to conventional building methods. Generally this is due to the industrialised off-site construction method of modular containers which has resulted in the production of cheaper buildings. It should also be noted that additional costs for transportation, foundations and connection to utilities will be added; however, these costs are marginal compared with the savings achieved through container modularisation. Transportation options are flexible and are more affordable when compared with other manufactured homes. The size of a shipping container module is limited by ISO standard dimensions, so it may be shipped internationally as an ordinary shipping container or transported locally using a semi-trailer, requiring limited resources and minimal logistical costs.

Environmental benefits

From an environmental perspective, the process of container construction possesses an advantage compared with conventional construction methods. As container buildings can be installed on temporary foundations, they are demountable and removable. For example a temporary structure can be removed and relocated with almost no impact on the environment. Once the planned service life of the building has expired, it may be disassembled back into individual modules and re-used, allowing modular buildings to be extended comparatively easily. This is because modular units can be set up on-site relatively more quickly than conventional renovations and therefore it is possible to increase the size and area of a building quickly and with greater flexibility to suit spatial requirements. For example, a smaller affordable container modular home can be built for a young couple and as their family structure grows with additional children, their home can have modules added to extend it to accommodate their demand for additional space. With this method the amount of inefficient and unoccupied building space can be reduced.

Global economic considerations

Economic influences are a major consideration in assessing the use of containers. In particular, the type and amount of trade (import/export) plays an important role. For example, at times the USA imports a large volume of goods with comparatively fewer exports, resulting in an excess of shipping containers that are sold relatively cheaply at container terminals. Adding to this excess supply, the global financial crisis reduced the overall global level of exporting and importing activities, leading to a worldwide oversupply of containers. This means that shipping containers can be purchased at very economical prices.

Regional economic considerations

Promoting cost-effective construction approaches can be a significant factor in both less developed regions and highly developed nations. Container modules are imported or locally manufactured and address urgent spatial requirements or infrastructure, when the local market has a lack of skilled trades and an absence of construction material. In addition this building system can be erected at short notice, given volatile economic circumstances. In highly developed countries that enjoy prosperity, it has been argued that economic development can strain the local construction industry's capacity by increasing demand. This is the case in Australia where the demand for housing cannot be matched by the supply capabilities of the national building industry. This increase in demand causes an increase in labour and material costs, partly decreasing the affordability of homes. Container modules built overseas or locally, with efficient production methods and with minimum wastage, reduce construction costs.

Container modularisation and design

There are two major problems associated with container modular homes when compared with housing produced by traditional on-site construction methods: (1) limited floor plan design flexibility; and (2) restricted external presentation. Container modules are limited spatially to dimensions governed by ISO standards. Modules are built within these governing parameters and arranged on site to create a preferred floor plan configuration. The external inherited box-like shell of steel container is constructed using 2 millimetre thick trapezoidal sheets that are welded into the container frame. If this type of finish is left exposed without any external cladding or facade treatment, the home will be negatively perceived as a temporary and poor quality home.

The floor plan design of a home can be highly personal and influenced by multiple factors. These factors include the allocation and spatial distribution of areas, on-going running costs, sustainability and the adaptability of the floor plan. When designing an affordable home it is important to consider its on-going running costs as being equally important as initial construction costs. With rising energy costs and increasing market emphasis placed on sustainable living, an affordable home must also be a sustainable home. Therefore, the interior of the home must be designed using the principles of passive design with the underlying aim of maximising the level of comfort for its residents while minimising energy use and promoting low-cost living. This can be achieved by utilising free, natural sources of energy, such as the sun and the wind to provide heating, cooling, ventilation and lighting to contribute to a reduction in energy usage.

While designing an interior layout it is important to ensure that the floor design is adaptable. An adaptable house accommodates lifestyle changes without the need to demolish or substantially modify the existing structure and services, and is one of the key advantages of building a modular home. Dedicated purpose rooms should be avoided with the emphasis placed on large multi-purpose rooms, like living rooms with open kitchens, with access to outdoor areas. Such flexible floor plans allows family-orientated central areas to be used to host guests during formal gatherings instead of having formal entertaining areas that are rarely used. Based on this logic and with the amalgamation of outdoor and indoor areas, the regularly used areas of the home increases and allows a smaller and more affordable home to offer large living spaces.

The geometric planning of floor plans is governed by the spatial parameters of the different container modules and current market trends. Although container modules dictate the design of the floor plan, it does not limit the number of possible design combinations. Container modules can be built within different size containers and can be modified with any required number of openings. Therefore, with such an array of choices, the possible spatial distribution of the floor plan is governed by current market requirements rather than modularisation limitations.

Cost-benefit analysis

- Material cost saving: Modularisation represents the standardisation of all aspects of the construction process including materials. In contrast, conventional on-site building is often subject to customisation and material wastage, resulting in frequent quantity variability due to non-standardised building practices. Therefore, it can be conservatively assumed that, on average, there is a 10% saving in the cost of materials used with the modularised off-site construction process.

- Labour cost saving: Subcontracted labour costs represent a large proportion of the cost of construction in conventional on-site building. Modular construction minimises the use of such costly labour by hiring skilled labour to support the manufacturing process. Two types of skilled labour are required: (1) skilled labour and (2) licensed skilled labour. Skilled labour is used for work that does not require a certification of compliance upon completion. For example insulation installation, cabinetry and carpentry are tasks that can be completed by skilled labour without the requirement of formal certification of work. Plumbing and electrical work cannot be carried out by a non-licensed skilled labourer. These tasks require formal certification from the relevant government bodies and therefore must be completed by licensed personnel.

- Common expenses incurred: Common expenses include all expenses incurred irrespective of the selected construction process. These expenses include earthworks for site preparation, and foundation and concreting work. These tasks are usually subcontracted out to specialist contractors and their costs are not affected by the method of construction chosen.

- Hybrid modular construction – container module construction cost: When analysing the cost and benefits associated with hybrid constructions the container module was treated as an additional cost, where the area each module covered was deducted from the net total area of the home. This allowed conventional building rates to be applied to the area of the home built on site, while discounted modular construction rates were only applied to the portion of the home built off-site. Costing the two differentiated areas separately, then combining them resulted in the total cost of the hybrid construction.
- Project management charge: Project management charges include expenses a construction firm will incur on a monthly basis to manage the delivery process of a construction project. These expenses include administrative costs, site management costs, occupational health and safety management costs, temporary site provisions (toilets, fencing, etc.) and other fixed costs such as office rent and other relevant running costs. This expense is variable and will decrease as the number of homes constructed increases, and vice versa.
- Time saving: Time saving represents the number of weeks of site work reduced by using an alternative construction process in comparison to the time taken if the same product were built using conventional on-site building methods. In this case it was established that a traditional home can take an average of twenty-four weeks to be completed on-site. A fully modularised home can be built off-site while its foundation is prepared, then completed modules can be delivered onto the site and the remaining construction work can be completed in as little as ten weeks. When using hybrid constructions the modules can be built off-site and delivered onto the completed foundation and then the remaining sections of the structure will be constructed. Although this method utilises a greater on-site construction duration when compared with full-modularised constructions, it is still more time efficient that conventional on-site construction. Saving on-site construction time usually results in a quicker project delivery time. This can potentially result in a multitude of cost savings for the average consumer such as savings on paying rent or another mortgage while their home is being constructed.

This case study has highlighted an entrepreneurial approach to property development by considering alternative design and construction approaches. Care needs to be taken to ensure there is adequate demand for the completed product when released onto the market. In such cases the use of careful market research is essential and will ensure any benefits gained (e.g. savings in time and costs) are not exceeded by an extended selling period. This places the emphasis on ensuring the business case is viable and a holistic evaluation approach from 'cradle to grave' (i.e. total life cycle of the development) is undertaken and documented.

Company overview

Since its corporate inception in 2001, UDAYA has grown to become one of Australia's most reputed mid-tier residential design and construction firms (UDAYA Pty Ltd, 2012). The firm is a credit to its founder, Suda Udaya's leadership and his endeavours to build a design and construction firm that would offer its clients an honest, headache free and cost effective development solution (UDAYA Pty Ltd 2012). UDAYA consists of a group of subsidiary entities that offer residential town planning works, architectural design work, engineering, construction, project marketing and project finance (UDAYA Pty Ltd 2012). The collection of these complementary services managed by respective industrial experts, ensures potential UDAYA clients are offered an efficient design process, construction, marketing and finance (UDAYA Pty Ltd 2012).

Note

1 RICS 2012 draft guidance note – Developing a building procurement strategy and selecting an appropriate procurement route. Available at https://consultations.rics.org/consult. ti/procurement/viewCompoundDoc?docid=2704532&partid=2704628&sessio nid=&voteid= (accessed 3/4/2013).

Chapter 8

Market research

8.1 Introduction

In a similar manner to professionals in other disciplines, although property developers are experts in the specific area of commencing and completing physical property development, they may lack the same high level of commitment in other related areas. Undertaking adequate market research is often acknowledged as such a shortfall. Unfortunately this conclusion is often reached with the benefit of hindsight, especially if a property development has not been as successful as planned. As a consequence, too many property developers expose themselves to high levels of risk which would be almost entirely avoidable had the importance of market research been realised prior to commencing the project, as well as during the entire construction phase (including disposal through sale or lease).

It is unclear exactly why the importance of sound market research is often under-estimated in the property development process. A possible reason could be a lack of formal training undertaken by most property developers who may have evolved into their role mainly via industry experience. Another possible reason is that when a property developer has previously operated in a rising market (i.e. demand exceeds supply), the impact of failing to carry out adequate market research may be hidden by higher prices. Any downturn in the property cycle requires the developer to be more cautious because the demand for new or refurbished property may be reduced due to an oversupply of similar property, or property values may fall. Market research is critical at all times but more so in a market downturn.

Conducting thorough market research has the potential to make or break a successful property development. Market research itself is a specific discipline and its importance cannot be under-estimated in the property development process. Every successful property developer acknowledges that undertaking well-planned and carefully executed market research will substantially increase the likelihood of success although there will always remain some information that is either unknown or unknowable. The rapidly growing area of property market research continues to embraces different specialisms, including information and database services, strategic and site-specific analysis, forecasting and portfolio analysis.

There are varying approaches to undertaking market research. Both site-specific and large-scale strategic analysis are used by developers and investors to help assess the viability of individual projects and/or to assist in making decisions which affect the company's long-term real estate strategy. For example research analysts can provide invaluable insights about the fundamental demand and supply factors, at any given time throughout the development process, as well as the relevance of underlying market conditions. In addition, market research can identify previously invisible risks and opportunities which might not be readily evident from viewing a list of current transactions in the market.

This chapter highlights the importance of market research, emphasises why it always needs to be incorporated in every project and explains how it can assist the success rate of property developers. In addition there is a discussion about the differences between land types that further complicate direct market comparisons. Prior to commencing any form of market research it is essential for the nominated land type (e.g. retail, industrial, residential) to be identified early in the analysis stage. A simple approach such as this in turn will assist the overall market research process; on the other hand, omitting this step will increase the likelihood of failure due to higher exposure to risk.

8.2 Undertaking market research

The concept of market research means different things in different circumstances to different stakeholders. For example, the term 'market research' or 'market analysis' is used broadly in economics but has a more specific meaning when related to the context of property development. A good starting point is to consider market research or analysis as the identification and study of the market for a particular economic good or service. Accordingly it can be considered at two different levels as follows:

a. Based on a relatively broad market viewpoint without a specific individual property as the primary focus of the study. For example this type of market research could be applicable for multiple properties located in the same marketplace.

b. Based on the perspective of the actual real estate market in which a given property competes. In this scenario the relevance of the market research is limited to an individual property.

With most forms of research (see Figure 8.1) the focus can range from (a) specific research, and (b) to broad research where the decision for the property developers will be based on what levels of research are relevant to the project at hand. The amount of market research possible is practically infinite and in a constant state of change. Resource limitations, especially financial, will partly dictate how much market research can be undertaken for a particular property. It is this balance between broad and specific information which must be closely thought through for each property development.

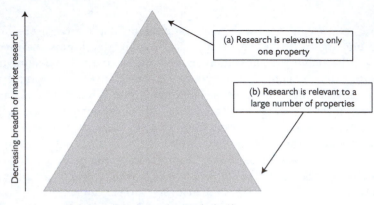

Relevance via land use or location (or both)

Figure 8.1 Market research approach from broad to specific

Figure 8.2 highlights the relationship between the overall aggregate real estate market and the subsectors within, based on classification via different land uses. This replicates the operation of the market where each land use has a set of unique characteristics. For example a prospective purchaser (or tenant) of an industrial property will have completely different requirements and therefore buyer/occupier characteristics, than a prospective purchaser of a residential property.

Building upon this discussion, the next stage is to model the relationship between land use and geographic location to determine which sector of the real estate market should be the subject of the market research. Using the simplified model in Figure 8.3 it is possible to enter four different locations (*y*-axis) and four different land uses (*x*-axis), resulting in 16 different sectors. A typical market could be broken into additional land uses (say 10) and also additional geographical locations (say 15 for ease of discussion), now resulting in approximately 150 different sectors which need individual market research. This example highlights the underlying complexities when determining where boundaries commence and end. Of course real estate developers

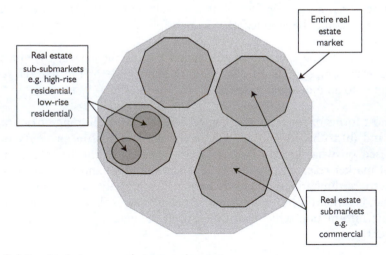

Figure 8.2 Relationship between market research areas

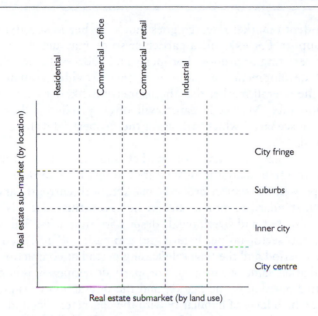

Figure 8.3 Cross-tabulation – land use by geographical location

operating in the global market will be examining an extremely large number of locational sub-markets.

During the actual task of undertaking market research and gathering data it is essential to consult with many market stakeholders and participants. Direct sources of information are those involved or likely to be involved in property transactions that can be relied upon for comparison purposes. There are also secondary sources of information which may provide only general or broad information about the local economy as supporting information. It is important to understand how the interaction of supply and demand affects the property's value. Through the investigation of recent property transactions, properties for sale in a competing market with the same land use (although not necessarily in the same location subsector) and the behaviour of market participants, it is possible to examine supply and demand relationships and investigate the current and future status of values. This approach will assist in the interpretation of market attitudes toward current trends and anticipated changes. If current market conditions do not indicate adequate demand for a proposed development, the market research may identify the point in time when adequate demand for a project will likely emerge. It can be observed in this example that market research helps to forecast the timing of a proposed property development and also the amount of demand anticipated over a specified time.

As the determination of the best use of a parcel of land should be the ultimate goal of a successful property developer, market research provides a sound basis for determining what this actually is. For example an existing or proposed improvement may be considered to be suitable as a particular land use – therefore it must also be put to the test of maximum productivity which will require confirmation via market research that an appropriate level of market support exists for that specific use at the time when the real estate development is released onto the market. When undertaken, in-depth market research goes much further in specifying the type and level of that support. For example a particular study may determine key marketing strategies for an existing or proposed property and address the design characteristics of a proposed development. In addition it may provide estimates of the actual proportion of the overall market that the property is likely to capture and its most likely absorption rate. Market research will often provide a 'best case scenario' and a 'worst case scenario' which will assist the property developer to undertake a sensitivity analysis.

In order to estimate the level of market support for a proposed property development, a research analyst must identify the relationship between demand and competitive supply in the subject property market, both currently and in the future. This is based on standard economic theory; however, this relatively basic approach is often underestimated and overlooked, despite its simplicity. The supply–demand relationship for this sector (as per Figure 5.6) can easily be modelled for the market over the forecast period and the final sale/leasing period at completion.

The overall market value of a proposed property development is largely determined by its competitive position in the market and the interaction of supply and demand levels. Ensuring a high level of familiarity with the characteristics and attributes of the subject property (divided into (a) land component, (b) building component, and (c) completed development) will enhance the ability to identify competitive properties (*supply*) and to examine the comparative advantages and disadvantages that the subject offers potential buyers or tenants (*demand*). Only with a comprehensive understanding of economic conditions can the effect on property markets and the momentum of these markets (and various subsectors) be accurately reliably understood. Only at this stage will it then be possible to quantify and better understand the externalities affecting the proposed property development.

8.3 Sourcing real estate market information

The most common reasons for undertaking market research are as follows:

1   To ensure informed decisions are made about individual developments or investments. This may include validation of preliminary unsupported estimates of a developer or investor.
2   To assist in the formulation of long-term development and/or investment strategies.
3   To analyse and predict the performance of property investment portfolios.

Each of these uses will bring together a mix of different research skills and approaches. In addition to these established uses, property market research is being increasingly used for decision support systems for a wide range of real estate related disciplines including valuers, investors, lenders, purchasers and real estate developers. For companies which are publically listed on the share market there is a greater need for higher levels of transparency in dealings and decisions relating to financial decisions about property. Some of the areas where market research can help include questions about:

1   how variables and decisions on the local economy affect value;
2   how the property development is affected by trends (e.g. demographic trends by varying cohorts) in the market;
3   the level of risk associated with a development at a particular point in time in the future.

There has always been a need to forecast the future which is designed to reduce risk. However, such models are never risk free and longer timeframes are associated with substantially higher levels of risk. For example it is possible to forecast the weather pattern for the next hour with a high degree of certainty, although less certainty is associated with tomorrow's or next week's weather. How much certainty is associated with the weather when looking one year ahead or five years ahead? Some decisions (e.g. which government will be in power in five years' time) are practically impossible to estimate with a high degree of accuracy so there will always be a certain exposure to risk. However, real estate developers must keep up to date with constantly changing forecasts about future trends through the use of property market research. Ignorance and the 'benefit of hindsight' are not valid excuses if a development has failed to reach its full potential due to reliance on poor or inadequate market research. At times the research may indicate a proposed development is not a viable proposition to commence in the current climate and should be delayed until the market recovers or demand returns. This results in a positive outcome for the developer since it will reduce exposure to risk. All real estate developers entering the market need to be conversant with the full range of research services available, if only to become sophisticated purchasers of them when required. The next section provides a guide to the main types of property research.

8.4 Types of research

This section discusses information, site-specific and strategic market analysis, forecasting and portfolio analysis.

8.4.1 Information

Generally speaking there are two different types of information: (a) property market information and (b) supporting information.

(a) Real estate market information

Reliable accurate statistics about all transactions in the real estate market are not freely available in all markets, especially with reference to purchaser's and seller's names for example. This is partly due to privacy laws and also to the lack of a centralised marketplace. While some partial information is available on development activity, very little exists about transactions or prices. Detailed information on development activity at a local level is rarely available on a consistent basis over long time periods, especially with reference to sensitive financial information. In the commercial market, data about recent leases and development completions usually have to be pieced together painstakingly from a number of different sources. Some property market reports contain information about rents and capital values in selected locations and cities; however, sufficient information is only provided to gain a reasonable view of market activity, either at the national level or for individual markets.

In the absence of detailed inexpensive comprehensive statistics, it is clear that successful real estate developers have collected their own records of transactions and prices. For example larger developers monitor information about most aspects of the market, including building availability, transactions (take-up and rents) and performance (yields and capital values). This information is often provided by larger valuation and appraisal firms via regularly produced bulletins or in periodic reports on sectors or topics of special interest.

One of the core challenges when conducting property research, especially across national, international or global markets, is the lack of benchmarks or commonly agreed standards. For example there are many challenges when comparing the level of sustainability of commercial office buildings as the sustainability tools used in one country may not be compatible with sustainability tools used in a different country (Reed and Krajinovic-Bilos 2013). Indexes for monitoring the real estate market are becoming available (e.g. the IPD Index produced by Investment Property Databank) although they remain expensive and lack detail at the local level.

Property market information is available via many free online websites and also is aligned with Geographical Information Systems (GIS) which greatly assist with monitoring the market. It can be argued that there is at times too much information available at this level which may confuse the market and potential purchasers/tenants. Caution must be exercised with free information with regard to reliability and accuracy of data.

(b) Supporting information

As real estate market analysis has evolved and become more sophisticated, it has drawn increasingly more on 'supporting evidence'. For instance, if demand is analysed solely on the basis of transactions completed on individual properties then part of the picture may be missed; the number and type of completed transactions, for example, is often restricted by the availability of suitable property. Attention should be placed on the underlying level and character of demand by looking at an area's demographic characteristics (especially for retail developments) or the structure and performance of its local economy, among other things. Official statistics of this type are available from the relevant government bodies which conduct a regular census and also from consultancies which provide an additional analysis of census data.

As well as indicators of local economic well-being, the analyst may also be required to advise on the overall state of the economy or particular occupier types, e.g. manufacturers or retailers. Many such macro-economic indicators are employed in a real estate market analysis including, for example, statistics about Gross Domestic Product (GDP), retail sales and manufacturing output. Information about employment and unemployment is relevant as well as national employment statistics and selected data for regions, counties and local authorities, often available at no cost.

The list of data sources (real estate and supporting) is not exhaustive and the range of useful publications is increasing all the time and updated regularly via internet access. Nonetheless, substantial gaps persist. All too often, researchers and analysts are spending their (and clients') time identifying and collecting information which might already be more efficiently provided via official sources. Moreover, the fact that property information has developed in a piecemeal fashion, led by competing private-sector interests, means that it is often inconsistent between sources (though usually consistent within any given source) and has focused on areas which have attracted the most investment – for example inner-city commercial property. This latter aspect is not altogether bad so long as the market remains focused in the same direction. A market downturn in a particular area often shifts investor and developer interests elsewhere and into locations and property types on which existing information is poor quality. As the next development cycle then gathers pace, errors of judgement are often made in these poorly researched and understood markets.

8.4.2 Strategic and site-specific analysis

In certain circumstances it is often sufficient for a researcher simply to supply his/her client with information, here the real estate developer may be the external client or the research is conducted in-house. As information has become more widely available, clients now expect interpretation of the information in the light of specific questions about their company or individual development plans. Two different types of analysis are commonly required, namely (a) a strategic analysis and (b) a site-specific research.

(a) Strategic analysis

Strategic analysis is either conducted by a company's own research team or commissioned from external consultants. It typically examines questions which are long term in nature and rarely relate to individual development projects. This type of research is often employed where a development or investment company is either reviewing its current strategy or evaluating moving into a new direction. For example the question might be whether the company should be moving into a different property type, land use or location or whether it should diversify its activities. Reports of this nature typically compare performance across sectors and/or locations over the long term. The analysis will usually examine real estate market performance within the context of underlying economic forces where analysts may also draw on economic and property market forecasts to predict future performance. This type of strategic research is rarely published and it is difficult to measure the impact on decisions made by major real estate developers.

Strategic research is also used to keep companies informed of emerging and future trends in occupier, development and investment markets. Research of this nature is sometimes funded by a group of developers and/ or investors joining together to commission analysis on a particular topic. While little of this research is published, we know that large property companies and investors are considering issues such as demographic change (e.g. the retirement of the 'baby boom' generation) and how it affects demand from a retail perspective. Some developers also use this type of research to spot new market niches and opportunities to maintain a competitive advantage.

(b) Site-specific analysis

Site-specific analysis is employed where a developer or investor needs answers to questions about an individual scheme. These questions might include, for example, whether the developer's predictions about an under-supply scenario are correct, whether to refurbish an existing property or whether an investor should fund a particular scheme. The purpose of the site-specific study is to look beyond the immediate state of the market (on which agents can advise) and to identify risks and opportunities which might not otherwise be apparent. Among other things, the analyst should be able to place perceived market shortages within the context of the property cycle and to identify how the market might move during the course of the development project. An important role for real estate market research is to provide the developer with information to assist informed decision-making and identify sections of the market which will be in demand in the future.

Site-specific studies typically examine some or all of the following:

DEMAND: BOTH QUANTITY AND TYPE

The first step in any site-specific study is to define the scheme's most likely market area. This is often referred to as 'the catchment area', especially in research relating to retail developments. The size of the retail catchment area will normally depend on the size of the scheme or the population of the town on which it is dependent and the expected area of 'draw' in terms of demand. Generally, the larger the retail real estate development, the more extensive is the catchment area. Where a scheme is designed for occupiers who are likely to be long distance relocations (e.g. a large business park), the analyst may research demand levels on a much broader scale such as nationwide or global, as well as with reference to the local market area. Demand is analysed both in terms of underlying economic drivers and recent property market transactions within a specific catchment area.

An understanding of the local economy is critical when identifying any potential long-term threats or opportunities, as well as the potential size of the market. For example, since employment has a direct influence on consumers' spending potential and is directly related to how much floorspace is needed in commercial, retail and industrial property, an evaluation of the current and future jobs situation is crucial in any study. An analysis of economic structure and recent market performance may reveal that the economy is over-dependent on a particular industry even though the overall outlook is poor. Changes in the size and composition of the local population will also directly affect the level of disposable income, spending patterns and the available workforce. This local economy research will draw on the various sources of 'supporting information' and will normally also include visits to the area and contact with local government bodies, organisations and the main employers.

Different methodologies have been developed to translate actual and forecast volumes of spending and employment in a particular area into the quantity of floorspace needed or the capacity of the market to absorb this demand. This capacity can be compared against existing floorspace provision, together with any additional development proposals, to assess whether there is likely to be a quantitative shortfall in supply, now or in the future. Establishing the quantity of demand is, however, rarely adequate to provide a case for or against a particular development. After the level of demand has been estimated, an evaluation needs to be undertaken of its qualitative characteristics and how these align with the developer's proposed design. It must be noted that all retailers will not experience demand for the same identical goods in the same stores, in the same manner that all office tenants will not want the same configuration and specification of space. The purpose of looking at the 'shape' of demand can be twofold: first, to estimate what proportion of the total quantity of demand is likely to be for a proposed property development; second, to assess critically the design of a proposed (or existing) property development. Qualitative aspects of demand can be assessed either via further desk-based research or via direct research into tenants. Examples of desk research include: the socio-demographic composition of the population; the number of banking, legal or accountancy offices; or the size distribution of manufacturing firms. The methods are many and will depend on the research question(s) being asked. Data on property transactions can also be useful indicators of the 'shape' of demand, e.g. the size, location and rents achieved in recent office developments or the profile of retailer enquiries. While recent lettings can be a useful indicator of demand, a crucial distinction must be made between effective demand (completed transactions) and latent demand (requirements yet to be satisfied). In certain circumstances, such as in very active markets, the level of demand may be adversely influenced by what developers are currently providing rather than what occupiers will actually want over the long term.

Types of qualitative surveys commonly conducted are direct individual interviews, either face to face, via telephone, email or post, depending on the complexity of the questions being asked. Surveys are either carried out by property researchers or subcontracted to specialist market research firms, especially in the case of retail surveys for which the required sample size is normally large. Surveys reduce the need to infer occupier behaviour from sometimes weak secondary data sources, such as reported transactions. These surveys can provide excellent first-hand information and insights into occupiers' real requirements and the reasons behind their property decisions. The principal disadvantages of surveys are that they can be relatively expensive and poor survey design can produce misleading results.

SUPPLY: QUANTITY AND TYPE

An analysis of supply can similarly be divided into quantitative and qualitative components. Quantitative analysis will typically establish the amount of existing floorspace in the market area, and any development proposals and when they are likely to be built and add additional floorspace. This analysis should be placed within the context of the development cycle to indicate whether increasing or decreasing building activity is probable in the next time period. Qualitative analysis may assess how much of the total floorspace is likely to be competing directly with a proposed real estate development, given its characteristics and intended occupier profile. With this approach it is often possible to identify a shortfall in an apparently oversupplied market and vice versa.

Often detailed information and related data on floorspace and development activity are not readily available, especially at the local level. This requires the researcher to piece together evidence from various sources to provide a best estimate of supply.

Sources include local government authorities, commercial databases, collection of information in the 'field', local real estate agents, industry bodies and other developers via networking.

The balance between demand and supply should be reflected in current real estate market conditions, especially in current rental information. Property researchers frequently use the prevailing level of vacancies as an indicator of balance in the existing market. Where available, the data about trends in area take-up and availability are also used to indicate the direction in which market balance is moving and why. A falling vacancy rate, for example, may not always indicate an improving level of demand as it can also indicate the withdrawal of vacant stock from the market, perhaps to be redeveloped later. Unfortunately a lack of reliable data at times means that detailed analysis of take-up and availability is not always possible and the analyst may need to devote considerable effort to collecting and analysing data about rental agreements and vacancies to gain a reliable picture of the market or future trends.

A real estate developer must refer to a range of indicators to present a view of future market conditions and the likely outcome for an individual development. Where extensive data is available then projections of take-up, availability and vacancies can be used alongside projections of underlying demand and supply conditions to produce a formal forecast of rents. Otherwise the rental levels may be projected simply on the basis of underlying economic demand conditions and the supply of new developments, i.e. at one step behind market indicators about take-up and availability. Assumptions about future rents contained in the formal development appraisal can then be tested against rental projections, further helping the developer to come to a view about the viability of the development project.

Research can also provide a more qualitative view of the viability of a scheme, pointing to intrinsic 'strengths' and 'weaknesses', any unforeseen 'opportunities' and any future 'threats' being commonly referred to as a SWOT analysis. This type of analysis is especially useful in situations where formal forecasting is not possible and there is a need to address questions about the optimum size and configuration of proposed development. Conclusions from a qualitative analysis of this type can also begin to identify the characteristics of the target market. For example occupier profiles thus identified can be used to nominate companies for inclusion in a marketing contacts database.

8.4.3 Forecasting

As they have become more familiar with property research, real estate developers and investors have become increasingly interested in the possibility of predicting future market conditions with direct reference to the level of rents, yields, capital values and returns. This is designed to reduce direct and indirect exposure to risk. This research service is provided by specialised consultancies who produce regular forecasts of the property market, usually on a confidential basis to developers and investors at a cost. Not all forecasts are published in a high level detail as they involve many assumptions. Forecasts are typically available for the office, retail and industrial sectors as well as the investment market at national, regional and increasingly at a local level. Most national and global real estate companies employ in-house research teams which produce property market forecasts for use by their own staff.

Forecasts are used for two principal purposes. First, they can provide an indication of the likely future operating environment through forecasts of the real estate market as an aggregate. In this respect the forecasts can indicate a required change in direction for a company's strategy, e.g. by highlighting future risks in an under-supplied market. Second, forecasts are sometimes employed in site-specific studies to provide rental projections (and, more rarely, yield forecasts) for the local market into which a real estate development will be undertaken. An aim of forecasts is to predict and model future changes in rents and yields on the basis that the relationships between each of these variables and the factors which drive them – namely supply and demand – will remain the same in the future as they have been in the past. Most forecasters employ formal econometric modelling techniques based on various forms of regression analysis or hedonic modelling.

It is commonly accepted that the direct measurement of demand and supply in the real estate market is extremely difficult. Forecasting models, therefore, normally employ proxies for demand and supply. On the demand side the economic factors which drive property demand are often used as the demand variable, e.g. retail sales or employment in service industries. On the supply side the amount of space being released onto a specific market and also building completions are sometimes known but often they have to be inferred from data such as building approvals and planning permission.

The models used and associated formulae for forecasting are closely guarded secrets but the aim is to minimise the number of assumptions as much as possible. Reliance on facts and actual information should be incorporated into the model as much as possible. For example with retail property the level of current rents is often strongly related to rents in the recent past. In addition to recent rental agreements, a number of other factors 'explain' retail market rents, including consumer expenditure on the demand side, together with floorspace, stock and construction starts on the supply side and interest rates. A common feature of forecasting models is that some explanatory variables are 'lagged' where events that occurred one or a number of years ago may be reflected in rents this year. These types of explanatory model form the basic building blocks of property market forecasts. Once the fundamental relationships have been established, forecasts of the independent variables (e.g. rents, yields) can be produced if projections are available for the explanatory variables.

On the demand side the economic forecasts are typically obtained from specialist economic and regional forecasters. On the supply side a real estate developer may provide their own estimates of future building and stock from analysis of the current development pipeline. In some circumstances it is also possible to use data on take-up and availability in forecasting models. As the use of property market forecasting has become more widespread, so the dangers of overly mechanistic analysis have grown, as has the reliance on 'black box' models, where the user may be unfamiliar with how the calculations actually work. A number of simple rules can be applied to the process of forecasting to avoid some of the potential pitfalls. First, the cliché 'garbage in and garbage out' can be applied rigorously to property market forecasts. This remains an issue in spite of the vast improvements in the quality of property data which have taken place in recent years. Second, and reflecting the restricted availability of suitable data, models should be kept simple and robust. Third, 'answers' generated by forecasts should be assessed critically in the light of the experience and knowledge of the real estate developer. Fourth, the users of the forecasts must be made aware of the uncertainties surrounding forecasts and the limits imposed by the data. Forecasting is not a substitute for interpretation as it is only a decision-support tool for the real estate developer to work with.

8.4.4 Portfolio analysis

A further application of property research techniques is in the area of portfolio analysis. Portfolio analysis was originally developed for the other capital markets in the 1950s but has only recently been applied to property. It is used to analyse the comparative return and risk performance of:

1   property as a whole against other competing investment types, e.g. equities, cash at bank;
2   an individual property type against the other land uses (office, retail, industrial).

In a portfolio analysis the expected relative returns for different property types (or asset classes) and their volatility on an inter-relationship basis are examined based on long time-series. The analysis identifies the optimal combination of property types (or asset classes) which will provide the maximum return for any given level of risk and therefore the optimal mix within a portfolio. Forecasts are increasingly being used alongside portfolio analysis to predict the optimum real estate holdings in given future market scenarios.

8.5 Impact of research

The focus has been placed on what property research is, how it is undertaken and what it refers to. The reliance on property research has increased very rapidly and subsequently it has become an important part of the development and investment process. However, the usefulness and limitations need to be acknowledged. Real estate developers and investors find research to be helpful for individual projects if the rapid expansion of and demand for such research can be taken as a measure of its usefulness. At the level of the property market as a whole, however, this is less clear. For example with reference to economic theory, the wider availability of information should arguably assist to make the market more efficient. This was evidently not the case with the Global Financial Crisis and the widespread availability of unprecedented amounts of information via the internet. It could be argued that the growth and availability of property information and research over the past decade, especially information freely available instantly over the internet, has increased speculative decisions and interpretations. In other words the amount of information appears to have helped to make responses more acute. This may not be a problem in markets for other asset classes (e.g. equity market) in which the supply side can respond quickly to changes in demand – either by increasing or decreasing output – but in the real estate market the consequences can be disastrous because of the long lead times in the development cycle. Against this background it must be acknowledged that there are limits to the usefulness of research into the future operation of real estate markets. If used correctly, research can potentially reduce uncertainty but not eliminate risk completely.

As real estate developers seek to undertake successful developments in the next cyclical upturn in a market which operates in a relatively uncertain economic climate, the use of research will undoubtedly continue to expand. Improved databases and more consistent analysis, however, will not provide all the answers to the real estate market. The real estate developer should never lose sight of the all-important final consumers of property – the occupiers – where decisions are made by humans rather than mathematical formulae. Perhaps in future more research should be directed to understanding exactly what is required from the development industry if we are to avoid a repeat of the mistakes observed in every financial downturn.

8.6 Reflective summary

This chapter discussed the importance of incorporating market research as an integral component in every successful real estate development. A lack of reliable and appropriate market research can be directly linked to higher risk and uncertainty, as well as reducing the potential profit margin and associated lost opportunities in ensuring the highest and best use is achieved.

It is essential to undertake appropriate market segmentation to accurately identify the most appropriate market for the proposed property development, with careful attention paid to varying levels and direct/indirect drivers behind demand. With reference to office development the drivers will predominantly be businesses seeking office space. Retail development will be largely dependent upon the location and characteristics of nearby households. Industrial development success will be associated with a range of variables including the strength of the local economy and government decisions about international trade for example. An analysis of the competing supply of similar land use will ascertain the total amount of stock available to the market, as well as the relevance and potential to achieve equilibrium.

Different types of market analysis are possible including economic base analysis, marketability studies, investment analysis and feasibility analysis. The rapidly growing area of property market research embraces a number of different specialisms including information and database services, strategic and site-specific analysis, forecasting, and portfolio analysis. This is a very specialised area which has increased in importance due to its recognised potential contribution towards reducing exposure to future risk for the real estate developer. While significant advances have been made in the availability of property market information, a number of major gaps still exist and the benefits of conducting detailed (and often expensive) research are often under-estimated. For example the application of GIS mapping in the real estate market allows the analysis of increasingly sophisticated and widely available data; however, relatively few developers take the time to model the locational attributes and thematic mapping implications associated with their potential real estate development.

When required, both a site-specific and a strategic analysis are used by developers and investors to help assess the viability of individual projects and/ or to better inform the company's long-term property strategy with regards to exposure to risk. Analysts provide critical insights into the fundamental demand and supply factors underlying market conditions. Effective timely research can also identify potential indirect risks and opportunities which might not be readily evident from current transactions in the market. Forecasts and portfolio analysis are increasingly being used in these types of analysis to provide predictions of future rents and yields and the optimal portfolio mix of properties as a result. A successful real estate developer will understand the importance of relying on high quality relevant research to minimise their exposure to risk. This may be informal risk evaluation (e.g. networking) but most often this will involve a substantial capital outlay for in-house researchers or an external consultancy to ensure today's decisions about the future are as fully informed as possible and based on minimal assumptions and unknowns.

8.7 Case study – due diligence and contaminated sites

This case study overviews a property development in Melbourne, Australia, constructed on a site which adjoined a highly contaminated site which was not adequately decontaminated prior to commencing the property development. After considerable financial outlay over ten years, including construction of a high quality property development to completion stage, the new buildings were demolished and the development site is currently vacant pending decontamination and a new development. The underlying failure can be traced back to the previous use of the site (a dry cleaning business) where the site was not suitable for residential use in its current condition. Clearly the initial property development should not have proceeded and was always flawed. It is worthwhile to examine the timeline of events.

- 1992. A large dry-cleaning business which had operated for more than twenty years ceased operation on a site which was adjoined the actual development site. The dry-cleaning business potentially contaminated its own site via chlorinated hydrocarbons and degraded white spirit or naphthalene, which are both used in the dry-cleaning business. With the benefit of hindsight it was evident that the adjoining site was also contaminated.
- 2001. Planning permit issued to allow construction of 49 units. Construction commences.
- 2003. Deposits paid by individual purchasers to the property developer subject to completion. Construction completed (Plate 8.1) but final approval to occupy not granted by government body.
- 2006. Developer attempts to sell entire development regardless of contamination (Plate 8.2).
- 2007. Development falls into a state of disrepair (Plates 8.3 and 8.4).

Plate 8.1 Property development completed with 49 units ready for sale (Source: Richard Reed)

Plate 8.2 Entire property development placed on the market in 2006 (Source: Richard Reed)

Plate 8.3 After an extended period of uncertainty due to the contamination issues the property was vandalised with graffiti (Source: Richard Reed)

Plate 8.4 After an extended period of uncertainty due to the contamination issues the property was vandalised, with internal fittings and white goods removed (Source: Richard Reed)

Plate 8.5 A deep excavation to remove the decontaminated soil (Source: Richard Reed)

Plate 8.6 Eventual decontamination of the site finally took place in 2013 (Source: Richard Reed)

Plate 8.7 The promotion of a proposed property development after the excavation of the site with removal of the previously failed development (Source: Richard Reed)

- 2012. The original buildings are demolished and decontamination commences (Plates 8.5 and 8.6). This followed legal action and the conclusion that the site must be decontaminated.
- 2013. A new property development is promoted following the decontamination (Plate 8.7).

This case study highlights the challenges associated with not undertaking a thorough due diligence process. This statement applies to the actual site but also to the risks attached to neighbouring properties. Clearly this site was not suitable for development without (a) the previous owner paying for decontamination of the site or (b) the property developer allocating sufficient funds for the decontamination. While potential recourse is available via the legal system, there are substantial indirect costs which can adversely affect the property developer's reputation. Another downside is the overall uncertainty associated with the development when seeking closure. This period may take years or decades, resulting in a detrimental effect on cashflow and holding costs. The saying 'fools rush in' may be linked to the inability to undertake adequate risk analysis to determine why the market has placed such a low value on a property.

A contaminated property may have a negative value and may not be viable as a potential development site without substantial external stakeholder (e.g. government) involvement. In the future it appears that companies operating businesses which contaminate sites may be held accountable to some extent to decontaminate or 'make good' the site. It should be remembered that most sites have a previous alternative use and due diligence must be undertaken to ensure that the site has been fully decontaminated and is suitable for the potential property development. The engagement of specialised advice to examine the soil below the surface is essential and outside the expertise of most property developers.

Chapter 9

Computer technology

9.1 Introduction

Until recently many property companies, with the exception of estate agents and realtors, have been relatively slow to capitalise on the opportunities offered by the computer industry. The recent emergence of software packages designed to meet the full spectrum of property company requirements, together with the increased portability of this technology through the use of laptops, tablets and smart phones, means that the developer now has many tools available both 'in office' and 'in the field' to assist in the complex task of managing a development.

The purpose of this chapter is to introduce examples of the type of development software available, from the spreadsheet to the more specialised bespoke property development software. Although most development appraisals start off as a fairly simple cost/benefit analysis they inevitably become more complex due to the interaction between the individual elements within the scheme (e.g. build costs, land/acquisition costs, rental values). As previous chapters have discussed, the individual elements which contribute to a development scheme from inception to disposal are rarely stable. Property development software provides the developer with the ability to model the consequences of foreseeable changes in different variables to enable contingencies to be built into the plan, thus reducing risk by ensuring that any 'knock-on' effect is taken into account. It also allows a developer to adapt to changes during the development process (e.g. unavailability of specific build materials) and make adjustments for price increases/decreases. Over the period of a development many of the variables are subject to change outside the control of the developer. This especially applies for macro-economic variables such as borrowing interest rates and inflation. The largest single benefit from using computer software is the straightforward ability to conduct a sensitivity or scenario analysis to assess the impact on a development due to changes; for example, a change in one or more of the following:

- higher or lower initial purchase price for the land;
- delayed construction time period, e.g. due to poor weather;
- market downturn and therefore lower final sale/leasing value;
- increased competition resulting in extended sale/leasing period.

In addition to the scenario or sensitivity analysis, the use of property development software has many benefits including ease of use, standardised output (i.e. reports) and reduced potential for errors. The successful property developer will keep up to date with industry standards including the development software which best suits their purposes. As the software industry experiences a rapid rate of change with version updates and hot fixes (for bugs) for example, reference needs to be made to the current version of the industry accepted real estate development software.

The financial analysis of a real estate development was substantially advanced in the late 20th century with the widespread availability of computer hardware and software. Prior to these advancements, complex calculations over extended time value of money were undertaken using pen and paper. Although there are numerous advantages from using computer software, arguably the single largest benefit is the ease of conducting a sensitivity adjustment if there is a change in a variable. Such a change is a relatively common occurrence when considering the large number of financial variables and related assumptions in a property development financial analysis. For example a variation (i.e. increase or decrease) in the interest rate for borrowed funds would affect calculations throughout the entire development timeframe and the final profit/risk breakdown. Undertaking such a modification without a computer would be very time consuming, as well as increasing the likelihood of a calculation error.

The three most commonly used tools are Microsoft Excel, Argus Developer and Estate Master DF (Havard 2013). However, there are so many products currently on the market (for example Caldes, Prodeveloper and Kel) that it would be impossible to review them all here. Instead this chapter focuses on the typical features associated with the many types of software packages available. The reader is encouraged to search for (e.g. via the internet) and become aware of all available real estate development software, as well as keeping up to date with constant and rapidly changing developments in this area.

9.2 Spreadsheets

For many real estate developers, the spreadsheet still remains the most suitable software package available for examining a real estate development. This is perhaps due to the widespread use of electronic spreadsheets (such as Microsoft Excel) and that in general they are very simple to use – the ease of use of any software is a critical consideration.

Spreadsheets were developed for use in accounting and have largely replaced paper-based systems throughout the business world. This type of program allows a user to organise and analyse data in tabular form on a worksheet. By adding formulae to the spreadsheet, each value can have a predefined relationship to other values so that a change in one value will automatically recalculate and display a new value based on the contents of other cells.

This ability to link any number of cells together makes the spreadsheet especially useful for 'what-if' scenario analysis, since changes in any value can be observed without the need to manually recalculate the figures. Modern software has most of the common financial and statistical functions built into the program, which makes calculations such as standard deviation or net present value easy to compute.

The property profession now has a choice of spreadsheet-based software specifically tailored towards the property industry, depending on whether the focus is on management or development. For the developer some computer software offers a spreadsheet with a standardised format that can be adapted for a range of different development projects. These programs are designed for ease of use and require minimal user training. The flexibility of such programs means that they are ideal tools for tracking changes or scoping out the viability of a development. One example of a spreadsheet program often used by developers is Microsoft Excel which is part of a suit of programs belonging to Microsoft Office. It features calculations, graphing tools, pivot tables, and a macro programming language called Visual Basic for Applications (Microsoft 2013).

9.3 Property development appraisal software

There are specialised software packages available which are specifically designed for development appraisal. These programs can be used for valuing residential, commercial and mixed use building plots in addition to managing the project costs once a development has commenced. Among other features, such as phasing costs and incomes, this type of program allows the user to residualise the land value or add a land value and calculate the likely profit. This type of software can be used in obtaining bank finance or managing development costs, income and profit. Examples of available programs capable of performing these tasks include: (1) Estate Master DF, and (2) Caldes.

1   Estate Master DF (Estate Master Development Feasibility) is a powerful real estate valuation tool. This program calculates key investment performance indicators, such as Residual Land Value, Profit, Margin, Net Present Value (NPV) and Internal Rate of Return (IRR). The user can also conduct multi-option scenario analysis to achieve the best financial outcome for their development site (Estate Master 2013).

2   Caldes Development Valuer is a Microsoft Windows based property development appraisal software program. The software is used by developers, banks, funds, agents and surveyors for valuing developments, land and building plots. The software can be used for appraising residential, commercial and mixed use building plots/sites. Once a project is on site the program can be used to manage the on-going project and view budget against actual figures (Caldes 2013).

9.4 Project management software

Many developers now use a specialised project management software package. Some software packages (depending on how sophisticated they are) have the ability to support the management of the entire development process by helping to plan, organise and manage resource pools and develop resource estimates for any given development. In addition this type of software can aid decision-making, aid communication between suppliers, contractors and other stakeholders and also help with documentation and administration.

Project management software packages require the developer to break down the individual stages of the development process (specifically including detailed information on the actual build) and enter this information into a spreadsheet or interactive Gantt chart. This allows the developer (and other stakeholders) to view the entire process on a timeline with financial outlays and processes visible at any given point throughout the entire development process. The critical stages of the development or processes which must be completed before another stage can proceed (i.e. build the foundations before the walls), can easily be identified. This type of software is typically used throughout the scheme because it can adapt the entire development schedule in response to change (for example an unforeseen delay in the delivery of materials for a particular process). It also enables a developer to adapt to change so as to avoid unnecessary additional costs and to control budgets and resource allocation. Many project management software packages are available; their features generally fall into two categories: those which focus on scheduling and those which provide information to the user.

Scheduling tools

These tools are designed to sequence project activities by assigning dates and resources to each task. The level of detail and sophistication of a development schedule produced by a scheduling tool may vary considerably depending on the software or methodology employed by the developer. Typically, scheduling tools may support the following activities:

- identifying and controlling multiple dependency relationships between scheduled activities
- managing resource assignment and resolving conflicts and levelling
- undertaking critical path analysis
- calculating duration estimation and probability-based simulation for any given activity
- cost accounting.

Information tools

This software can also provide information to the developer, stakeholder or anyone else involved with the project. It can be used, for example, to provide information on workload planning, timing for delivery of equipment and materials or to measure the level of effort required to complete a development project. The type of information this software might be required to produce might include:

- accounting
- risks to the project
- projected timescales for the completion of specific tasks
- workforce planning and availability (planning holidays, etc.)

- detailed record of activities throughout the development process (i.e. historical evidence in relation to progress or performance)
- utilisation of available resources
- collaboration with stakeholders or others.

9.5 (Real) estate agent, realtor or letting agent software

If a developer intends to manage the sale of the completed development, he should be familiar with at least one of the software systems that have been introduced into the property market. These are available as web or internet packages, often on a monthly subscription, or can be installed onto individual PCs or servers. Most packages offer similar features; however, some have the added benefit of being linked to social media sites such as Facebook, Twitter and LinkedIn. Modern social trends are geared towards the use of the internet, and it is essential that every Agent starts to build relationships with future clients by sharing news and posting frequent updates. This is likely to increase your brand awareness and secure future clients based on internet based social strength.

9.6 3D modelling/design software

Other recent software developments have focused on 3D modelling and it is now possible for a developer or project manager, with no prior knowledge of 3D drawing software, to create a model of a prospective development very easily. There are a number of very intuitive programs available (for example SketchUp) which usually have an option to interface with aerial imaging software (via the internet). This allows the developer to upload an ordnance survey map or aerial image of the site they wish to develop and produce an accurate 3D model of the proposed development. The benefit to the developer is that any number development options can be explored before the architect is called in to draw up the plans.

New developments in computer software for the property industry are building on the technology used for 3D models. For example building information modelling and 3D laser scanning.

9.7 4D modelling

This type of modelling is now widely used in the construction industry (e.g. linking a project portfolio management program to a BIM model to facilitate construction scheduling) and it is likely that the use of 4D (four-dimensional) modelling will significantly increase over the next few years, especially in geographic information systems (GIS), first, to understand change that has already occurred and second to enable predictive modelling of future trends (Zeiss 2013a).

9.8 Mobile computing

The benefits from the use of mobile computing have already been noticed by estate agents who can now work either at home or out in the field. The use of laptops, tablets and smart phones and associated high speed technology allows them to be more productive. They can obtain current real estate information by accessing multiple listing services, either from home, office or car when out with a client. They can provide their clients with immediate information on a specific property such as local neighbourhood details including proximity to schools and local crime rates, which in turn saves time, improves efficiency and enables the agent to provide a more complete service to the client.

9.9 Emerging technologies for real estate development

Building information model (BIM)

BIM is a digital representation of physical and functional characteristics of a development or building. A BIM provides a shared knowledge resource for information about a development and forms a reliable basis for decisions during its life cycle (from cradle to grave) (National Institute of Building Sciences 2013).

BIMs are now being used to produce virtual 3D models from 2D architect's drawings to help understand how a development will function when complete. Current BIM software is used by individuals, businesses and government authorities who plan, design, construct, operate and maintain diverse properties (e.g. infrastructure, residential units, offices, factories, warehouses, etc.).

Some of the benefits from using BIM include:

- Energy use: The ability to significantly reduce the amount of energy that is consumed each year as building design and thermal calculations can be analysed, enabling improvements to be made to the design of the building fabric, services, use of renewable energy and building orientation to benefit from solar gain.
- Building maintenance: The provision of digital information on completion of the project (rather than large paper files) can help with maintenance scheduling by providing quick access to information.
- Informed decision-making: A virtual model allows the client to see how the building and fixtures/fitting will function when complete. Adjustments can easily be made to improve functionality and reduce long-term costs from replacement materials or alterations once a building is complete.

Benefits to the developers during the construction phase include:

- Purchasing: BIM enables a digital list of all construction materials to be made which allows the developer to quickly select products that comply with proprietary, prescriptive or performance specifications.
- Clash detection: 'Clash Detection enables effective identification, inspection, and reporting of interference clash in a 3D project model between various 3D solid objects. Using Clash Detection can help you to reduce the risk of human

error during model inspections' (Revit Services, no date). Although the cost of correcting a clash can be significant, the cost of detecting and fixing a clash between structural elements and services at design time is insignificant.

• Reduced information requests: The use of a digital model reduces the need to supply repeat information during the construction phase. There can be many requests for information due to poor or incomplete documentation. Every alteration to documentation or design has a cost. Using BIM in effect allows the building to be constructed twice, once digitally at design time and a second time as the physical building.

While the use of BIM may increase design fees, the ability to run simulations, design out clashes and waste can significantly reduce costs during the construction phase, with the added benefit that the client is handed usable asset data once the building is complete. The value added from using BIM may offer significantly more value to a client than can be achieved by traditional design methods.

BIM + 3D laser scanning

BIM (discussed above) and 3D laser scanning offer new opportunities for capturing, mapping and analysing building information. While the application of this technology has tended to focus on the design and construction of buildings, the benefits of this technology are beginning to be recognised by property surveyors. 3D laser scanning can be used to record data on the spatial layout of a building, although this technology is currently unable to record information as to the material scanned or its condition. A major benefit of 3D laser scanning for surveyors, valuers, purchasers and tenants is the ability to capture detailing with high precision, including building characteristics normally only captured by a photograph (Mahjoubi, Moobela and Laing 2013). The integration of BIM and 3D laser scanning is likely to provide a significant change in the accuracy and speed of building surveying in contrast to conventional surveying techniques. Currently BIM and 3D laser scanning is undertaken by specialist consultants and external companies and the cost of using this technology may be prohibitive to some smaller companies until it becomes more widely adopted. That said the ability to scan buildings, create an accurate 3D replica and provide a faster and more efficient service to clients suggest that the use of this technology will become widespread in the near future.

9.10 Future trends

Emerging technology is focusing on developing ways to link data from different sources together so that it becomes more useful. Since the adoption of BIM by the construction industry, new programs are being developed to combine BIM with Linked Data and the Semantic Web.[1] For example, ontologies[2] can be used to enhance the interoperability[3] of linked data[4] which could be used, for example, to link a virtual building (created by BIM) with other external building data 'to perform some building decision analysis such as energy performance and environmental assessments' (Abanda et al. 2013). However, combining such data presents significant challenges that need to be overcome. According to Zeiss (2013b), 'surveying/civil/geospatial interoperability is part of a longstanding set of challenges that includes BIM interoperability and 3D modeling interoperability'. This technology is still in its infancy and it is likely to be some time before it becomes widely available and adopted by the property industry.

9.11 Software package options

The type of software we have discussed here can be expensive, although arguably any outlay in purchasing or using this type of software is recovered through improvements in efficiency, thus reducing financial outlay during the development process. Each developer will have their own system requirements and a number of different options are available from access to a web-based program to installation onto a desktop PC.

Installation

- Desktop option: If the developer choses this option then he will purchase the program and install it onto a desktop/laptop. This type of software is designed to be a single user application. The user will typically be the project manager or another expert who may be in charge of scheduling or managing the risk associated with the development.
- Web-based application: Some project management packages are available on a subscription basis and the user will access this through a web browser. This may appeal to many users due to the increase in the use of smart phones and tablets, although they are limited to some degree in that they can only be used when the user has live internet access.

Program users

Once the system requirements are chosen in relation to access, there are other options based on the desired number of program users.

- Individual (single) user system: An individual or single user application is typically used by small companies or individuals. The software is based on the assumption that only one person will be tasked with editing the project plan. Desktop applications generally fall into this category. A well-known example of this software is:
  - Microsoft Excel (part of the Microsoft Office suite).
- Multi-user or collaborative system: This type of system is designed to give multiple users access to the system at the same time and allow them to modifying different sections of the plan at once. This means that users can update sections they are personally responsible for and these changes instantly update the overall plan. This is a typical feature associated with web-based tools but does rely on the user having access to the internet. Some desktop software overcomes this limitation by providing a fat client (also called a heavy, rich or thick client; a computer network that typically provides rich functionality independent of the central server) that runs on a user's desktop computer and both receives and sends updated information to the other project team members through a central server when users connect periodically to the network. Any changes to

scheduling can also be synchronised with the other schedules as soon as the user logs on to the network. One example of this type of software is:

- Revit Building Design Suite (designed for Building Information Modelling (BIM)).

- Combined (integrated) system: This system will allow a developer to combine project planning or management with many other aspects of company life, for example calendars, messaging, task lists and managing customer relationships. One example of this type of software is produced by the company Argus. Argus products include a range of software aimed at the property industry for example:
  - ARGUS Developer
  - ARGUS Valuation – Capitalisation.

9.12 Reflective summary

This chapter has discussed examples of computer software which will assist real estate developers to manage a development project from inception to disposal. The type of computer program a developer will choose to invest in will depend on factors such as whether the development is a one-off or a relatively straightforward simple project (e.g. with one parcel of land or one building over a short period of time) or a complex multi-stage development over an extended timeframe. The widespread availability of inexpensive computer technology including laptop ('in the field') portability has increased the reliance on property development software throughout all phases of the development. Complete solution packages are typically used for managing and growing a commercial real estate portfolio. This type of software offers tools which include asset management, asset valuation, portfolio management, budgeting, forecasting, reporting, lease management, collaboration and knowledge management and will appeal to professionals managing REITs, institutional investors and developers. The use of property development software has many benefits including ease of use, standardised output (i.e. reports) and the reduced potential for errors. With the increased use of BIM and on-going developments in computer technology, programs and software, there will be some exciting and extremely useful tools to aid property professionals in the future.

Relevant websites

Argus Software http://www.argussoftware.com,
Caldes Software http://www.caldes.co.uk/home.asp
Estatemaster Software http://www.estatemaster.net/index.html
Feastudy Software http://www.devfeas.com.au/
Microsoft Excel http://office.microsoft.com
SketchUp http://www.sketchup.com/

Notes

1 'The Semantic Web is a collaborative movement led by international standards body the World Wide Web Consortium (W3C). It promotes a standard or common data format on the World Wide Web. By encouraging the inclusion of semantic content in webpages, the Semantic Web aims at converting the current web, dominated by unstructured and semi-structured documents into a "web of data".' (Wikipedia 'Semantic Web'. Retrieved from http://en.wikipedia.org/wiki/Semantic_Web#cite_note-W3C-SWA-2 [viewed 15/12/13]). See also 'W3C Semantic Web Activity'. World Wide Web Consortium (W3C). November 7, 2011. Retrieved 26/11/11.

2 'An ontology, in computer science, formally represents knowledge as a set of concepts within a domain, using a shared vocabulary to denote the types, properties and interrelationships of those concepts'. (Wikipedia 'Ontology'. Retrieved from http://en.wikipedia.org/wiki/Ontology_%28information_science%29#cite_note-TRG93-1 [viewed 15/12/13]). See also Gruber, T.R. (1993). 'A translation approach to portable ontology specifications' (PDF). *Knowledge Acquisition* 5 (2): 199–220.

3 'Interoperability is the ability of making systems and organizations to work together (inter-operate). ... the term was initially defined for information technology or systems engineering services to allow for information exchange'. (Wikipedia 'Interoperability'. Retrieved from http://en.wikipedia.org/wiki/Interoperability#cite_note-1 [viewed 15/12/13]). See also, Institute of Electrical and Electronics Engineers. IEEE Standard Computer Dictionary: A Compilation of IEEE Standard Computer Glossaries. New York: 1990.

4 'In computing, linked data (often capitalized as Linked Data) describes a method of publishing structured data so that it can be interlinked and become more useful. It builds upon standard Web technologies such as HTTP, RDF and URIs, but rather than using them to serve webpages for human readers, it extends them to share information in a way that can be read automatically by computers. This enables data from different sources to be connected and queried.' (Wikipedia 'Linked Data'. Retrieved from http://en.wikipedia.org/wiki/Linked_Data [viewed 15/12/13]). See also Bizer, Christian; Heath, Tom; Berners-Lee, Tim (2009). 'Linked Data—The Story So Far'. *International Journal on Semantic Web and Information Systems* 5 (3): 1–22. doi:10.4018/jswis.2009081901. ISSN 1552-6283. Retrieved 18/12/2010.

Chapter 10

Marketing and sales

10.1 Introduction

As with all businesses, one of the developer's primary objectives is to 'increase shareholder's wealth'. This is regardless of whether it is a one-person property development company or a multi-national organisation. Hence the underlying objective of undertaking a property development is to complete the development and then commence another development project. However, the final stage of the development, and hence the reason for undertaking many of the stages in the development process, is the eventual sale (or lease) of the completed project to a third party. To facilitate this process, marketing is a key yet frequently misunderstood component. Often this third party only becomes aware of the development via a structured marketing programme which may include relatively simple marketing strategies – for example placing a site board on the property boundary. In today's competitive marketplace there are relatively few prospective new owners or tenants who are not subject to strategic marketing initiatives from competing developers. Maintaining a comprehensive email database as assembled from online enquiries is an example of accepted marketing practice. Therefore, developing and implementing an effective marketing plan is an integral part of a successful property development.

Where possible the dual objectives of the marketing plan will be to (a) sell or lease the completed project as well as (b) achieve the predicted sale price or rental level as originally forecast prior to commencing construction. The result would theoretically be a lower exposure to future risk coupled with increased future certainty; however, this can only be achieved by adopting a sound strategic marketing plan incorporating proven and reliable promotion and selling strategies. Otherwise, who will know about a completed property development unless they visually see it from the road in the construction phase? This 'free' marketing option would result in exposure to an extremely small sector of the potential level of demand in the marketplace, primarily based on pure luck – it is a high risk approach. Although marketing has a financial cost it has been proven that you have to 'spend money to make money'.

Most importantly this part of the development process must not be a task which occurs only at the end of the process and therefore receives relatively little attention. Marketing must be viewed as a priority that potentially will adversely affect the success (or lack thereof) of every property development. In isolation a marketing strategy has the potential to 'make or break' a property development. This aspect has also affected lending criteria since the global financial crisis where banks and financiers required a high level of pre-commitment (i.e. usually at least 50% of the development assigned to dedicated purchasers) prior to lending any funds for the development. In such instances the developer actually commences marketing prior to undertaking the development, as well as during the development up until final completion (if required). It is clear the strategic marketing programme ends when 100% commitment to the completed development by prospective purchasers/tenants has been achieved.

The significance of the marketing phase has often been under-estimated by property developers who may focus too much on the preceding steps of property development, including land acquisition and construction, rather than maximising the eventual return on disposal. Some developers appear to be more focused on identifying the next potential development, as their current development is nearly completed, rather than on the marketing phase prior to completion. In addition, certain developers may be unfamiliar with the optimal marketing strategies for their project – for example in the 21st century an 'online' presence has replaced the newspapers as the most up-to-date source of property space available for sale/lease. Furthermore an 'online' presence has the widest instantaneous global reach and can be instantly updated. The primary goal of the property developer should be to achieve 'full market value' for their completed development, based on the highest and best (legal) use of the site. The nature of the real estate market with individual buyers and sellers ensures there is an absence of a centralised marketplace. This provides additional challenges for the developer as any advertising must target individual purchasers and tenants. Overall this places additional importance on the marketing phase yet again.

The widespread use of computer technology has radically altered the methods used to market completed developments. For example it is commonplace to use computer simulations of a completed development to replace a typical display unit. This allows a prospective tenant to view a development (finished or unfinished) remotely. In addition, prospective buyers and tenants may be contacted via a variety of mediums including mobile or cell phone text messages and emails. Although traditional communication mediums such as telephone and direct mail-out are still in use, they only represent a relatively small proportion of the total strategic marketing campaign. This chapter discusses the methods and options available for marketing the completed property development with the emphasis placed on the marketing and selling aspects.

10.2 Marketing approaches

As much as there are different land uses, there are a large number of different marketing approaches which are tailor-made for each development. Nevertheless the developer must be fluent with all marketing options and also be fully conversant with all aspects of the market. This includes using resources to identify and accurately define the target market using specified parameters such as by geographical location (e.g. for a retail shopping centre development) or by prospective investor size (e.g. such as for a major office building). This should be followed by an organised campaign to actively promote the product (i.e. the completed property development or part thereof) to the nominated sector. The underlying aim is to promote the development to potential purchasers and tenants in the target market based on the highest level of efficiency, i.e. not promoting to sectors outside the target market where possible. For example if the property development is a completed retirement village and the developer is accordingly only targeting a market aged 50 years and over, then a television marketing plan would be aligned with television programmes (e.g. mid-morning lifestyle) which are watched by this cohort and not typically viewed by younger generations.

It is accepted that a developer has limited resources, especially financial, in every stage of the development including the marketing phase. Therefore, the developer must establish a realistic budget to actively market the completed development and be prepared to increase this allocation if deemed necessary. In a perfect world every additional dollar spent on marketing should result in a higher return on marketing investment, although in reality this relationship is often hard to measure accurately. The size of the marketing budget will depend on the nature of the development scheme and the state of the property market prevailing at a particular point in time. The rationale behind any decision about the marketing budget will be affected by the prevailing level of demand in the marketplace – for example the developer should consider the impact of other completing developments which have been (or will be) released onto the market. For a developer it can be frustrating that successful marketing is not an 'exact science' and therefore the actual proportion of the development budget spent on marketing will vary between projects. While some property developments will practically 'walk out the door' there will be other developments which need a concentrated and at times relatively expensive marketing campaign to ensure the completed project is successful. The alternative option, an ill-informed marketplace which is unaware of the product available for sale, should be avoided where possible.

The amount of financial resources that a developer commits to promotion is often based on their judgement, partly influenced by their past experiences and expert advice received, e.g. from their selling/leasing agent. Developers who specialise in a particular type of development or who concentrate their development activities in a particular geographical area will tend to have established personal contacts and a good reputation with potential occupiers. They should also have a good grasp of the supply and demand interplay. In such scenarios the amount of promotion (and associated financial resources) required will be minimal. Often this type of developer will be able promote the project by personal contact either directly or through appointed agents. Another example is the potential for free exposure via the media, especially for landmark or interesting developments, which may create sufficient publicity to reduce reliance on marketing campaigns. In these situations it is a good idea to forward a press release to the media rather than wait for their interest first. In the case of large-scale developments the promotional campaigns tend to be very extensive and may have to extend over considerable periods of time to be effective. Associated costs can also be expensive and long-running campaigns may be cost prohibitive.

Regardless of the size of the promotional budget and available resources it is essential for the various components of the marketing campaign to be closely monitored. The level of success with the marketing campaign should be observed where possible – for example it should be possible to record the number (and timing, e.g. which day/time) of enquiries which will help to gauge the effectiveness of a marketing strategy in attracting potential occupiers. This is a common feedback mechanism and frequently used by major retailers who note customer postcodes to assess consumer behaviour. When monitoring buyer/tenant enquiry levels it is important to optimise maximum value for money; therefore, a careful analysis of all enquiries in terms of both numbers and 'quality' will help to identify the most effective methods of promotion. Careful recording and analysis of results will provide a useful database for the planning and direction of future promotional activities. There is large difference between attracting an enquiry from a potentially interested future occupier and converting this enquiry into an occupier (e.g. sale or tenant). The conversion rate is just as important from a marketing perspective.

It is essential for the initial planning phase of the promotional campaign to be carried out in consultation with the sales/leasing team who will be involved in selling/letting the scheme, whether they are appointed agents or an in-house team employed as part of the entire development project. For large-scale developments the sales/leasing team may include an advertising agency and a public relations consultant who are experienced in the promotion of property and real estate, especially with reference to a newly completed property development. Again, this will include being skilfully able to identify a genuine enquiry rather than spending time with non-genuine customers. Prior to commencing promotional activity the developer, agent and any consultants must identify the relevant strengths and weaknesses of the particular property development, in the context of the competition, so the advantages may be highlighted. This may also include reference to the unique selling points (USP) of the development. During the evaluation and design process it is critical that the developer defines the target market for the property, either through their own knowledge and experience or, where necessary, via market research. The property market is very fragmented, partly due to the lack of a centralised marketplace. For example potential occupiers may vary from small businesses to major companies with their own retained agents or internal property departments, and they have different needs and perceptions. In order to be cost-effective, all aspects of the promotion must be specifically designed to attract the attention of the target market.

Discussion points

- Why is marketing and selling as important as any other part of the property development process?
- How has technology changed the way that property is marketed?

A relatively early decision in the overall property development process is to confirm the name of the particular building or development, where applicable to create an identity which the promotion can attach itself to. It should be noted that the individual identity/name of the building, together with an associated design/ logo, will last throughout the promotional campaign and can be a major marketing aspect. Most importantly the 'name' of the development will distinguish it from the competition in the same marketplace. The name identifies the building or development, which is particularly important in relation to the perception of a large shopping centre development as the name will endure well beyond the letting of the individual shop units. Most importantly the name will be used in promoting the centre to the shopping public and wider society. With an office building the building's name will tend to change from time to time when let to a large individual firm who may name it after their own organisation. For example sports stadia are typically named after an organisation or team. This approach has spread to all land uses upon which a major development has or is being undertaken including major residential land subdivisions. Also the name of a completed property development may reflect its former historical or locational significance, often for many years after the development has been completed. It is important that the developer, before eventually deciding on a suitable name, checks with the local authority and also relevant stakeholders that the name is acceptable to them and broader society. Due diligence should also include an analysis of other developments with the same or a similar name (even from a global perspective), which will help to mitigate any confusion when the development is searched for using a typical internet search engine.

We will now discuss some typical approaches to marketing a property development in more detail.

10.2.1 Site boards and site hoardings

Still today this remains one of the oldest and most cost-effective means of marketing a property based on the location attributes. Other than the cost of manufacturing, there are no costs associated with advertising due to the 'free' benefits attached to the site. Also the advertising will ensure there is a connection between the broader marketplace and the site. It is human nature to read signs which are designed to attract attention – property and real estate professionals also have a trained eye to observe sites with signs erected.

The erection of a site board is usually one of the first methods of promotion which can be arranged as soon as the developer has acquired the site. In addition site boards can be a very effective and inexpensive means of communicating news about an upcoming property development scheme, even before there is any physical sign of land clearing or construction. Multiple site boards may be erected during the course of the development depending on the information available at any particular time. For example before the final details of the scheme are known a simple 'all enquiries' or 'register your interest' board may be erected, providing the name of the developer or an agent and their contact details. It is now standard procedure to include all communication mediums including a website address, phone numbers (i.e. at times a mobile number only) and email addresses. It is also acceptable to include a barcode for scanning by a mobile phone which will then lead the prospective purchaser/tenant to a third party website. For larger developments the contact details for several agents are usually listed. A site board with brief details will often encourage preliminary interest but must be replaced at the earliest opportunity by a board providing detailed information on the nature and size of the scheme. Practically all development sites will have a board giving details of the development, both during the course of construction and from the moment the construction phase has been completed until sales or lettings have taken place.

The positioning of the board and its actual wording (i.e. both spelling and grammar) are most important to ensure maximum exposure value. For example if the site is in a prominent location (e.g. a busy major road) then a board may be the primary method of promotion for passing traffic and may actually generate more enquiries than other advertising mediums by having the highest profile. The site board should be clean, lit by lighting (i.e. so it is visible at night and in dull weather conditions) and regularly maintained. A site board in a poor condition gives a poor first impression and reflects badly upon the quality of the development itself. Purchasers/tenants will quickly form their first opinion of the quality of the development which may be hard to change. Care should be taken to locate the board away from obstacles such as trees which may partially or fully obscure the board when they grow. Also the developer should ensure that the advertising boards of contractors (including subcontractors) and the professional team are kept to a minimum and do not adversely distract from the site board. If the developer sells advertising space on the site hoardings to advertising agencies then care must be taken to ensure that their positioning does not distract from the main site board. The benefit of the on-site board may be lost if the perimeter is too 'busy' with advertising material.

It should be remembered that a site board is actually an advertisement and in many jurisdictions will require planning permission prior to its erection. Obtaining planning permission is not usually a large hurdle since it is a matter that can often be dealt with under a planning officer's delegated powers. However, some planning authorities may impose size restrictions or dictate the maximum site board size (i.e. height and width) and even the actual colours. This also includes the maximum height of the letters (e.g. 25 mm or 1 inch) which is the main reason why some site boards appear to contain 'squashed' writing. Producing a board which does not comply with planning regulations is a waste of resources.

Above all, the message on the board must be very simple and uncluttered, yet include the relevant information. The board should inform the passer-by (i.e. in a car) at a glance of the following information:

1   the name of the scheme (incorporating any logo)
2   the type and size (or range of sizes) of the accommodation available
3   whether the accommodation is 'to let' or 'for sale'
4   the name and contact details (business phone, mobile phone, email address) of a contact who can provide further information, either the developer or the appointed agents
5   the website for the property (where applicable)
6   the band on FM radio which broadcasts details about the property only in close proximity to the site board (where applicable); and
7   the location and inspection hours of the display unit (where applicable).

It is critical to ensure the site board is highly visible and well positioned for passing traffic (e.g. at eye level) although not overly detailed or busy with too much information. The address of a simple URL for internet access is often a good tactic which directs interested person/s to the site where detailed information is available. At times additional information could be added to the site board including the anticipated completion date and what proportion of the development has already been sold (e.g. 70%); this approach has the added advantage of encouraging a sense of urgency. This could be accompanied by a brief specification highlighting the existence of other influencing factors including sustainability features, potential views (e.g. water, city) and internal finishes. Other options are possible, such as a coloured 'high visibility' written message placed diagonally across one of the top corners of the board or alternatively bunting could be used. In certain instances a video/LCD monitor may play (on a continual loop) an internal view of the proposed development and/or a simulated fly-by, although clearly this is for up-market or large developments. Many signs are also lit up at night, often by a single light, which increases the profile of the site over extended hours and also catches the eye of passing traffic as the remainder of the site is typically pitch black. Often these lights are powered by a solar charger on top of the sign.

A trend in some areas is to use decorative site perimeter hoarding to surround the development site during construction to assist promotion of the scheme. Decorative site hoardings have two advantages: first, they distract from the unsightly chaos of building sites, providing an attractive outlook to passing traffic; second, they can provide information about the scheme in an imaginative and informative manner to help the awareness of the development. Often passers-by will recall a property development with decorative eye-catching hoardings rather than the usual contractor's painted plywood effort. Humans will always be observant, especially potential purchasers or tenants who are already on the look-out for potential property and real estate acquisitions or tenancies. On the other hand, decorative site hoardings can be costly when compared with ordinary site hoardings and they also lose the benefit to the developer of revenue from advertising hoardings. In the case of a refurbishment, where protection of the building during construction is usually in the form of mesh hung from the scaffolding, in high profile locations it has become increasingly popular to reproduce a life-size image (or even a promotional message) of the proposed building onto the protective mesh positioned on the side of the building.

10.2.2 Particulars and brochures

In today's increasingly competitive marketing environment practically every property development is supported by particulars and/or brochures of the scheme, providing the potential occupier with further information in response to their enquiries. The advent of internet marketing, desires to reduce costs and the push for sustainability has decreased the importance of hard copy material (i.e. 'it's all on the web'); however, prospective purchasers/tenants may still demand a hard copy flyer/information brochure outlining information about the development. Importantly, these particulars need to be aimed at potential occupiers and agents, where the nature of the particulars will depend upon the actual nature of the property and the target market identified by market research. The fundamental question that needs to be asked is: what information will the potential purchaser/tenant require in order to take the next step, which is usually arranging to view the property? This approach is based on the marketing strategy AIDA, referring to creating (A) attention, (I) interest, (D) desire and (A) action by the prospective purchaser/tenant. The content and nature of the flyers and brochures need to be targeted directly at the likely readership and arouse interest in the project and eventually action. For example, with reference to a new shopping centre development, the target market may be residents visiting a nearby older shopping centre in an adjoining area. For a new office building, the target market may be existing tenants in other nearby office buildings.

Usually for a smaller property development, the information about the available individual components in the development should be set out on a single sheet of paper (i.e. A4) with a photograph or on a double-sided printed colour glossy sheet. For larger schemes, more detailed information is required and therefore particulars need to take the form of glossy colour brochures or booklets. Property is all about presentation and perception; therefore, it is not appropriate to hand out a blurry photocopy of black-and-white text, which is hard to read and contains poor quality images, although it is cheap to produce. The prospective purchaser/tenant (or 'prospect' in marketing terms) will not receive the impression of a high quality development, regardless of the size of the project. The objective here is to convert a 'prospect' into a purchaser, tenant or user by presenting a professional image and showing the development in its best light. The 'prospect' is more likely to be impressed with the development scheme and act immediately to benefit from this opportunity.

Basic particulars on the flyers, which may be supported by photographs or illustrations depending on the size of the project, should contain the following:

1   a description of the location of the property, including a small map;
2   a description of the accommodation, giving dimensions/areas and a brief specification;
3   a description of the services supplied to the property, such as gas and electricity, as well as other considerations such as sustainable attributes;
4   the nature of the interest that is being sold if it is freehold, the nature of any leases or restrictions to which it is subject, and, if leasehold, the length and terms of the lease.

The details must give the name, address and contact details of the agent and developer, and the name of the person in the agent's office or on the developer's staff who is dealing with the property. This is aligned with the 'action' phase of the AIDA principle so that a direct contact can be made by the potential occupier and more information can be supplied as requested. Hopefully this will take place when the prospect is standing at the sign and contact will be made via an email, calling a mobile/cell phone or visiting the website listed. Care and attention must be taken to ensure that all information given in the particulars is accurate and not 'false or misleading', and consequently all information needs to be carefully checked to ensure its accuracy as far as can be reasonably ascertained. This will rule out the use of inaccurate or subjective language in the description of the property. Legislation varies between different jurisdictions but nevertheless has the common goal of protecting a prospective purchaser. This legislation also applies to any information supplied or to statements made by agents and developers in promoting the property, and therefore applies to practically all forms of promotion.

Usually it is necessary to prepare a multi-page colour brochure describing the property, giving more details than the brief particulars referred to in a one-page flyer. As part of the marketing campaign the brochure is sent to prospects who have actually registered interest and replied to any advertising, site boards, mail shots or preliminary particulars, or may be sent directly to people who are known to be interested. This type of target marketing can also including emailing a soft copy of the brochure to a database of prospects which have been identified by an external party, e.g. the developer may purchase a database of prospective purchasers/tenants from an industry provider. Accordingly it is important that the brochure is integrated with the rest of the promotional activities with all information error free and up to date. The appearance to a potential purchaser is of great importance as it reflects the quality of the development that it describes.

Most developers or their agents employ a specialist outside designer or agency with experience in property brochures, since they can provide professional designers and copywriters, ensuring the most effective use of words, typography, colour and shape. Furthermore, an external design is usually more cost-effective considering the time delay that often occurs between property developments. It is critical to demand and receive extremely high standards of design and production. All photographs need to be taken professionally; many photographers specialise in property and real estate. Photographs which are hypothetical, such as a concept drawing, must be noted as such and clearly 'subject to change upon completion'.

The brochure should provide clear directions and maps to enable the reader to find the property's address easily. It should also provide information about the locality which will be helpful to a potential occupier. For example, in promoting a particular office location information should be provided on drive-times to nearby access roads, towns, airports and cities; transport routes and major cities/towns nearby; and the availability of public transport services. Also, particularly if the office is in a regional or rural town and the likely occupier is a firm decentralising from a major city, information should be provided on local housing, shopping facilities, availability of local labour and so forth.

For an extensive long-term promotional campaign it may be necessary to produce more than one brochure at different stages in the development, especially when there are multiple high-rise buildings in the development. At times substantial financial resources can be spent on brochures; in addition to being expensive to produce, it should be remembered that they can quickly become dated. To reduce wasted expenditure, simple particulars or what is known as a 'flyer brochure' or 'flyer' (i.e. a single-sheet brochure) may be produced showing an artist's perspective of the proposed building and including very basic details. These can be replaced with photographs of the completed building and additional details as soon as they become available. At times it may be worthwhile producing a tenant's guide to the building for prospective occupiers who show continued interest in the building. The guide should provide a detailed description of the building and specification, together with information on operation, maintenance and energy-saving data. Due to the relatively high expense associated with glossy brochures, care should be taken not to hand out detailed brochures to prospective purchasers/lessees who are only showing a mild interest. The underlying aim should be to provide basic information (i.e. a flyer) to parties on their initial enquiry, and then to forward a detailed brochure to parties who show a genuine interest.

For most development sites a dedicated internet web address is standard policy, and often accompanied by a three-dimensional 'walk through' or a 'fly by'. On particularly large and complex development schemes there may be a computer simulation and/or graphics combining video and voice in a multi-media presentation. The internet profile can also be hosted on a real estate agent's site rather than a stand-alone site for the property development. Even though there is a reduced cost for being listed on a real estate agent's internet site without the added cost of a dedicated URL for the development, the downside is the loss of identity for the development among all of the other competing properties on the agent's site. The initial attraction of the development could lead the prospective occupier to a website where other competing developments are listed. This provides strong encouragement for a developer to set up their own individual website for each development – a process which is fortunately becoming less expensive over time and can be undertaken in-house by larger real estate developers.

For interested potential purchasers it is possible to show three-dimensional images of the building proposed if the architect is using a CAD (computer aided design) architectural design system to produce drawings of the scheme. This technology enables the viewer to be effectively taken round the building from every angle and perspective. Photographic stills of the three-dimensional images produced can also be used in any promotional material.

10.2.3 Advertisements

As a long-established method of raising the profile of a new development and attracting interest, advertisements are still considered a cost-effective medium. Usually the advertisements are either aimed directly at potential occupiers or indirectly through their agents depending on the target market. Traditionally such advertisements were usually placed in the property sections of local and national/international newspapers, as well as various industry property/real estate journals. However, the internet has quickly overtaken the printing press as the preferred medium for advertising a development due to: (a) the lower cost per advertisement, (b) the wider audience coverage, (c) the ability to update the advertisement instantly, (d) the higher level of professionalism including both visual and audio responses, and (e) the ability to identify the number of views of the website, including the viewer's computer location (i.e. country) and their actual search terms. If the potential occupier is likely to be represented by an agent, then a combined presence on the internet and in the newspaper property pages is often the most effective approach for spending money on an advertising campaign. While the content of an advertisement is undeniably important, the design and layout are crucial to project the desired image. It is also essential to reach a balance between too much detail (i.e. too 'busy') or insufficient detail to promote interest. The goal of the advertisement should ultimately be to encourage action.

Good design and layout increase the impact of the advertisement, whereas poor design not only wastes money but can sometimes create a negative impression. Thus, poor design should be avoided at all costs. It must be remembered, particularly on the global internet with millions of websites and in the property press, that the advertisement will compete with others of a similar nature and therefore must be designed to make an impression and illicit the desired response.. It must simply be 'eye catching'. The cost will vary with size and position on the webpage, in the newspaper or magazine. An advertising link from a high profile industry website or a front page advertisement in the newspaper, while costly, may provide the opportunity to create a bold, imaginative advert which will make a lasting impression. The cost of the website link or a one-off advertisement on the front page has to be weighed up against the cost of a series of smaller advertisements in less prestigious positions. The style and advertising media should be varied in a constant attempt to improve results.

Care needs to be taken over the way in which the advertisement is worded and presented; in nearly all cases it is worth employing an experienced advertising agency to ensure the maximum impact, although some large firms of agents have an in-house advertising department. The advertisement should contain sufficient information to attract the potential occupier, but it should not be so cluttered that it is difficult to read. Information should include:

1   the type of property, e.g. office, industrial;
2   approximate size of the building area (and land area where applicable);
3   services available, e.g. electricity, lift;
4   location and proximity to transport and amenities, e.g. a train station for an office building;
5   whether the property is for sale or to let;
6   the telephone number/email address of the agent or the developer, or both, so that the potential occupier can make a direct contact.

At times it may be extremely difficult to judge the short-term or long-term effectiveness and worth of an advertising campaign, but some attempt should be made to identify the source of any enquiry. For example this could simply be recording the postcode of the home address of prospective occupiers. It must be remembered that some readers might see an advert and take no immediate action, but the information will be retained in their memory and subsequently when looking for accommodation, they will recall the advertisement and make enquiries. This is referred to as 'top of the mind' advertising and used by billboard advertisers on the side of motorways.

Internet, newspaper and magazine advertising is a common form of advertising where costs can vary depending the size of the readership (i.e. distribution) and the competition by other advertisers for space in that medium. Other factors, such as the number of words in the advertisement and the type of photographs (e.g. size of videos to be hosted on a website, alternatively the actual page number or photos in black and white versus colour in a newspaper), will adversely affect the overall advertising costs.

It is accepted that relatively little use is usually made of television and radio for advertising individual properties, primarily due to cost restrictions, including the expense associated with making a high quality professional 15 or 30 second advertisement. Other negatives include the 'one-off' nature of an advertisement (i.e. you either see/hear it or miss it) and the limited audience (i.e. off-peak periods are cheaper with a smaller audience). Some developers argue that radio and TV should only be used when all else has failed, although advertisements on local commercial radio during 'drive-time' are often effective to spark interest in certain projects. The decision to advertise here will be made on a case-by-case basis and based on criteria including the marketing budget of the developer and the perceived effectiveness of this approach.

At times simple poster advertising can be used to promote larger properties, employing prominent locations such as railway stations, roadside hoardings, and even on or inside public transport (buses, trains) and taxis. As an example of its effective use, consider the promotion of an out-of-town office development with ample car parking. Posters displayed in appropriate railway stations, in the underground, on buses or in taxis, can be an effective and often overlooked means of advertising since they are viewed by commuters every day. A poster advertisement, like a site board, should convey its message instantaneously as people will only have time to glance at it and rarely have time to study it carefully. The downside is that poster advertising is expensive and it is difficult to target specific groups. This type of advertising should always highlight the web address where more detailed information can be accessed.

10.2.4 Internet presence

As the property market is part of the information society it is important to appreciate the wider role of stakeholders (i.e. people, government and business) with regards to information and communications technology (ICT). From an ICT perspective a successful property development definitely must have a presence on the internet in some form or another. Many developments have their own domain name or web address, which provides detailed up-to-date information about the project and saves the cost of printing expensive colour brochures. Other sites have live web cam feed of construction on the site so that interest is maintained. In addition, many newspapers/ television/radio advertisements highlight a web address that prospective purchasers/ lessees can instantly access for more information and contact details.

It is essential that a professional webpage designer is engaged to design and maintain the webpage. Although practically anyone can construct a simple webpage in Microsoft Word (e.g. in HTML format), a professionally designed webpage is worth the additional expense. There are other benefits such as the time taken to load the website (prospective purchasers/lessees using the internet are used to fast loading webpages and will not wait more than a few seconds), the use of multiple links to other pages and photos/videos, and the use of high quality sound (where applicable) which collectively contribute to an impressive window to the international world. It is recommended that examples of impressive webpages are studied and noted through the course of daily business to suggest as examples to the web designer. Finally, the costs outlaid for web design and maintenance must be in proportion to the size of the property development and in the context of the overall advertising budget. There are no hard and fast allocation budget models, but generally a project with a broad global purchaser base would be reached more efficiently via the internet, as opposed to advertising in multiple newspapers in other regions/countries.

Discussion point

• Why has the internet become the most popular advertising medium?

10.2.5 Direct mail

Direct marketing via post (i.e. mail shots) are commonly used as they are targeted and relatively cost-effective. Mail shots are used on their own or to support an advertising campaign. However, because it has become such a widespread promotional method, great care has to be taken if it is to be really effective. The mail shot should be aimed at a very carefully selected list of potential occupiers; the success of the shot will depend very much on the compilation of the mailing list and any follow-up procedures. There are a number of specialist direct mail organisations capable of producing mailing lists with a degree of specialisation and accuracy. They maintain general lists of industrial or commercial firms which can then be broken down into particular categories such as company size, trade and location. There is a limit to the frequency with which direct mail shots can be used and the employment of one of the specialist firms will often be found to be advantageous. Care must be taken in selecting a firm; they must routinely and frequently update their mailing lists. To be effective the mail shot must reach the right person within the organisation, who is responsible for property matters, and at the right address. To overcome the almost automatic tendency of the recipient to throw the contents of the direct mail into the wastepaper basket, the message should be seen at a glance. A very long letter which has to be read through to the end before the message is fully understood, or a brochure or leaflet without any covering message, will often fail. By contrast, a short sharp covering message attached to a brochure or leaflet will often be absorbed by the recipient before they have time to throw it away. The letters should be prepared in such a way that each one appears to have an original signature.

With direct mail, unlike broad advertising on the internet or in a newspaper, the results and effectiveness of any promotion can be quantified. It can be also followed up by a telephone call to obtain a reaction or information on general property requirements. Such telephone canvassing may be carried out by the agents or a representative employed in the show office/suite. Mail shots, provided that the target market is accurately identified, can be cost-effective.

10.2.6 Email marketing

In this age of technology, email has become one of the primary communication mediums. Email has many benefits including low delivery cost, low maintenance/upkeep costs and the ability to send out bulk emails. As many prospective purchasers/lessees are busy and do not have the time to answer the phone (either they are at work, at home or elsewhere), for many email has emerged as a viable and convenient alternative. To send out the same information via postal mail is comparatively expensive with an accompanying time delay, as well as the occasional incorrect or old address. Detailed information can be accessed via the email using links (rather than as email attachments) as opposed to sending out emails that have a large size.

For larger projects it is possible to subscribe to regular information emails from the developers that keep the reader up to date, as well as allowing the developer to gauge the interest in the project. In addition it is possible to email special offers or the date of the final completion. Be wary, a balance must be struck between saturating the reader with too many emails and not keeping them up to date with progress. The core aim should be to encourage action by visiting the property development and/or committing to a detailed discussion. There is a danger that emails could disseminate too much information and therefore not give the recipient any reason to seek more.

10.2.7 Opening ceremonies

For large-scale property developments (e.g. a major office building) it may be appropriate to launch the development by conducting a ceremony during the course of construction or by having a well-publicised opening ceremony when the building is completed and ready for occupation. It is normal practice to invite to such functions the local and national press, local and national agents, and sometimes potential occupiers. Local councillors or their officers may also be invited to help with the development company's public relations. For example the opening of a shopping centre may be a high profile event with a celebrity to perform the opening ceremony. Often this will result in media coverage which will be priceless from a developer's perspective. It is important to pay great attention to detail, ensuring not only that the buildings are ready for occupation with working services and landscaping but also that the ceremony itself runs smoothly. An unsuccessful opening ceremony, perhaps due to overcrowding, may result in negative perceptions of the development itself. Planning here is key and should be based on past experiences of similar events and the accompanying media 'hits' in the press or on the internet, e.g. via Twitter feeds.

In the case of a shopping centre it is standard practice for the major retail spaces to be let and occupied, where the opening of the development will be linked to the opening of those units. In an increasingly competitive environment in the real estate market, developers have had to make their opening event both increasingly imaginative and innovative in order to encourage agents to attend. In other words, agents will receive many invitations to ceremonies and they have to make a choice about which to attend. Accordingly, such incentives as free gifts, a prize draw to win a holiday or a night in a major hotel may have to be offered. The main advantage of this type of promotion is that it attracts stakeholders including leasing agents, particularly those retained by a potential occupier, to the property and gains/retains their attention.

10.2.8 Show suites and offices

A promotional campaign must be timed to have its maximum impact when the building is completed. For example it is essential for photographs to be included in the material so stakeholders including potential occupiers can view the final product, e.g. on the website and on promotional advertisements (such as via search engines). Due to the nature of property development itself and the large time period prior to completion, the developer can be creative in marketing the final product. To ensure the completion of a successful project it is essential that the developer places importance on selling/letting the building before completion to maximise the cashflow as soon as possible after the end of construction. During the construction period any promotional activity can only be supported by the use of conceptual drawings, plans, artist's perspectives, 2D and 3D models, as well as CAD images to help build up a picture of what the completed project will look like. For large developments including mixed-use or phased developments (e.g. a business park) it is standard practice to set up a show (display) office on site during the course of construction. This display office will be fitted out to the highest quality and display plans, models and promotional material about the scheme. This will often include a video simulation of the completed project and at times this display office will be elevated to give the prospective occupants an idea of the view and location on completion.

The office, often a temporary structure which has been appropriately decorated, painted and fitted out inside, should be staffed by an on-site representative of the developer or agent who is able to talk knowledgeably to leasing agents and prospective occupiers who make the effort to visit. Refreshment facilities should be available to offer to the prospective purchaser/tenant. A 'finishes' board should be on display, showing samples of the internal finishes proposed for the scheme. It is important to make sure that the show suite or office is clearly signposted and, if possible, the landscaping should be brought forward to enhance the environment. In any event, any landscaping in a scheme should be planted as soon as possible, allowing for planting seasons, so that it can mature slightly before completion of the whole scheme.

When the development has been completed, its appearance is of vital importance. In the case of an office scheme or an industrial scheme which has a high proportion of office space, it will often be advisable to completely fit-out a floor, or a part of a floor (including partitions, carpet and furniture) so potential occupiers can confidently see how the offices will look when they are occupied. This may be supported by plans for all floors showing alternative open-plan or cellular floor plans. In addition, it is usually advisable to fit out the main entrance hall and/or foyer and it is important to ensure that the common areas, bathrooms and toilets are kept clean. In most buildings the carpet flooring throughout the building will be provided by the developer. Incentives will become more commonplace in the market if there is limited demand, with the developer offering to pay partially or wholly for the occupier's fit-out. In the case of a shopping scheme or industrial warehouse scheme where units are in shell form, a sales office located on site is usually recommended to attract and welcome potential occupiers.

10.2.9 Public relations

Predominant larger developers (particularly public quoted companies and REITs) place a large amount of focus on ensuring they undertake effective public relations. Usually they either retain external public relation (PR) consultants or employ their own in-house staff responsible for liaising with stakeholders, the general public and the media. Everyone involved with public relations should be linked to a particular development scheme right from the beginning and be kept informed at all times on all aspects of the development, especially the progress status. One of the main priorities of the public relations team is to promote and enhance the reputation of the development company as a corporate entity; however, there will be spin-offs in relation to the promotion of individual development schemes. Where a scheme is particularly sensitive – for example due to historical reasons or political reasons within the local community – it is practically essential to employ a public relations consultant specifically for that scheme to ensure it all runs as smoothly as possible.

Through their press contacts, public relations consultants can achieve effective media press editorial at significant stages in the development programme. For example, carefully controlled and timed media or press releases may be circulated to the property press on one or more of the following occasions: initial acquisition of the site, obtaining planning permission or consent, the completion of the scheme, and on the leasing of all or part of the scheme. Editorial coverage is often offered by a publication or on a website if advertising space has been booked to coincide with coverage of a particular geographical area or a particular type of property. It should be noted that editorial coverage is much more likely to be read than an advertisement; furthermore the editorial coverage can be achieved at no cost.

Promotional material should be on display at every public relations event where either the developer or the development itself is being promoted. Leasing agents should have hard copy promotional material in their office and should promote the development, where possible, at corporate hospitality events and exhibitions.

Discussion point

- List the various types of promotion available and the positive and negative aspects of each option.

10.3 Role of the agent

A real estate agent is usually directly responsible for selling/leasing a particular property and they are engaged by the developer. Although some developers may have their own experienced in-house marketing team, it is not always possible to rely 100% on their efforts alone. The appointment of an external agent will depend upon the circumstances of a particular development project and the potential market for the scheme. In addition, many agents offer certain advantages over an in-house agent and this is an important consideration.

One of the major benefits of enlisting the services of a local agent is that they are often strategically positioned in a better location for attracting business. For example, for an out-of-town developer undertaking an industrial development in a provincial town it makes more sense to employ a local agent with established offices in the town. A good firm of agents should have extensive networks and a detailed knowledge of the particular market in which they operate, being thoroughly familiar with current and future levels of demand and supply. They should possess up-to-date accurate information about this data. Furthermore they may have additional specialised knowledge and experience, such as specialising in certain types of property use and developments. The agents will also have personal contacts with other agents retained by a specific client or decision-makers within the property department of a potential occupier. For example many agents closely monitor the status of leases for existing clients, therefore knowing exactly when a particular lease in any buildings is coming up for renewal or expiry. The agent will then be able to contact the occupier prior to the lease renewal/expiry at a time when they are in a decision-making mode. A successful agent must be active and continuously involved in the marketplace so they are aware of all changes in market conditions.

Agents can be retained on a 'sole/exclusive agency' or 'joint/multiple agency' basis. As sole/exclusive agent they alone are responsible for disposing of the property and are entitled to a financial remuneration or fee as a reward for each new lease or sale. The actual level of remuneration will vary due to a number of factors – however, a normal range of fees might be 10% of the first year's rent for a lease (allowances will be made to take account of any rent frees or inducements given to the tenant on a letting) or 2% of the sale price if sold, although again this can be negotiated depending on the size of the development and the relationship between the developer and the agent. A joint agency arises where two or more agents are instructed to sell the same property, which can often happen when both a national and local firm of agents are instructed together. In such a case it is commonplace with each new lease or sale for the developer to pay a larger fee, perhaps one and half times the normal amount, where the agents share the fee between themselves on some agreed basis. For example, joint agents would each receive 7.5% of the fee from an agreed total fee of 15% of the first year's rent for a new lease. The appointment of a second or third agent (if two are already appointed) is often not the decision of the developer; however, it may be a condition of a funding agreement where the fund requests that their own agents are to be involved.

Often a developer may employ a national or regional firm of agents operating either with or without the input of a local firm. There is no general rule to apply and this depends largely on the size of the development and the target market. Each scenario must be examined on its own merits. However, a national agent (or at times, an international agent) normally has a greater understanding of the larger and more complex schemes and has more direct and frequent contacts with the larger companies and multiple retailers attracted to them. A local agent typically does not have this advantage. Also a national agent can usually offer additional advantages to include investment advice if the scheme is to be forward-funded or potentially sold on completion. On the other hand, a local agent will often have a better understanding of the particular characteristics of the local market and of local occupiers and retailers.

Regardless of the final decision regarding the type of agent or combination of agents, it is very important that they are not an 'afterthought' and have the opportunity of contributing to the planning, design and evaluation stages of the project. They should be able to draw attention to key features of the design that add to or detract from the overall marketability of the property. Also the agent/s must be able to accurately identify the market levels of rent (on final completion), the nature of competitive development and the most effective timeframe for leasing or selling the development. Much time and resources can be wasted by undertaking a premature marketing campaign too early before completion. On the other hand, commencing the campaign too late will result in additional holding costs for an extended period post-completion. Nevertheless it is recommended that the agent/s are engaged at an early stage, which will allow them to become thoroughly familiar with the property they are going to sell or lease. In other words, it can be a disincentive for a potential purchaser or tenant to find that the selling agent cannot provide full details of the property they are promoting. In addition the agent/s should attend regular meetings with the client to be kept informed of progress on the development and to plan the promotional activities.

It is essential to distinguish between (a) the appointment of agents and (b) offers to pay commission to an agent introducing purchasers or tenants. There is a contractual relationship with (a), the terms of which (both expressed and implied) need careful forethought and consideration. The binding agency agreement should clearly detail in simple layman's terms the length of time the appointment will remain in existence and how the agreement may be terminated, specifying any retainer payable and setting out the rate of commission, including in what circumstances and when it will be payable. In addition, it should specify whether the agreement is for a sole or joint (multiple) agency and what the developer's rights are concerning their discretion to employ additional agents. Also it should be made clear whether agents are entitled to expenses in the event that they do not succeed in disposing of the property.

At times offers will be made to pay commission to agents on the signing of legal agreements for sales or lettings for purchasers or tenants introduced by the agent. No formal agency appointment is made in these circumstances, and this type of arrangement is substantially different from the binding formal appointment of an agent to act for the developer for the disposal of a particular scheme. Sometimes the offer is open to any agent/s formally introducing a purchaser or tenant, or such an offer is only made to specific agents. Although a developer might argue that an offer to pay introductory commission to any agent would be likely to result in wide exposure to the market and attract the most prospective buyers or tenants, unfortunately this is not necessarily so. For example if too many different agents are handling a particular property there is a danger that the property may lose its attractiveness as a unique location, which in turn may create an unfavourable impression on the ground. In other words, after an initial flood of enthusiasm many agents tend to lose interest if they know that the property is in a large number of hands, where each agent is competing with many other agents and their hard work may not be rewarded. Therefore, some of the best agents might be reluctant to be involved with properties which are widely and indiscriminately offered in such a way. The smarter agents will devote their resources to potential sales/lettings where they have the highest likelihood of success, i.e. not shared with a large body of other agents with associated risk. Sole or joint agents, or perhaps three or four agents working on an agency basis, are much more likely to have a sense of involvement. Whatever the agreed arrangements are, it is important to keep all the agents up to date at all times. The worst case scenario is for the completed development to be perceived as being poor quality and not in high demand in the market. This type of stigma can potentially affect any development and should be avoided at all costs since overcoming negative perception is difficult to accomplish.

Communication and accurate record keeping of all enquiries is essential so that it can be seen at a glance which prospective purchasers or tenants have been introduced by which agents and precisely when. The question of disclosure is important. If an agent has already been retained to find accommodation for the prospective purchaser or tenant, they will not be able to accept an introductory commission from the developer. Furthermore, when agents refer to 'clients', they are normally retained and will not seek a commission; but when they refer to 'applicants', they will normally expect a commission because they are not retained by applicants. In all cases these details must be clarified explicitly at the outset.

When an agent is appointed by a developer they should be involved at the commencement of the development process. In some cases an agent may even have been responsible for initiating the development process by introducing the site to the developer and/or arranging the necessary development finance. Often the agent has a valuable input, ranging from their in-depth knowledge of the both the occupier and investment market through to the final market evaluation and design processes. Developers should note that agents often tend to be cautious when advising on design and the specification of the scheme from a letting/selling point of view, since they wish to ensure that the scheme has the widest tenant/purchaser appeal. Furthermore, developers should make their own judgements about the advice received and undertake their own due diligence as supported by their own experience and research. In addition both developers and agents should obtain feedback from tenants about similar schemes.

Regular meetings and communication between the developer and agent/s is essential and should normally take place on a monthly or fortnightly basis, largely depending on the stage reached in the development process. At these meetings the agent/s should provide status reports about the leasing/selling of the scheme. The report will normally take the form of an update detailing all enquiries received from interested parties and include any known requirements for accommodation that are to be followed up. The agent/s should advise on the availability of competing developments in the same marketplace and what terms are being quoted on those buildings, and concluded sale or leasing transactions on similar accommodation should be listed and discussed in light of the current development at hand. The source of each enquiry should be noted so that the developer can judge the effectiveness of a particular method of promotion, e.g. internet, direct email, newspaper advertisement. These promotional activities will also be analysed to ensure the highest levels of efficiency are achieved for the marketing budget. The agent should have a full input in any decision on the content and design of all promotional material within a marketing efficiency context. In many cases the agent will be responsible for the booking and scheduling of advertising space, unless a separate advertising agency has been appointed. As both agents and advertising agencies book space on a regular basis they should be able to pass on to the developer the benefits of any discounts and should be able to obtain good positions within the relevant publications or on a particular website.

10.4 Sales/Lettings

After a potential occupier or an applicant has expressed their initial interest in inspecting a property, the next step is for the agent to show them round at a mutually agreed time. Regardless of how many telephone and email enquiries are received, the objective should be to convert the remote enquiry into a physical initial inspection. If continued interest is shown (e.g. a second inspection), then the developer may be involved in all future viewings and ensuing negotiations with the interested party. There is clearly no substitute for personal contact with the applicant although there may be times, particularly early on in the negotiations, where the agent should lead the discussion. The developer should make a judgement as to when it is appropriate to become involved.

During the on-going regular meetings, the agent/s will be able to advise the developer when it is appropriate to quote terms and financial information (i.e. relating to the final purchase or leasing price). The agent/s should be familiar with the level of flexibility they actually have when negotiating terms, bearing in mind the underlying requirements of the financiers and/or the investment market. This information will vary between each development; however, the terms of each sale/lease will be determined by both the method of funding the scheme and the intentions of the developer as to whether it will be retained as an investment or alternatively on-sold to a third party investor. If a scheme has been forward-funded then the fund will need to approve the terms of every lease and the actual type of tenant. If the intention of the developer is to fully on-sell the completed development to an investor, it is essential for the leasing terms for all tenants (i.e. tenant risk) to be acceptable to the investment market at which it is aimed. This aspect will largely depend on to what extent the property is considered to be 'prime' in terms of both its quality and location. The developer should be advised on this matter by the agent/s who will be responsible for handling the investment sale. Furthermore the developer must acknowledge that the terms of every lease will directly affect the investment value of the completed development. Even though a property may be initially retained by the developer, thought must be given to any eventual sale in structuring the terms of a lease so it is generally acceptable to investors and not perceived as being poorly let.

It is important to focus on the various terms which need to be agreed when negotiating with a potential tenant. If negotiations are being held directly with an owner-occupier then the only terms to be agreed upon are the leasing/sale price and any additional work to be carried out by the developer.

10.4.1 Demise

The accommodation space being let needs to be clearly defined and at times this is known as the 'demise'. A decision needs to be made, in the case of an office development, as to whether it is possible to lease/sell only whole individual floors or alternatively if it is acceptable to the financier and/or the developer to lease/sell a portion of a floor. In many instances when seeking to achieve the maximum investment value of the completed development, agents will be instructed to pursue a single letting of a building. In comparison a multi-let building will narrow the potential investment market appeal due to the additional management and risk involved which will be reflected in any investor's required yield. Immediately following each downturn in the property market cycle there are often examples of recently completed property development schemes that remain vacant – this scenario is then further complicated when a developer has strictly adhered to a policy of accepting only single lettings. However, the actual decision as to whether to accept multiple lettings will usually need to be made only if the level of demand in the market for a completed building proves difficult. Other factors come into play here, for example the level of flexibility with this decision will largely depend on the financial status of the developer and/or the additional influence of any financier or development partner. The following questions should be asked: Is it better to maintain cashflow at the expense of maximising the capital value? Or alternatively is it better to have a building 100% let at a lower rate/m^2 or 50% let at a higher rate/m^2? In other words, the final capital value of a building is linked to the rate/m^2 at which the building can be leased/sold, a lower rate/m^2 will ensure cashflow but may result in a lower overall capital value.

If each unit in the completed development is not self-contained then consideration needs to be given to adapting the premises to suit the needs of other tenants. For example, access may need to be reserved for the tenant/s in respect of the use of common areas such as foyers, staircases, lifts, bathrooms and toilets. Arrangements will also need to be made in respect of separating services for metering purposes (e.g. water, electricity and heating/air-conditioning).

10.4.2 Rent

The developer will usually quote an advertised rental value for the accommodation in consultation with the agent, based on a current assessment of the hypothetical market value of the property for lease/sale. The developer will also calculate the rent required, which when capitalised at the appropriate market-derived yield will provide the developer with a satisfactory return on investment (ROI). This return should compensate the developer with profit for their exposure to risk (i.e. profit/risk) for the entire project. At the same time it is important for the developer and the agent to be aware of all recent leasing transactions and the advertised rents quoted for other properties in order to be familiar with the interaction of market supply and demand. The real estate agent/s should be able to assist with interpreting and monitoring overall levels of demand via the number of enquires for the development. On the other hand the level of supply is also critical, quantified by examining the number of buildings under construction or planned as well as recent sales of similar property in the surrounding area. When analysing lease deals and quoting rents, allowances must be made for any differences in the specification and location of the comparable properties.

The actual rent agreed between the developer and a potential tenant will depend on how the negotiation is conducted and the willingness of each party (refer to the International Valuation Standards Committee (IVSC) definition of market lease). Also it will be affected by the strength of their respective bargaining positions in current market conditions. For example, in a difficult market where supply exceeds demand and tenants have considerable choice of competing developments in the marketplace, the developer may consider offering incentives in order to maintain the required rental value for the property. However, such incentives should be kept within reasonable limits otherwise the developer will be losing income in the form of incentives in return for a rent in excess of a market rent. In other words a developer may be 'over-renting' the building in the view of any potential investor who will anticipate an incentive being included in the negotiations. At times reasonable inducements, such as six months rent free (to allow time for the tenant to fit-out the building), or a capital contribution towards specified fitting-out works, are considered in many markets to be an accepted practice by potential investors. A fit-out contribution will usually be related to fixtures and fittings which are referred to as 'landlord's fixtures and fittings' for rent review purposes. Accordingly, when the rent is reviewed in accordance with the terms of the lease, account will be taken of the benefit of those landlord's fixtures and fittings in determining the rent for that particular property.

From a historical perspective, during periods of severe oversupply in the property market substantial inducements or incentives (e.g. from twelve months to three years rent-free periods) have been offered in order to maintain a certain rental level on office schemes. Even though this would ensure the building is 100% let at market rates, the inducement may not be so transparent to a potential purchaser of the building. In other words this would appear to maximise the investment value for the developer, especially if the rent review clause in the lease is drafted on the basis of an upward-only rent clause (this type of clause is illegal in some markets), and in the process completely disregarding the existence of incentives being offered in the open market at the time of the rent review. The downside of achieving a higher rent is that in a period of low rental growth, potential investors will discount the value of the slice of the rental income considered to be in excess of market rent, as they perceive there will be an absence of growth until after the first review.

10.4.3 Lease terms

The circumstances surrounding the actual length of a lease and also the existence of any lease options for many property developments will depend largely on whether the property is viewed as a potential investment for an institutional purchaser. In other words, the strength and value of the lease (i.e. the income flow) will be linked to the level of perceived risk affecting a reliable source of future cash flow. Note that most institutional investors prefer long-term leases to decrease the risk of a tenant void and loss of income, as well as avoiding the associated leasing fees. Over time the length of the lease term has been changing in the marketplace in response to changing tenant demands – for example shorter leases have become increasingly acceptable since tenants have often been in a better position to negotiate such terms to suit their particular business plans. At times tenants have also been able to negotiate break clauses at certain specified times within their leases to allow flexibility as their business changes. Some institutions may, depending on the circumstances, agree to such clauses provided the tenants also agree to the lease terms. When necessary a developer may be called to negotiate a landlord's break clause to coincide with a major redevelopment or refurbishment opportunity.

With forward-funded arrangements the institution will dictate the length of the lease in the funding agreement and any flexibility will need to be strongly argued by the developer in the light of market conditions. For schemes where the likely tenants are small businesses or sole traders with little or no financial track record, such as in a specialist shopping scheme or small industrial units, the developer will often be more flexible due to the limited market demand. Due to the type of tenant, such schemes will not be considered 'prime' and, therefore, institutionally acceptable lease terms are not so important.

10.4.4 The tenant

The final value of the development will be linked to its ability to earn money and the strength (i.e. reliability and security) of the tenant. The financial status and background of all potential tenants will need to be checked by the developer to ensure they have the ability to pay not only the agreed rent but also all outgoings on the property. For example an institutional investor may prefer that the tenant's profits meet a certain criteria, such as exceeding a sum three times the rent. In most cases the developer will insist on being provided with at least the last three years' accounts of the tenant to confirm the tenant's financial standing and reduce the developer's exposure to tenant risk. In the case of new businesses or those without sufficient track record, the developer should obtain bank references and trade references. In addition, the business plan and cashflow projections of the tenant should be examined to confirm viable long-term projections over the timeframe of the proposed lease length.

If the developer or financier considers the financial covenant of a tenant is insufficient and therefore too risky, then bank, financier and/or parent company guarantees may be requested in order to reduce perceived risk and provide some form of guarantee in the event of non-payment of rent during the lease period. In the case of private companies and sole traders, directors' personal guarantees will often be sought as well.

In a retail scheme the developer and/or the financier may wish to influence the tenant mix within the scheme to ensure a variety of retail uses for the shopping public, which will in turn determine the success of the scheme and reduce exposure to risk. The diversification of different tenants will ensure the shopping centre can cater to a wide variety of shopping needs in the market, which in turn may alter as seasons change or shopper preferences vary over time. Having tenancies of different areas (m²) will be more attractive to the varying demands of multiple tenants than if only one size was available.

10.4.5 Repair obligations

The responsibility for repairs and maintenance during the term of the lease will depend largely on whether the property unit is self-contained. For example, in the case of a single building the sole tenant will usually be responsible for all internal and external repairs. On the other hand, where the unit is only part of a building that is shared with other tenants and/or the owner, the tenant will usually be responsible for internal repairs and the landlord will be responsible for the external and common parts. Usually this will depend on how well each stakeholder (i.e. the landlord and the tenant) negotiated the original tenancy agreement and what was agreed at that point in time. The landlord's expenses associated with repairing, servicing and maintaining the common areas will usually be recovered by a 'service charge' directly levied on all tenants, most often in proportion to the area of their demise. This is often referred to as 'recoverable outgoings'. The developer/s must provide prospective tenants with estimates of the service charge.

10.4.6 Other considerations

There are many other related factors to be considered during the process of promoting and selling/letting a completed property development. The information listed in this chapter has presented a broad overview of the basic principles – however, every development is unique and the circumstances will vary much in the same way that each property itself differs. Nonetheless there is by no means an exhaustive list of the matters to be considered by the developer when negotiating with tenants.

In order to be contractually binding, it is essential for every lease and/or sales transaction to be legally documented in the form of (a) a lease or (b) a contract and transfer of title. It is essential for the developer to instruct a solicitor as soon as the promotional campaign has commenced so that all the necessary draft documentation can be prepared. Once negotiations are concluded with a tenant or purchaser the legal work can then proceed quickly. A developer will also need to ensure all the necessary draft deeds of collateral warranty (guarantees against faults in a new/refurbished building for a period of time) have been agreed with the professional team and contractors (with design responsibility), should either the tenant or the purchaser wish to benefit from them. A tenant may require warranties on schemes where the repairing obligations could be considerable if something goes wrong and they wish to be able to pursue the relevant professional or contractor for a remedy. In such an example the remedies under a warranty are limited and a tenant may wish to benefit from decennial insurance to protect against latent defects. However, the developer will need to judge the likelihood of such insurance being required by either a tenant or purchaser as the option of cover must be arranged before construction starts.

Prospective tenants and purchasers will typically make enquiries through their solicitors as to the existence of services and would like to see copies of all planning and statutory consents. A potential tenant or purchaser will also require full information about the building in the form of 'as-built' plans and maintenance manuals. The developer should therefore ensure that all the necessary supporting paperwork is in place *before* draft contracts and leases are issued to reduce the risk of delays in the legal process. The developer should be prepared to answer all queries and supply any documentation related to the development.

Most importantly the development process does not end there once the documentation has been completed and the keys are eventually handed over to the tenant or purchaser. The developer will have on-going responsibilities in the post-completion phase, such as ensuring the satisfactory completion of the defects identified at practical completion and assisting the tenant with any 'teething' problems. The developer will have a continuing relationship with the tenant in their role as 'landlord', unless the property is sold either for occupation or as an investment. Even if there is no continuing contractual relationship, the developer should maintain contact with all their tenants or occupiers and provide an after-sale service. This will pay future dividends and help the reputation of the developer in the market, as failure to protect a reputation can adversely affect future proposals, although the developer may never find out the reason. For example, occupiers may not speak to each other but they may talk to their agents if they have cause to complain about a particular development. In today's increasingly competitive economic environment a tenant/purchaser will be much more vocal when things are going wrong than when things are going very well. The advent of social media has assisted this process and many prospective tenants/owners would Google a developer to find out about their track record. Thus there is no substitute for direct feedback from occupiers of completed projects from which invaluable lessons can be learnt for future developments. Overall there is a strong argument that the development industry as a whole should increase their level of research into occupiers' needs to a much greater degree than currently undertaken. At all times the occupier is the customer and they should come first.

10.5 Reflective summary

In order to reduce their underlying exposure to risk, the task of securing an occupier should be at the forefront of a developer's plans at the start of the development process and at all stages thereafter. Marketing should not be an afterthought. During the evaluation stage it is critical to research and accurately quantify the level of occupier demand for the development in the proposed location and in the prevailing market, and further research should be carried out to identify the target occupier market and their exact requirements in terms of the design and specification. At the same time consideration should be given to the level of competitive supply in the market and when competing developments are scheduled for completion. However, such research is often overlooked by developers, which may result in a building being difficult to dispose of. Unfortunately, this cannot be solved by simply throwing money at promotion. The aim of promotion is to make potential occupiers aware of the development scheme and for a campaign to be effective it must be carefully targeted at and tailored to its audience. Leasing and selling agents, with their integral knowledge of the market, have an important role to play in disposing of the development to an occupier whether through a letting or sale, together with any subsequent investment sale. Any letting should be secured on terms acceptable to the investor market to maximise its value and reduce perceived tenant and property risk.

Chapter 11

Sustainable development

11.1 Introduction

There is little dispute that buildings are substantial carbon dioxide (CO_2) emitters and contribute substantially to climate change (Reed et al. 2008). This argument is based on the environmental footprint of buildings, especially the high reliance on resources due to increased use of air conditioning and heating. At the same time it has been demonstrated that the value of a building can be linked to the building's perceived level of sustainability (Myers et al. 2008; Warren-Myers 2013). The underlying subject of this chapter is the meaning of 'sustainability' and 'sustainable development' in the context of property development, where both concepts have now become an important if not an essential consideration in most developments. In addition, the concept of corporate social responsibility (CSR) is discussed and the relationship between sustainability and CSR is explained. Then consideration is given as to why property and related development is an important sector in relation to both sustainability and CSR, with the emphasis placed on the key areas of impact. This is followed by a discussion about how the impacts are addressed by property development stakeholders including developers, investors and occupiers.

11.2 What is sustainability and sustainable development?

A commonly used definition of sustainability was established by the United Nations World Commission on Environment and Development report in 1987 entitled *Our Common Future* (WCED 1987). This definition states:

> Sustainable development is development that meets the needs of the present without compromising the ability of future generations to meet their own needs.

Conducting a development in a sustainable manner is arguably already being undertaken by all developers as they re-use some of their resources (e.g. computers, scaffolding, tools) for subsequent developments. The focus has moved from considering 'which individual property development is sustainable?' to 'how sustainable is each property development?' The challenge at hand is to increase the level of sustainability in the built environment during (a) the task of undertaking the property development and (b) the operation of the developed property after completion. This situation is further complicated as there are as many definitions and interpretations of the terms 'sustainability' and 'sustainable development' as there are groups trying to define it. To a large extent all definitions relating to sustainability are concerned with aspects of the following broad parameters:

1   an understanding of the relationships between the economy, the environment and society
2   equitable distribution of resources and opportunities, and
3   living within limitations.

Even though they create a level of confusion in society, the different ways of defining sustainability are useful for different situations and varying circumstances. For example there are varying levels of development in and between individual countries which range from advanced western civilisations to third world developing countries. Furthermore there needs to be a holistic approach to sustainability from a global climate change perspective where all countries are fully committed to sustainable practices – any less than 100% commitment by all countries will be inefficient because the actions of one country also affect all other countries. For these reasons various groups have created different definitions of sustainability. The commonly used definition of sustainability which has stood the test of time is the one established by the United Nations World Commission on Environment and Development report in 1987 (cited above).

This landmark definition, also commonly referred to as being drawn from the 'Bruntland Report' (WCED 1987) after the chair of the UN Commission, established two key principles which over time have substantially influenced the sustainability debate. These principles relate to both 'inter-generational' and 'intra-generational equity' and an understanding of both terms are important in the context of property development. *Inter-generational equity* is based on the principle of equity between people in society who are alive today and those generations in the future, although yet unborn. The inference here is that unsustainable production and consumption occurring today (e.g. over-consumption) will diminish and degrade the environmental, social, and economic resources which would otherwise be available for tomorrow's society. For example the widespread consumption of fossil fuels, which is a finite resource taking millions of years to produce, is unsustainable. Therefore living a sustainable existence is ensuring future generations will have the ability and means to achieve a quality of life equal to or better than today's life. On the other hand *intra-generational equity* is the principle of equity between different groups of people who are alive today. The emphasis is placed on social justice issues, therefore it manifests itself in areas such as alleviating the debt and improving the health of developing countries. There is direct relevance to inequity here and attention should also be placed on assisting developing countries to be sustainable, rather than focusing only on reducing the resource consumption of developed countries.

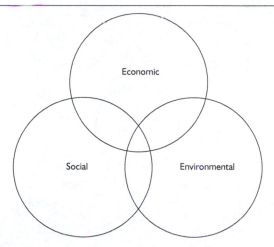

Figure 11.1 Three-pillar model of sustainable development based on triple bottom line accounting

An alternative definition of sustainability was produced by the World Business Council for Sustainable Development (2000):

> Sustainable development involves the simultaneous pursuit of economic prosperity, environmental quality and social equity. Companies aiming for sustainability need to perform not against a single, financial bottom line but against the triple bottom line.

There are similarities between the earlier definition provided by the Bruntland Report (WCED 1987) and the later WBCSD definition where both refer directly or indirectly to social, economic and environmental aspects of society. In the WBCSD definition there is direct reference to the 'triple bottom line' approach as introduced by John Elkington in 1994, which has since been widely acknowledged as the initial recognition when seeking to incorporate sustainability in business (Figure 11.1). Triple bottom line (TBL) accounting requires moving from the traditional reporting only framework to also accounting for environmental and social performance, as well as the standard economic performance. When adopting a TBL approach, the company is broadening its level of responsibility to also include stakeholders, as well as the traditional responsibility to shareholders. In this context the term 'stakeholder' refers to those influenced, either directly or indirectly, by the company's activities.

The argument behind the use of TBL is that all three aspects must be accounted for when undertaking sustainable development. Over time it has been observed that social values have gradually influenced these ideas and some argue that TBL is now out-dated. The latest model (Figure 11.2) includes a fourth pillar to reflect the importance of cultural aspects in creating lively communities in which people want to live, work and visit.

Nevertheless it is important to recognise the significance of TBL where these ideas have also evolved into the concept of corporate responsibility (CR) or corporate social responsibility (CSR).

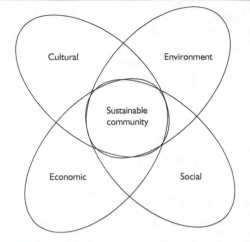

Figure 11.2 Four-pillar model of sustainability (Source: adapted from http://sustainableantigonish.
ca/4-pillars/ 2013)

11.3 Corporate social responsibility

Corporate social responsibility (CSR) is at times referred to using other terms including
'corporate citizenship', 'responsible business' or simply 'corporate responsibility'.
Defining CSR is difficult since many definitions exist. According to Van Aaken et
al. (2013), CSR is linked to companies which incorporate 'the actions of corporate
actors that address social and ethical values beyond legal requirements'. This
definition covers part of what CSR is about. A more detailed explanation is that CSR
is linked to a company's obligation to be accountable in its operations and activities
to all stakeholders, with the aim of achieving sustainable development not only in
the economic dimension but also in the social and environmental dimensions. This
definition also incorporates the social, economic and environmental elements of
sustainability as referred to in the WBCSD definition of sustainability in section 11.2. It
is worth noting that the WBCSD defines CSR as being 'the continuing commitment by
business to behave ethically and contribute to economic development while improving
the quality of life of the workforce and their families as well as of the local community
and society at large'.

So in essence CSR is generally about taking a broader approach to business activity.
Historically, based on standard economic theory, most freehold businesses are profit-
seeking, motivated solely by financial or economic gain and pay little or, in some cases,
no regard to environmental issues, such as pollution resulting from their activities or
social issues such as the exploitation of labour in developing countries. Clearly then a
business that adopts CSR is committed to following 'best practice' for all stakeholders
and to making all their activities open and transparent.

The foundations of CSR can be traced back to the philanthropy of Victorian
industrialists, such as Joseph Rowntree, Titus Salt and George Peabody and, in the
1920s, American industrialists Andrew Carnegie and Henry Ford. Then CSR re-
emerged in the late 1970s and early 1980s as organisations became increasingly
concerned about their public image (Weber 2008). Since then CSR has mainly been
driven and promoted by large companies and corporations, many of which voluntarily

issue CSR disclosures in their annual reports and on their websites (Rodriguez and LeMaster 2007; Vurro and Perrini 2011).

Today CSR has moved 'from ideology to reality' (Lindgreen and Swaen 2010: 1) and become firmly entrenched in the business community. Although many organisations demonstrate their commitment to CSR, 'many struggle with this effort' (Lindgreen et al. 2009), perhaps due to the continuing debate about what CSR actually is (Lindgreen and Swaen 2010). Although there is still no universally accepted definition of CSR it has been argued that it assumes the role of acting ethically in business dealings at all times (Moir 2001). This is based on the belief that business organisations have a moral obligation 'to pursue those policies, to make those decisions, or to follow those lines of actions which are desirable in terms of the objectives and values of society' (Bowen 1953).

An alternative perspective about CSR relates to the concept being a pragmatic public relations exercise. This viewpoint considers CSR as an attempt to gain a competitive economic advantage by creating a favourable impression on employees, clients and consumers (Rangan, Chase and Karim 2012). In between these two CSR perspectives is another view that CSR relates to the obligations of a company to its stakeholders, being 'the people and groups who can affect or who are affected by corporate policies and practices' (Lantos 2001). It appears that most organisations would consider this is an appropriate concept of CSR. Since organisations exist within networks of stakeholders, one issue is their potentially conflicting demands based on what individual stakeholders believe to be the 'societal expectations of corporate behaviour; a behaviour that is alleged by a stakeholder to be expected by society or morally required and is therefore justifiably demanded of a business' (Whetton et al. 2002). An landmark global study undertaken by PricewaterhouseCoopers as early as 2003 concluded that the majority of companies considered that being socially responsible involved the following activities:

* providing a healthy and safe working environment for employees
* creating value for shareholders
* supporting community projects, and
* good environmental performance.

The same survey also showed that with CSR the concept of 'one size' does not fit all, and this view seems unchanged. It was also noted that the perception of what CSR was varied between different sectors in the same market and also between different countries. In the USA for example it is common in some areas for CSR to be perceived more as a form of philanthropy, where businesses make profits, unhindered, and then they donate to a charitable cause. On the other hand CSR is seen as the link between corporate profitability and social improvements in many African countries. In Europe and North America it can be argued that CSR is at times viewed as a combination of environmental responsibility, ethical economics and charitable giving. However the European concept has been targeted on operating the core business in a socially responsible way, complemented by investment in communities for solid business case reasons. It could be argued that this approach is more sustainable as CSR becomes an integral component of wealth creation, adds to business competition and also increases the value of wealth creation to society. Furthermore, in difficult periods there is reason to exercise CSR further, whereas if CSR is philanthropic outside the core business mission and objectives, it will be dropped when times are tough.

At present most of the information disclosed by businesses in their voluntary social disclosures relates to employees, the environment and the community. There is no standardisation in the type of content included or the way this information is reported (Reverte 2009). However, it has previously been demonstrated (Williams and Pei 1999), after examining companies' CSR disclosure practices, that:

1   There are variations in the way that each company approaches CSR. For example some companies have internal individuals or groups dedicated to ensuring the company fulfils its obligations, whereas others use CSR as a form of public relations.
2   There are differences in the amount and type of information disclosed. For example, some companies provide detailed information of past performance in areas such as health and safety, environment and community relations, where other companies simply present their overall CSR strategy.

This indicates that the drivers for CSR vary across sectors and between companies, that perceptions of CSR differ and that a valid indicator of the real value a company attaches to CSR is when, if at all, they locate this function within their organisational structure (Frankental 2001). For most companies the primary driver behind their CSR efforts is not the belief that they can be a positive force for social change but the 'multi-faceted business returns' that they can reap. Engaging in CSR generates favourable stakeholder attitudes and better support behaviours from purchasers, investors and those seeking employment. Research (both academic and marketplace) indicates that key stakeholders (consumers, employees and investors) are 'likely to reward good corporate citizens and punish bad ones' (Frankental 2001: 8). One example can be seen from the results of a 2013 study of American consumers which found that 91% of consumers would 'switch brands' in favour of one that supports a good cause, given there is similar quality and price (Cone Communications 2013). 'The basic belief that CSR can be good for business clearly drives corporate interest in CSR' (Lingreen and Swaen 2010). Organisations can gain a competitive advantage and differentiate themselves by integrating non-economic factors which improve their corporate image and reputation, and in turn increase consumer goodwill and engender positive employee attitudes (Brammer et al. 2004).

Maximising business returns from CSR policies requires some form of measurement of its impact. Studies in this area have primarily used one of three methods to measure CSR as follows: (a) the use of expert evaluations, (b) content analysis of voluntary social disclosures and other corporate documents, and (c) analysis of performance in controlling pollution. Analysing a company's financial performance can also help to explain variations in approaches to CSR and disclosure practices. For example the number of philanthropic activities is traditionally affected by a company's overall profitability and return on investment. The underlying argument here is that CSR can pay off, however it really depends on how an individual company approaches the issue and a variety of other relevant variables. For example companies exist in many different formats with varying degrees of complexity, ranging from those that are highly profit orientated in the private sector to those in the public sector that are not-for-profit orientated. The activities in which companies are engaged varies considerably and it is not surprising that this translates to varying approaches to incorporating CSR. Efforts to maximise business returns from CSR activities are often impeded by stakeholders' low awareness and scepticism towards such activities. The key to overcoming this is effective communication about CSR to stakeholders.

11.4 Different types of CSR

There has been a long-standing view about the existence of different types of CSR (Lantos 2001) where most types have evolved from economic, legal, ethical and philanthropic responsibilities which in turn also relate to Lantos's three classifications of ethical CSR, altruistic CSR and strategic CSR. According to Lantos (2001) ethical CSR can be described as the moral mandatory fulfilment of a corporation's economic, legal and ethical responsibilities. The concept of altruistic CSR is generally linked to philanthropic CSR and involves a contribution to the good of society, even if it compromises the level of profit. For example altruistic CSR might be the funding of schools or supporting local community initiatives. The justification is based on the premise that corporations, through their wealth and profit making capacity, have the indirect power to affect parties beyond participants directly involved in their transactions. It can be argued that altruistic CSR is not really legitimate for corporations because it involuntarily affects shareholders (through lower share prices), consumers (via higher purchase prices) and workers (receiving lower pay). The arguments against company involvement in altruistic CSR are that firms are not competent to involve themselves in public welfare issues and that, through taxation, corporations already contribute to society.

However, strategic CSR can been seen as a legitimate activity which can truly enhance a company's business performance while benefiting society. Strategic CSR can be broadly defined as undertaking caring corporate community services that accomplish strategic business goals. Corporations contribute not only because it is morally right and proper but also because it is in their best financial interests to do so, hence fulfilling their underlying duties to stakeholders (Brammer and Millington 2004). Within strategic CSR the voluntary aspect leads to increased employee morale and higher productivity, attracts customers to a perceived caring corporation, as well as contributing to the local community and making it easier to attract good employees. Overall an effective strategic CSR will present a win–win situation benefiting both the community and the firm.

In order for strategic CSR to operate effectively, corporations must to be able to measure and benchmark their activities. Without a form of reliable measurement, strategic CSR is reduced to a public relations exercise and little else. It is essential for strategic CSR to embrace the stakeholders, be rewarded by financial markets, be defined in relation to goals of ecological and social sustainability, be benchmarked, be audited and also be open to public scrutiny when necessary. Finally it has to be embedded vertically and horizontally within the business and must be embedded in the corporate planning function.

When discussing CSR there are other considerations which may affect a company's approach to it. For example a company's approach to CSR may be complicated and adversely affected by its legal position. In most legal frameworks companies are legal entities with certain rights and privileges, but also have specific liabilities. Another factor which may affect a company's overall approach to CSR is its position in a certain business sector and its level of power. For example some companies are so large they have a turnover exceeding the gross national product of some small individual countries, which effectively means their power and influence can be immense. Therefore the perception of larger corporations can vary on an individual basis and these corporations may have a completely different perception of the role of CSR. Therefore, given the variation in conceptual understanding and the different types of CSR, each company will generally incorporate CSR in a different way. With direct reference to property development, each developer is likely to observe some variation both within and across the sectors, based on parameters such as company size.

11.5 Sustainability reporting

When undertaking CSR a fundamental aspect is to maintain a high level of accountability to all stakeholders in relation to social, economic and environmental impacts. In this context, sustainability reporting is the method of demonstrating accountability. In addition the role of reporting enables businesses to develop new targets and goals in terms of relevant social, economic and environmental impacts, thus enabling performance gains to be made. Over time protocols have been developed for non-government organisations (NGOs) to adopt when undertaking sustainability reporting, for example when involving stakeholder consultation to identify the goals and targets relevant to the business. It is imperative to benchmark activities to industry standards where possible – for example a relevant measurement may be the amount of waste products produced by a company or the amount of water or electricity consumed per annum. After benchmarks are established, targets can be established, such as reducing electricity consumption in a company's offices by at least 15% per annum. It is necessary to set up a strategy whereby accurate reporting and documentation of the policy can be recorded within a specific time frame. At the same time a responsible person has to be identified within a senior management role and given the power to ensure targets are implemented within the business. The actual number of people engaged in CSR depends on the size of the business and its activities. An annual report is prepared by the CSR manager and made available to the public, often via the company's website. Within the annual report new targets will be set out for the forthcoming period.

There are other considerations with sustainability reporting which relate to an international need to address climate change. For example the Global Reporting Initiative (GRI) was established in 1997 to provide business with a reporting framework to ensure consistency across and within business sectors with reference to sustainability reporting. The GRI is a non-profit coalition of over fifty groups focusing on investment, environmental issues, religion, labour issues and social justice considerations. The GRI elevated the importance of sustainability reporting to a somewhat similar level as financial reporting by identifying metrics applicable to all businesses, also sets of sector-specific metrics and a uniform format for reporting information relating to a company's sustainability performance. The GRI Sustainability Reporting Guidelines recommend specific information related to environmental, social and economic performance and are structured around a CEO statement, key environmental, social and economic indicators, descriptions of relevant policies and management systems, stakeholder relationships, and management, operational and product performance as well as a sustainability overview. For further information see http://www.globalreporting.org/Home.

There is a close affiliation between CSR and GRI, referred to as ISO 14001, which specifies a framework of control for an Environmental Management System (EMS) against which an organisation can be certified by a third party. This was established in 1996 to set out a framework for businesses to manage their operations in a more environmentally aware manner. For example ISO 14001 established a system of record keeping, auditing, reporting and managing a business to identify environmental impacts and to set targets for reductions. Such reporting has to be publicly available and therefore is typically found on company websites and also within an annual CSR or Environmental Report. Companies are required to identify key personnel within their organisation who have responsibility for managing the EMS. There now exists a whole series of ISO 14000 international standards on environmental management which provide a framework for the development of an environmental management system and the supporting audit programme. For further information see: www.iso14000-iso14001-environmental-management.com.

Discussion points

- How does the concept of sustainability relate to property 'development'?
- Consider the process of development and the direct relationship, in most instances, with the planet as the land component.

11.6 Real estate development, sustainability and CSR

CSR and property development interface in various ways. Given that CSR is related to the quality of a company's governance and strategy (including people and processes) as well as the nature and quantity of their impact on society, CSR may affect property development either in the direct employment of staff and how they operate and occupy their property or in the way the developer actually develops or redevelops property. This leads us onto the importance of developing 'sustainable property' in a sustainable manner.

One of the reasons why the property sector is directly related to sustainability is that buildings impact on the environment in many different ways, for example the built environment is responsible for approximately 40–50% of all energy use in many cities It is now generally accepted that climate change and global warming is linked to an increase of greenhouse gases in the atmosphere (EPA 2013; DECC 2013). CO_2 is the most talked about greenhouse gas which is produced during the consumption of fossil fuels such as oil and gas. Typically electricity has been generated by gas and coal fired power generation plants, therefore electricity consumption also releases CO_2 into the atmosphere. Some fossil fuels have a higher carbon content, for example black coal has a lower carbon content than brown coal and thus emissions will be lower when black coal is used to generate electricity. Global concern about the environmental impacts from burning fossil fuels has led to the development of alternative energy sources such as solar, wind and wave power. However, this technology is still relatively new and currently only supplies a relatively small amount of the energy consumed on a daily basis (i.e. for heating and light).

While developers can build a low or even zero carbon building it is the operational phase of a building's life-cycle which is crucial in its overall environmental impact; globally buildings are estimated to be responsible for approximately 40% of the total waste going to landfill sites, consume 16% of fresh water and 40% of raw materials. In addition about 25% of the global timber harvest is related to buildings. In the UK alone, 40% of the UK's energy consumption and carbon emissions come from the way our buildings are lit, heated and used (Department for Communities and Local Government 2013). In the USA, buildings are responsible for 36% of the total energy use, 65% of electricity consumption and 30% of greenhouse gas emissions (U.S. Green Building Council 2013).

These statistics clearly show that the overall environmental impact of buildings (irrespective of size) is substantial and needs to be reduced.

11.7 The case for sustainability

The environmental, social and economic arguments for sustainability in the built environment have been well proven. From an environmental perspective the Working Group II for the Fourth Assessment Report on climate change for the Inter-Governmental Panel on Climate Change (IPCC 2007) reported a high degree of confidence in concluding that mankind's, or anthropogenic, activity has led to changes in the planets' climate. The 2007 report followed on from three earlier reports which tracked changes reported in research worldwide since the 1970s. Accompanying the release of each report was the latest evidence which reaffirmed the established links between greenhouse gases and climate change. A number of working groups have continued to gather and review data in preparation for the Fifth Assessment Report to be completed in October 2014 (IPCC 2013). Such reports have encouraged many countries to adopt internationally agreed targets with regard to greenhouse gas emissions, such as the United Nations Framework Convention on Climate Change Kyoto Protocol, signed in 1997 (UNFCCC 2007). More recently (2008), the United Nations launched the UN-REDD Programme to reduce CO_2 emissions from deforestation and forest degradation (REDD) in developing countries. A broad-brush approach to tackling emissions was developed in 2009 when over 120 heads of state from at least 24 countries, who together are responsible for more than 80% of the world's global pollution, signed up to the Copenhagen Accord.[1] Under the Accord, both developed and developing nations agreed for the first time to reduce their emissions[2] and to register their national commitment by the end of January 2010. They also agreed to provide funding to assist developing countries reduce their CO_2 footprint, for example by deploying clean energy technologies, and adapting to the impacts of global warming such as flooding. The concern is that since this agreement is not legally binding it may not make a significant difference to the amount of CO_2 emissions produced worldwide and therefore global warming will continue to accelerate. Although the CO_2 targets each nation committed to varied, depending on their economic development, it is clear that there is broad acceptance of the environmental argument to support embracing a sustainable approach in society and the built environment. Many countries have also created new laws to ensure adoption of policies to reduce emissions, for example the Climate Change and Greenhouse Emissions Reduction Act 2007 (AUS) and the Climate Change Act 2008 (UK).

Of the social and environmental arguments put forward, sustainable buildings are increasingly promoted as being healthier buildings for occupants and users. One example is the use of natural materials in the design specification which leads to less off-gassing of volatile organic compounds (VOCs). Sustainable buildings tend to maximise natural daylight and fresh air, therefore improving both the internal air quality and the internal environment quality in their design which research has shown to be preferred by building users. However there is other evidence that a building which is 'too sustainable' may not be suitable for the occupiers, arguably due to features such as building management systems, which for example regulate temperature inside a building, giving individuals little or no control over the temperature of the environment they work in. A careful balance in property development needs to be struck here.

It has been shown that sustainable buildings lead to less absenteeism from workers, less churn or turnover of staff and increased productivity in workers, which in turn creates an economic argument for sustainable buildings. Another aspect of the economic argument is that sustainable buildings cost less to run. Similarly, if less waste is produced by building users and more waste is recycled, the overall running costs are reduced. Water consumption can be substantially lowered, allowing users or owners (depending on which party pays the bill) to enjoy lower water bills. However, there are claims that the initial capital outlay or construction costs of sustainable buildings have typically been higher than for non-sustainable buildings. Clearly any additional cost will vary depending on the range of sustainable measures incorporated into the design. However not all sustainable buildings are excessively more expensive and it is often possible to make some savings elsewhere in the specification to offset the cost of the sustainability features desired. In the future, when the property market recognises the added value from increased sustainability within a building in market valuations, the gap between the initial high capital cost and the current market value will lessen. There is another argument that sustainable buildings will attract higher rentals (often referred to as 'green leases') because of the perceived health and productivity gains for occupants attributed to sustainable buildings although long-term data to prove this theory is still being collected. In addition, some countries (i.e. the UK) have introduced legislation that will make it almost impossible to lease a commercial building with a low sustainability rating. This will undoubtedly result in some building stock becoming obsolete.

11.8 Sustainability and property development

It is clear that property development has direct links with sustainability issues which cannot be overlooked. We will now examine the different stages of the property development cycle and identify sustainability issues which affect each stage.

Developmental land and sustainability

The sustainability issues which affect land include loss of habitat and bio-diversity, and contamination of land either by natural causes or as a result of a previous use (e.g.

from industrial processes). Protection of natural resources is high on the agenda of many governments and is frequently linked to election pledges. The development of land with existing contamination issues can substantially add to costs and developers need to take appropriate steps to reduce their risks when acquiring land which has been previously used. In order to determine a viable solution it is often necessary to enlist the services of a contamination expert.

Development finance and sustainability

In addition to the traditional sources of finance for property development, new products are constantly emerging as financial institutions and lenders adopt and promote corporate social responsibility and risk management strategies. These developing forms of finance may require sustainable features to be a requirement of the finance package and/or offer incentives and discounts on finance for sustainable developments. For example 'eco finance' is a developing area as financial institutions integrate sustainability into their policies, practices, products and services. There is a clear awareness of the potential benefits for banks and financiers when integrating sustainability into their business strategy, where many financial institutions reported positive changes as a result of integrating social and environmental issues in their business.

Many banks and financiers have adopted environment-related policies and procedures when seeking to lend funds. An example of investment guidelines includes reference to lending-related activities guidelines such as regarding dangerous chemicals, freshwater infrastructure and forest products. Many financiers exercise caution when faced with environmentally sensitive transactions which in turn minimise the environmental, credit and goodwill risk associated with their investments. From a global perspective it is essential that banks and financiers fully consider the broader context of sustainability.

Some lenders actually screen environmental risks surrounding corporate loans to help clients improve their regulatory compliance and environmental management programmes. As environmental risks become more complex, the financial community has evolved its traditional approach. For example to reduce the exposure of commercial loans, increasingly lenders place more importance on businesses' ability to manage environmental liabilities. Environmental consultants estimate the nature and likelihood of risks and their advice informs the bank's decisions about whether to accept, avoid, manage or mitigate risks, or to seek insurance cover. This approach works when risks are quantifiable and there is a level of certainty, however the qualitative nature of many risks generates ambiguity. At times consultant panels are used on the development of contaminated land where the consultants recommend actions (i.e. conditions) to minimise risk and liability which may become a condition of the loan approval.

Certain sectors and industries are associated with high environmental risk, such as waste management, forestry and oil and gas, which sometimes involve developers. Banks and financiers rely on guidance to outline the environmental and social impacts of lending, covering risks, regulation and international best practice. The lender will identify a client's environmental risks and help reduce exposure. Where necessary loans are made conditional on the client taking measures to reduce exposure to environmental risks. Loan decisions are often informed by three risk considerations: (a) direct risk which can be affected by land contamination, (b) indirect risk related to regulatory impacts and changes, and (c) reputational risk. Larger development schemes may be considered by the banks under these provisions as well as smaller property development schemes situated on land previously used or contaminated. There is a strong argument that such policies relating to environmental considerations will continue to be developed and supported by an increasing number of banks and financiers.

Planning and sustainability

The key sustainability issues relating to planning are usually linked to:

- transport
- ecology and site issues, and
- zoning and land use issues.

Transport impacts on work, leisure and recreation patterns and on the environment in which society exists. Dependence on cars and road freight has continued to increase since the early 20th century, which has had a substantial environmental, social and economic impact. Carbon emissions produced by transport account for approximately a quarter of total carbon emissions for a developed nation. Social impacts revolve around the frequency and severity the health impacts indirectly related to inhaling emissions. Economic impacts relate to the costs of social and environmental impacts. For a sustainable environment it is important to provide a means of access in a way which has less impact on the environment. In planning terms this can be achieved by developing and implementing policies that:

1   improve and promote walking, cycling and public transport and change habits which reduce the amount of car use;
2   manage freight transport by moving to rail, canal or river, which in turn reduces reliance on heavy truck traffic;
3   make streets, bus and light rail/tram stops safer, including lower traffic speeds and better security;
4   reduce fossil fuel dependence and shift to alternative cleaner, renewable energy for transport;
5   ensure transport impacts are fully reflected in investment decisions and the costs that users actually pay; and
6   plan in an integrated manner to involve the community and link land use and transport together.

When property development schemes are considered in this framework, the typical transport issues include:

- Access to public transport access nodes and facilities. For example environmental assessment schemes advocate a proportion of a development has to be within a specific distance of peak and hourly off-peak public transport services.

- Provision of bicycling facilities and changing rooms for cyclists. Assessment schemes stipulate minimum requirements expressed as a percentage of the number of dwellings or occupants, depending on the property type.
- Proximity to local amenities such as banks, shops, pharmacy, school, medical centre, place of worship, parks, etc.

Ecology and site issues centre on the loss that results from the development of land and destroying or impacting on the local eco-systems such as flora and fauna (see for example some of the high profile media stories of developments which have impacted on ecology, such as wind farms, airport extensions or high speed railways). When an ecological feature exists on a development site then protection of this feature should be the objective of the development team who may be able to include it as a unique selling point (USP) for a particular development, e.g. rare bird habitat. Developers with proposals that will enhance the ecological value of sites are clearly welcomed by planning authorities. Therefore developers must consider the ecology and site issues at an early stage to avoid negative attention from environmental groups or unwanted media coverage. Ecological issues should also be benchmarked by measuring the ecological footprint of proposed building/s in the development.

On a regional and local level, sustainability itself can be affected by the zoning of the land and the different uses which have been envisaged by the planners. Clearly the property developer has less influence individually on the regional plans and should identify the prevailing trends in the locations or regions in which they operate. Consultation with the local authority plans will identify an authority's intentions with respect to zoning and land use issues in addition to the local authority's sustainability agenda.

Sustainable design and construction

The construction phase of property development can become environmentally friendly in several ways. First, there is the selection of the contractors on the tender list, second, environmental considerations can be incorporated into the procurement process and finally there are the activities undertaken during construction itself.

Increasingly, contractors are adopting CSR (see section 11.3) and are committed to reducing the environmental impact from development by including social responsibility in their business operations. They often post a vision statement on their website which sets out their goals and drivers in terms of environmental, social and economic sustainability factors such as the example from the Bovis Homes Group website[3] (see http://www.bovishomesgroup.co.uk/pdfs/CSR_2011.pdf). Such publicly available statements allow property developers to select contractors on the basis of their environmental credentials. In addition, these websites outline their progress towards embracing sustainability visions and goals. It is necessary to ensure that the organisations have targets that they monitor and audit. An annual sustainability report should detail relevant information about sustainability. Contractors will often illustrate their expertise by referring to case studies of sustainable building projects they have undertaken. Another approach to selecting environmentally aware contractors is based on their past performance in respect of the construction of sustainable buildings. Based on past experiences a developer can also build up a list of potential contractors with whom they work well on certain projects. Therefore this list may form the underlying basis of contractor selection for future projects.

In terms of design and build projects, the specific environmental issues will depend on variables such as the property type and specific location. Key sustainability areas that impact on the task of undertaking a property development include:

- reducing CO_2 emissions
- minimising pollution
- life cycle costing (LCC) or whole life costing (WLC), and
- using construction resources efficiently, including creating minimal waste.

Approximately 50% of all CO_2 emissions come from constructing or using buildings in the urban environment. Improving energy efficiency is an important way to reduce emissions. Relatively simple approaches include providing more insulation, more efficient glazing (e.g. double or triple glazing), introducing measures such as recovery of heat from waste water or air, and installing individual meters for heating and hot water which will also reduce the level of energy consumed, in turn leading to cost savings. In addition these measures also improve indoor air quality.

Embodied energy is generally described as the energy required to create or manufacture the materials used in the construction process. For example materials such as brick or concrete have a high embodied energy because the manufacturing process has a very high energy demand due, for example, to the heat required. However, offset against this must be the capacity for a material to retain heat which will then be released back into a building. It is necessary to adopt an environmentally friendly approach to determine the best combination of materials for a real estate development. The intended use of the property must be taken into account, along with services including the heating system and other installations to be used. It is essential to look holistically at the process rather than simply focusing only on the construction technique and materials.

Renewable energy has an important role in reducing CO_2 emissions. While many renewable technologies can be used, some are more appropriate than others for particular projects. Examples of renewable energy technologies include solar panels (photovoltaic and solar thermal), wind turbines, geothermal (for example ground source heat pumps), heat exchange systems and micro-scale hydro-generation. It must be noted the capital costs of the majority of renewable technologies have been relatively high, but as their use is becoming more widespread economies of scale have started to drive costs down. Some governments offer grants and rebates to encourage the use of renewable energy and also to offset the initial installation costs. Other incentives include feedback tariffs for any surplus electricity produced which is fed back into the main electricity grid for use by other consumers.

Pollution from the construction process can take many forms other than the pollution into the atmosphere of greenhouse gases; fuel spillages, fly-tipping, and mud/silt from sites or the wheels of transport vehicles are the most common. In addition, many construction materials can pose a pollution risk during their manufacture or in use.

Life-cycle or whole life costing is a technique which integrates the capital expenditure committed to a project with the operational costs involved in operating and maintaining the building. Clients procuring buildings will be adversely affected if their development requires expensive maintenance soon after completion. Well thought out designs which incur relatively little on-going maintenance costs are to be encouraged.

The concept of maximising efficiency by fully utilising resources is commonly viewed as the best approach for increasing levels of sustainability. However in reality this will not always deliver the best product for end-users. Therefore it is important to look to the end-users and then make decisions based on optimising not only efficiency but also the level of utility. The design phase and construction phase of development schemes are important here.

Reducing the amount of waste management during construction can both increase profitability and lower construction costs as wasted materials are an added expense. Sustainable property developers need to consider reducing waste to save money and reduce environmental impact. Developers pay to dispose of construction waste and since landfill taxes and disposal costs are increasing, minimising waste makes financial sense. A waste management strategy can help to design out waste, minimise waste creation on site and ensure any resulting waste is dealt with appropriately, which results in a tidier, safer site. It is estimated that an active use site waste management plan can produce savings approaching 3% of total construction costs.

As many developments involve demolition, a pre-demolition audit is essential for effective waste management. Contractors and or developers need to identify the type and amount of waste generated on the site. An increasingly common approach is to use a spatial mapping program incorporating a geographical information system (GIS) to locate the closest waste management site and therefore reduce costs and time associated with the transport of waste. With the same approach it is possible to locate nearby recycling sites, reclamation companies, composting facilities, manufacturer take-back schemes, transfer stations, landfill sites, and incinerators. This is an emerging area and it is vital to measure and benchmark the level of construction, refurbishment and demolition waste with industry standards. The construction industry must continue to develop minimum reporting requirements for construction, refurbishment and demolition waste and to generate performance indicators and benchmark figures. With increasing legislation being introduced in certain jurisdictions for site waste management plans (for projects exceeding a specified value), rising landfill costs and an increased focus on sustainability, it is practically essential to start measuring and monitoring waste generated. This practice is likely to become the accepted norm and developers seeking to procure sustainable construction in their projects will need to explicitly set out requirements in respect of waste management.

An alternative approach for a developer aiming to increase the sustainability of a development project is to require the designers to specify the re-use of materials, especially if the development involves the partial or complete demolition of an existing building. This is a common approach in a change of use site, e.g. from industrial to residential. Also the designer can specify the use of recycled materials such as recycled concrete for hardcore or recycled timber. In some locations there can be strict policies or legislation protecting the existing building, such as for heritage purposes, which may force the developer to keep or re-use some of the existing building, for example the façade. Re-use is better than recycling as no further embodied energy is used to physically transform the materials from one form to another. However, the use of recycled materials is clearly preferable to the consumption of raw resources and materials. Many environmental rating tools such as BREEAM give credits to projects which incorporate recycled materials.

In respect of materials the following questions should be asked during property development:

1   What is the environmental impact of the materials? All materials have varying levels of environmental impact so the choice of construction materials is important – for example excessive logging of timber can lead to deforestation and loss of the carbon sink.

2   Has responsible sourcing of materials been undertaken? For example has the timber been sourced from a sustainable re-forestation supplier?

3   Has the provision of recycling facilities for materials used within buildings during their life cycle been made? This includes internal storage and provision for external collection.

Embodied energy is directly related to sustainability and commonly defined as the total energy required to produce the construction material or product used in the development. It refers to the total energy involved in extraction or mining, transportation and manufacturing a material or product. Some materials such as concrete and steel have a high embodied energy, although the level of embodied energy in other material, for example stone and timber, is relatively low. If a developer seeks to undertake a project with low embodied energy, clearly they must avoid using large amounts of high embodied energy materials.

The amount of volatile organic compounds (VOCs) in the project is another consideration for the developer. Simply explained, VOCs are emitted as gases from certain solids or liquids and include a variety of chemicals which may have short and/or long-term adverse health effects. VOCs are emitted by a wide array of products including paints and lacquers, paint strippers, cleaning supplies, pesticides, building materials and furnishings, office equipment such as photocopiers and printers, correction fluids and carbonless copy paper, graphics and craft materials including glues and adhesives. At the very least the best practice is to avoid the use of VOCs where possible and limit exposure. Some individuals have a very low tolerance to these compounds and health effects include eye, nose and throat irritation; headaches; loss of coordination; nausea; damage to liver, kidneys and the central nervous system. Some organics are suspected or known to cause cancer in humans. Key signs or symptoms associated with exposure to VOCs include irritation, nose and throat discomfort, headache, allergic skin reaction, nausea, fatigue or dizziness.

Water

Water is a requirement for human survival yet increasing populations and growing water usage are causing water shortages throughout the world. This is obvious in densely populated areas where there is increased demand from a comparatively larger population for a reliable water supply. However in many locations a reduction in supply has coincided with increases in consumption. Water shortage also seems an obvious consequence of a drought, yet flooding, which is becoming increasingly common, impacts both water supply and wastewater services (Water UK 2011). Typically, reference to water in the context of sustainability refers either to (a) water shortage or (b) re-using existing water such as 'grey water' from washbasins or washing machines or 'black water' from toilets etc. A developer can influence occupiers and users to reduce water consumption and running costs by including such measures as water efficient appliances (e.g. low flush toilets), water metering, leak detection systems and water butts. Environmental assessment schemes measure what is considered a reasonable consumption level for a specific property type based on a benchmark amount.

Health and well-being

Increasingly the residents of many countries are spending more time each day inside buildings, either at home, in education, at work or enjoying social and leisure activities. In a global city it is estimated a typical resident may spend up to 90% of their time inside a building, which is their primary form of protection from the environment (Awbi 2013). Therefore the relevance of sustainable buildings to occupants is clear. Sustainable buildings are promoted on the basis of their benefits to health and consideration is given to optimising the health of occupants in the actual design of a project. These reasons are frequently cited as a good rationale for sustainability in buildings. With the needs of the occupiers in mind, developers should consider such features as:

1   maximising natural daylight
2   installing sound insulation to reduce the transmission of airborne and impact sound
3   providing adequate private space for occupants in residential property
4   avoiding the use of materials containing VOCs, and
5   reducing the amount of air conditioning and recycled air in buildings, to reduce the likelihood of sick building syndrome.

Market research and sustainability

There has been a significant change in attitudes towards sustainability since the early 2000s and there is now a commitment at local, national and global levels towards improving sustainability within the built environment (RICS 2013). A 2009 *Harvard Business Review* suggests that the drive for sustainability is forcing companies to adopt new approaches to business models, processes, the technologies they use and the products they produce, in other words it has become 'the key driver of innovation' (Nidumolu et al. 2009). The first global focus on sustainability in relation to property was placed on developing green buildings in the 1990s and this started a significant change in attitudes towards sustainability in the built environment. Surveys show that many people believe the environment to be increasingly important. Marketing firms also conduct primary consumer research to investigate the percentage of the population for whom environmental, social, and healthy lifestyle values play an important role in purchasing decisions. Findings from this type of research have observed a shift in consumer attitudes towards embracing sustainability and it is these general trends and perceptions which filter down to property development.

Property developers are an integral part of society and need to keep abreast of social and cultural trends especially relating to sustainability to ensure their properties meet market requirements, reduce environmental impacts and are future proofed to some extent. Although there are varying perceptions of sustainability, depending on which section of society is questioned, surveys show that society already expects a certain level of sustainability to be incorporated in the products we use – for example energy-efficiency in domestic appliances (e.g. refrigerators) and lower emission/more economical motor vehicles. An increasing focus on CSR has also raised the profile of environmentally friendly products and services. These changing perceptions have directly and indirectly affected the property market where the environmental impact is considered to be considerable. Clearly then a 21st-century developer needs to be aware of these changing attitudes and perceptions about sustainability in order to ensure their products meet market expectations and do not become victims to a new type of building obsolescence, namely environmental obsolescence.

Marketing a sustainable development

A developer needs to identify their target market and identify the optimal marketing strategy to reaching that specific market sector and it is essential to include features in the development which are attractive to that section of the market. It has become increasingly common to use a two-tier marketing strategy for a residential project, for example where one advertising campaign is targeted at a mature purchaser and a second campaign is focused on a younger or single household group. Different sustainability features might appeal more to different groups – for example an older resident might prefer a closer proximity to public transport or healthcare facilities, although a younger household might prefer access to recreational facilities such as gyms or fitness clubs. A marketing campaign focused on a sustainable property development will seek to highlight the social, economic and environmental benefits of buying or renting in the particular development. It has been increasingly observed in the marketplace that marketing campaigns now promote the features of sustainable buildings due to the higher level of purchaser demand for this product and the perceived benefits from occupation.

Promoting and selling sustainability

There is evidence that property developments are being promoted to the market and sold on their sustainability credentials as a core selling feature or advantage over a competing development. As property value is linked to perception, this approach has capitalised on the positive buyer perception of sustainability. For example, in the residential sector a higher volume residential property developer can promote a development as being sustainable because it is located closer to the city centre, as well as being energy and water efficient. With reference to the commercial sectors, office developments are commonly promoted on the basis of their sustainability rating. There are many regions where government departments will only commit to buildings which have minimum environmental standards (e.g. 4.5 star 'Greenstar'). Developers keen to promote their developments as complying with these standards will need to ensure they keep their current sustainability marketing information up to date.

In reality it has been argued that the task of 'selling sustainability' is an oxymoron. This is because sustainability is about conserving resources and reducing environmental impact, however selling is about increasing consumption. There are challenges with selling sustainability features like energy efficiency because consumers are often unable to see any physical benefit from their investment, unlike for example, a new kitchen. This barrier to the adoption of sustainable products or processes appears to be diminishing as levels of awareness are increased and benefits such as reduced running costs are accepted in society. Such arguments are easier to make when energy and fuel prices are increasing and sustainable options become more financially viable. So how can the two conflicting concepts be resolved? Clearly some degree of development is inevitable to cater for the changing needs of the community and changes in demographics and, vitally, to maintain economic activity (e.g. creating employment). In today's property market it is both appropriate and necessary to promote developments on the basis of their level of sustainability. It can also be argued that the promotion of sustainable buildings is highly desirable to further raise the awareness of sustainability in buildings to a wider audience.

As with any marketing and promotion campaign, it is necessary to engage specialists to identify target groups and markets for the product. Advertising campaigns in a range of media can then be developed to target each group effectively and create demand for the development. The key property development stages are summarised in Table 11.1 with the main potential sustainability issues also noted.

Table 11.1 Property development stages and key potential sustainability issues

Property development stage	Potential sustainability issue
Land for development	• Loss of habitat • Loss of bio-diversity • Contamination (naturally occurring or due to previous use)
Development finance	• Consider eco financing
Planning	• Transport • Ecology and site issues • Zoning and land use issues
Design and construction	• Selection of contractors – including their CSR performance • Reducing CO_2 emissions • Minimising pollution • Use of lifecycle costing or whole life costing techniques • Using resources efficiently • Waste management on site • Reuse of materials and recycling materials • Specification and selection of materials – health and embodied energy issues • Water • Health and well-being for users • Environmental assessment ratings
Market research	• Awareness of changing social and cultural perceptions towards sustainability
Promotion and selling	• Awareness of changing social and cultural perceptions towards sustainability

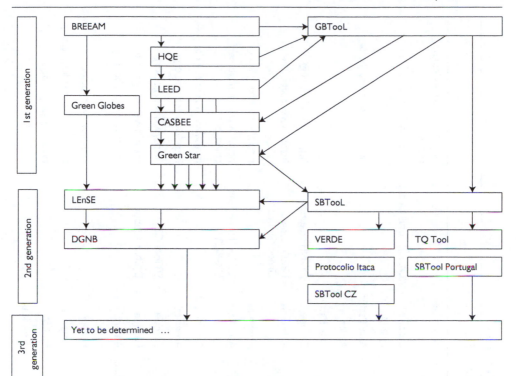

Figure 11.3 Evolution of sustainability rating tools (Source: adapted from 'Aufbau internationaler Bewertungsmethoden' in *Abhängigkeit voneinander, Zertifizierungssysteme für Gebäude*, p.25, Thilo Ebert, Nathalie Eßig, Gerd Hauser, DETAIL Green Books 2010)

11.9 Sustainability rating tools

Although there are no identical parcels of land in the world and every country is unique in its approach to land use, development and sustainability, there are common approaches to the valuation of land and buildings and an analysis of those values to identify trends etc. However, environmental rating tools have not followed this trend and generally different countries will have their own measurement tools and rating standards. Although it is possible to compare the value of an office building in New York, Berlin, London or Melbourne which may be valued on a ten-year discounted cash flow valuation approach (after allowing for exchange rates), making a similar direct comparison of the sustainability rating of the same building is complex. The first widely recognised environmental assessment method and rating system for buildings was BREEAM, which set a standard for best practice in building design, construction and operation. BREEAM was arguably the catalyst for sustainability changes within the property industry as well as the development of new sustainability rating systems like LEED, CASBEE and Greenstar. The second generation rating tools combined components from different systems of the first generation. The next step in this process is the arrival of third generation rating tools (Figure 11.3).

Table 11.2 Green Building Councils and rating tools (Source: Reed and Krajinovic-Bilos 2013)

	GBC	Membership Level	Country / Economic Zone	Region	Tools	
1	Argentina Green Building Council	Established	Argentina	Americas		http://www.argentinagbc.org.ar/
2	Austria Green Building Council	Prospective	Austria	Europe	OGNI, Total Quality	http://www.ogni.at/
3	Bahrain Green Building Council	Prospective	Bahrain	MENA		
4	Bangladesh Green Building Council	Associated Group	Bangladesh	Asia-Pacific		http://www.banglagbc.org/
5	Botswana Green Building Council	Associated Group	Botswana	Africa		
6	Brunei Green Building Council	Associated Group	Brunei	Asia-Pacific		
7	Bulgarian Green Building Council	Emerging	Bulgaria	Europe	DGNB adaption	http://www.bgbc.bg/
8	Canada Green Building Council	Established	Canada	Americas	LEED Canada, Green Globes	http://www.cagbc.org/
9	Chile Green Building Council	Established	Chile	Americas		http://www.chilegbc.cl/
10	China Green Building Council	Associated Group	China	Asia-Pacific	GBAS, Three Star, HK-BEAM	http://www.chinagb.net/chinagbc/
11	Colombia Green Building Council	Established	Colombia	Americas		http://www.cccs.org.co/
12	Costa Rica Green Building Council	Prospective	Costa Rica	Americas		http://www.crgbc.org/
13	Czech Green Building Council	Prospective	Czech Republic	Europe	SBTool CZ	http://www.czgbc.org/
14	Denmark Green Building Council	Associated Group	Denmark	Europe	DGNB adaption	http://www.dk-gbc.dk/
15	Dominican Republic Green Building Council	Prospective	Dominican Republic	Americas		http://www.drgbc.org/
16	Dutch Green Building Council	Established	Netherlands	Europe		http://www.dgbc.nl/
17	Ecuador Green Building Council	Associated Group	Ecuador	Americas		http://www.ecuadorgbc.org/
18	Egyptian Green Building Council	Associated Group	Egypt	MENA		http://egypt-gbc.org/
19	El Salvador Green Building Council	Associated Group	El Salvador	Americas		

No.	Council	Status	Country	Region	Schemes	Website
20	Emirates Green Building Council	Established	United Arab Emirates	MENA	LEED Emirates, BREEAM Gulfs	http://www.emiratesgbc.org/
21	Estonia Green Building Council	Associated Group	Estonia	Europe		
22	France Green Building Council	Established	France	Europe	HQE, Escale, BREEAM France	http://www.francegbc.fr/
23	German Sustainable Building Council	Established	Germany	Europe	DGNB, BNB, BREEAM Germany	http://www.dgnb.de/
24	Ghana Green Building Council	Prospective	Ghana	Africa		http://www. ghgbc. org/
25	Green Building Council Australia	Established	Australia	Asia-Pacific	NABERS, Green Star	http://www.gbca.org.au/
26	Green Building Council Bolivia	Prospective	Bolivia	Americas		http://www. gbcbolivia. org/
27	Green Building Council Brasil	Established	Brazil	Americas	LEED Brasil, BREEAM Brasil, AQUA	http://www. gbcbrasil. org. br/
28	Green Building Council Espana	Established	Spain	Europe	VERDE, BREEAM Spain	http://www.gbce.es/
29	Green Building Council Finland	Emerging	Finland	Europe	PromisE	http://www.figbc.fi/
30	Green Building Council Italia	Emerging	Italy	Europe	Protocollo Itaca	http://www.gbcitalia.org/
31	Green Building Council Nigeria	Prospective	Nigeria	Africa		
32	Green Building Council of Croatia	Emerging	Croatia	Europe		http://www.gbccroatia.org/en/
33	Green Building Council of Georgia	Prospective	Georgia	Europe		
34	Green Building Council of Sri Lanka	Prospective	Sri Lanka	Asia-Pacific		http://srilankagbc.org/
35	Green Building Council Slovenia	Prospective	Slovenia	Europe		http://www. gbc-slovenia. si/
36	Green Building Council South Africa	Established	South Africa	Africa		http://wwwgbcsa.org.za/

continued…

Table 11.2 continued

	GBC	Membership Level	Country / Economic Zone	Region	Tools	
37	Guatemala Green Building Council	Emerging	Guatemala	Americas		http://www.guatemalagbc.org/
38	Hellenic Green Building Council	Prospective	Greece	Europe		http://www.elgbc.gr/
39	Honduras Green Building Council	Associate Group	Honduras	Americas		
40	Hong Kong Green Building Council	Established	Hong Kong China	Asia-Pacific	HK-BEAM, Three Star	http://www. hkgbc. org. hk/
41	Hungary Green Building Council	Emerging	Hungary	Europe		http://www. hugbc. org/
42	Icelandic Green Building Council	Associated Group	Iceland	Europe		http://www.vbr. is/efni/english
43	Indian Green Building Council	Established	India	Asia-Pacific	LEED India, TGBRS India	http://www.igbc.in/
44	Indonesia Green Building Council	Emerging	Indonesia	Asia-Pacific		http://www.gbcindonesia.org/
45	Irish Green Building Council	Prospective	Ireland	Europe		http://www.igbc.ie/
46	Israel Green Building Council	Established	Israel	Europe		http://www.ilgbc.org/
47	Japan Sustainable Building Consortium	Established	Japan	Asia-Pacific	CASBEE	http://www.ibec.or.ip/CASBEE/english/index.htm
48	Jordan Green Building Council	Established	Jordan	MENA		http://www. jordangbc. org/
49	Kenya Green Building Council	Prospective	Kenya	Africa		
50	Korea Green Building Council	Prospective	South Korea	Asia-Pacific		http://www. koreasbc. org/
51	Kuwait Green Building Council	Prospective	Kuwait	MENA		
52	Latvian Sustainable Building Council	Prospective	Latvia	Europe		http://www.ibp.lv/
53	Lebanon Green Building Council	Prospective	Lebanon	MENA		http://www.lebanon-gbc.org/

	Council	Status	Country	Region	Tool	URL
54	Libya Green Building Council	Associated Group	Libya	MENA		
55	Luxembourg Green Building Council	Associated Group	Luxembourg	Europe		
56	Malaysia Green Building Confederation	Emerging	Malaysia	Asia-Pacific		http://www.mgbc.org.my/
57	Mauritania Green Building Council	Associated Group	Mauritania	Africa		
58	Mauritius Green Building Council	Prospective	Mauritius	Africa		http://www.gbcm.mu/
59	Mexico Green Building Council	Established	Mexico	Americas	LEED Mexiko, SICES	http://www. mexicogbc.org/
60	Montenegro Green Building Council	Prospective	Montenegro	Europe		http://www. greenbuildingcouncil.me/
61	Morocco Green Building Council	Prospective	Morocco	MENA		http://www.moroccogbc.org/
62	Namibia Green Building Council	Prospective	Namibia	Africa		
63	New Zealand Green Building Council	Established	New Zealand	Asia-Pacific	Green Star NZ	http://www.nzgbc.org.nz/
64	Norwegian Green Building Council	Associated Group	Norway	Europe	BREEAM Norway	
65	Oman Green Building Council	Prospective	Oman	MENA		
66	Pakistan Green Building Council	Prospective	Pakistan	Asia-Pacific		
67	Palestine Green Building Council	Prospective	Palestine	MENA		
68	Panama Green Building Council	Emerging	Panama	Americas		http://www.panamagbc.org
69	Paraguay Green Building Council	Associated Group	Paraguay	Americas		
70	Peru Green Building Council	Established	Peru	Americas		http://www.perugbc.org.pe/
71	Philippine Green Building Council	Prospective	Philippines	Asia-Pacific		http://philgbc.org/
72	Polish Green Building Council	Established	Poland	Europe	BREEAM Poland	http://www.plgbc.org.pl/
73	Portugal Green Building Council	Associate Group	Portugal	Europe	Lider A, SBTool Portugal	

continued….

Table 11.2 continued

	GBC	Membership Level	Country / Economic Zone	Region	Tools	
74	Qatar Green Building Council	Emerging	Qatar	MENA		http://www.gatargbc.org/
75	Romania Green Building Council	Established	Romania	Europe		http://www.rogbc.org/
76	Russia Green Building Council	Emerging	Russia	Europe	BREEAM Russia	http://www.rugbc.org/green-building-blog/lang/en/
77	Saudi Arabia Green Building Council	Prospective	Saudi Arabia	MENA		http://www.saudigbc.com/
78	Serbia Green Building Council	Emerging	Serbia	Europe		http://www.serbiagbc.org/
79	Singapore Green Building Council	Established	Singapore	Asia-Pacific	BCA Singaour Green Mark	http://www.sgbc.sg/
80	Slovak Green Building Council	Associate Group	Slovakia	Europe		http://www.skgbc.org/
81	Sweden Green Building Council	Established	Sweden	Europe	BREEAM Sweden	http://www.sgbc.se/
82	Swiss Sustainable Building Council	Prospective	Switzerland	Europe	MINERGIE; DGNB adaption	
83	Syrian Green Building Council	Prospective	Syria	MENA		http://www.syriangbc.org/
84	Taiwan Green Building Council	Established	Chinese Taipei	Asia-Pacific	ABRI	http://taiwangbc.org.tw/
85	Trinidad and Tobago Green Building Council	Prospective	Trinidad	Americas		
86	Tunisia Green Building Council	Associated Group	Tunisia	MENA		
87	Turkish Green Building Council	Established	Turkey	Europe		http://www.cedbik.org/
88	UK Green Building Council	Established	United Kingdom	Europe	BREEAM	http://www.ukgbc.org/
89	Ukraine Green Building Council	Prospective	Ukraine	Europe		
90	Uruguay Green Building Council	Prospective	Uruguay	Americas	LEED, Green Globes	http://www.uygbc.org/
91	US Green Building Council	Established	Usa	Americas		http://www.usgbc.org/
92	Venezuela Green Building Council	Associated Group	Venezuela	Americas		
93	Vietnam Green Building Council	Associate Group	Vietnam	Asia-Pacific		http://vgbc.org.vn/

Development of rating tools

The current era of rating tools commenced in 1990 with the introduction of the BREEAM rating tool, and this was followed five years later by the French system, HQE, and by LEED in 1998. The evolution and adoption of rating systems in different countries is largely based on the initial rating systems; for example see BREEAM (Netherlands), LEED (United Arab Emirates) and Green Star (South Africa). Some countries have more than one rating tool, however other countries have yet to develop or adopt their own – see Table 11.2.

In an era of international property development and investment where it is possible to compare valuations of buildings in different countries, it is unfortunate that rating tools do not exhibit the same level of comparability due to their unique characteristics and specific focus. Arguably this may potentially hinder the take-up rate of sustainable rating tools and be a barrier to increasing knowledge about the sustainable attributes of comparable office buildings located in different countries. A comparison of the uptake of the BREEAM Europe tool and Australia's Green Star showed that the impact of the global financial crisis was not apparent in the office sector for Europe in 2008 and 2009, however results for 2010 and beyond may possibly show lower growth and uptake levels in sustainable property. European countries with high population densities appear to have embraced the sustainable office rating tools. Uptake of the retail and industrial tools is lower, partly reflecting the later development of the tools and the overall number of new buildings constructed in the sector. As the Green Star tool does lead to lower levels of sustainability compared with BREEAM, one challenge is to increase the proportion of 6 star rated buildings, which in turn will contribute to addressing climate change in the built environment. This remains the single greatest challenge and continues to increase in importance at the same time as the new rating tools enter the market and contribute to the increasing complexity for all stakeholders.

Finally there is nothing to stop a property developer obtaining a bespoke assessment for a property development scheme. This can make it less easy to promote and convince the market that the bespoke building is as, or more, sustainable because the recognised benchmarks have not been adopted. Bespoke assessments may be more useful to those developers or owners wishing to retain a long-term interest in the building or scheme.

Discussion points

- What are the main environmental assessment tools available to property developers?
- Why would property developers use these tools?

11.10 Reflective summary

We are now in an environment where developers must embed sustainability in their development at some level. This chapter commenced with an overview of sustainability and how it relates to CSR. The question of why property is an important sector in relation to sustainability and CSR was addressed and the key areas of impact were identified. Finally the chapter discussed how these impacts are being addressed by governments and by developers, investors and occupiers. Rating tools were discussed, especially from an international perspective to highlight the increasing complexity of sustainable buildings.

It is impossible to ignore sustainability as some issues are embedded within legislation, such as planning and building regulations, for example energy efficiency and water economy. However developers should be aware that these regulations represent the minimum allowable standards and not best practice. Developers should adopt best practice wherever possible, including sustainability of course. The standards and quality of the built environment should be improved for both inter- and intra-generational benefits. Sustainability can be embedded throughout the development process from inception to site selection and acquisition to the financing of the scheme. The design and procurement phase is another key area where decisions will have a substantial impact on the sustainability of a project. Environmental assessment tools enable benchmarks to be set which the market recognises and acknowledges. The importance of market research is highlighted with the need to consider the different types of sustainability features that would appeal especially to the target groups that a particular project would be aimed at. In all areas of property development and at all stages in the process, the trends are for more sustainability. Over time the sustainability tools adopted by the industry will improve, thereby allowing developers to deliver more sustainable buildings to the market.

11.11 Case study – Library of Birmingham, UK

The public library in this case study was built for Birmingham City Council as a flagship project to form part of a wider city centre regeneration scheme and was completed in 2013. It is situated in Centenary Square and connects to the Birmingham Repertory Theatre which shares some of the new library facilities (Plate 11.1). Table 11.3 gives some statistics about the development.

Background

The decision to build a new library in Birmingham city centre was made over a decade ago. The first proposal, designed by Richard Rogers, was for a new library to be built in the emerging Eastside district next to the science museum and the city's new green park. However, due to financial constraints and concerns about the suitability of the location (Eastside is on the fringe of city centre) this plan was abandoned. The second proposal was to split the library between two buildings; the archives and special collections would be housed in a new building situated in Eastside, the main lending library would be built on the site of an existing car park in Centenary Square, close to the location of the old library; this plan was also shelved. The final decision was to create one building in Centenary Square which would be linked to the existing Repertory Theatre. This building would include a

Plate 11.1 View of the library from Centenary Square (Source: reproduced with kind permission from the Library of Birmingham, UK)

Table 11.3 Library of Birmingham vital statistics

- Client: Birmingham City Council
- Owner: Birmingham City Council
- Architect: Francine Houben from Mecanoo Architecten b.v.
- Project engineers: Buro Happold
- Project manager and cost consultant: Capita Symonds
- Contractor: Carillion
- Size: 30,000 m^2
- Floor area: 20,798 m^2. Additional floor area shared with the Birmingham Repertory Theatre: 6,804 m^2
- Number of floors: 10 floors including UG
- Building height: 60 m (200 ft)
- Predicted operational energy: 160 kWh/m^2/a
- Regulated carbon emissions: 19 kgCO$_2$/m^2/a
- EPC rating: B
- BREEAM rating: Excellent
- Construction period: January 2010 – September 2013
- Total build cost: £188.8 million

300-seat studio theatre, meeting and conference rooms, cafe and restaurant facilities that could be used by both library and theatre visitors. It was felt that with the two organisations working in partnership, the library could become home to a diverse range of activities which could combine literature, research, arts and community events.

Plate 11.2 Aerial view of the new library development in Centenary Square, Birmingham (Imagery © Google, DigitalGlobe, Getmapping PLC, Infoterra and Bluesky, Landsat, The GeoInformation Group)

Location

The new library building is located between the 1960s concrete-built Birmingham Repertory Theatre (the REP) and the 1936 Art Deco-style listed Baskerville House (Plate 11.2). The site is owned by the council and had long been earmarked for a special, high quality new development. It was used as a car park until July 2009.[4]

Choosing the design

The Royal Institution of British Architects held an international design competition with over 100 entrants. In August 2008, architects Mecanoo and multi-discipline engineers Buro Happold were announced as the winners. In 2010, Daden Ltd produced a virtual model of the building from architects' plans. Library staff used the model to plan the layout of the internal features and facilities of the new building and to engage with the community. It also enabled existing plans to be modified. According to Daden, the virtual model was also used by contractors to 'better understand the build and inform their bids',[5] and help them to determine the exact positioning of fixtures and fittings.

Planning permission

Ground clearing for the project started before planning permission had been granted. The initial design was modified slightly to reduce the height by one level in response to consultation and other elements of the design were refined, such as the amphitheatre and its relationship between the library and Centenary Square. Planning permission was granted in November 2009 subject to a number of conditions.

Site conditions

The development site adjoins a listed building with shallow foundations and close to an underground railway tunnel. The site was previously a canal basin and there was some soil contamination. The ground consisted of soft natural materials overlying rock. The basement level was set to be at rock surface level which allowed the contaminated canal infill material and soft ground material to be removed, eliminating any costly excavation into the rock to construct the building's foundations.

The foundations consisted of pad footings on the rock for smaller column loads which reduced cost, and bored cast in-place piles for larger loads, reducing the number of piles required thus speeding up this part of the development programme. The water table was below basement level and not predicted to rise within the life of the building. Hard landscaping surrounding the building meant that only a small amount of surface water was likely to permeate to the ground surrounding the basement level and a permeable contiguous piled retaining wall was constructed with draining cavity to collect and drain off any rainwater.

The design

The main building materials are steel, concrete and glass. The exterior of the building is constructed from 31,000 square miles (80,290 square kilometres) of glass and steel[6] which is covered in a dual overlay of interlocking metallic circles. Each metallic circle was hoisted onto the building by crane and affixed to a bespoke steel bracket.[7] The metallic circles were designed from the perspective of looking out from within the library building and create ever-changing shadows inside the glass building. The interior is formed from a sequence of eight circular, interlocking rotundas across several floors, each providing natural light and ventilation and connected by a series of travelators and escalators. The book rotunda is situated at the core of the building and is home to over 1.5 million books and 14 miles (24 kilometres) of shelves for archive and heritage collections. In addition to a number of conference, meeting or teaching rooms, there is a 300-seat auditorium, a flexible studio theatre, an outdoor amphitheatre (Plate 11.3) and a recording studio.

The roof garden reduces surface run off. The two terrace gardens (3rd floor and 7th floor), which overlook the city and the amphitheatre, have both been designed to be family-friendly, interactive outdoor spaces. The front entrance is covered by a cantilevered canopy that forms a shared entrance for both the library and the Repertory Theatre. This space is also used for outdoor performances.

The studio theatre, located between levels 0 and 2, needed to be acoustically sealed from the rest of the library to prevent noise transference. This was achieved by constructing a separate structure supported on acoustic bearings. Levels 3 to 7 of the main library structure which overhang the theatre void were supported by using hangers from transfer walls on levels 5 and 6. The building was designed to be an exemplar low energy and sustainable building, minimising the impact on the environment and reducing energy use by 50% over the previous library building; it

Plate 11.3 View of the library from the amphitheatre (Source: reproduced with kind permission from the Library of Birmingham, UK)

achieved a BREEAM 'Excellent' rating. The total reduction of CO_2 emissions for the development is around 40%. Environmentally friendly features include the use of low carbon technology for heating and cooling, low energy lighting and water conservation systems and a building management system (BMS) monitors and controls the temperature, air, lighting and energy consumption of the various systems throughout the building.

Sustainable features

- *Low energy lighting:* The library was fitted with a versatile lighting system to optimise energy efficiency. A lighting management system was required to meet a range of criteria, including optimising energy performance and integrating the management of general and emergency lighting. One of its more important features is that it provides the flexibility to control individual luminaires, as the library will host many different types of event, often with several running simultaneously. To meet these requirements the lighting control system has been configured to a multi-floor philosophy to deliver maximum flexibility. Energy issues are addressed through a range of control strategies, including timed operation, absence/presence detection and daylight linking, most of which can be controlled through a supervisory PC.
- *Ground source aquifer cooling system:* This is a free renewable resource that utilises ground water to provide chilled water to the air conditioning system, which supplements the energy required for cooling the building and further

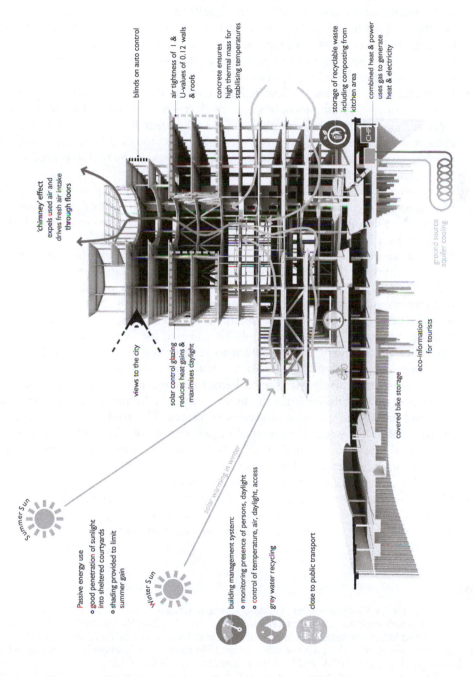

Summer Sun

Winter Sun

Passive energy use
- good penetration of sunlight into sheltered courtyards
- shading provided to limit summer gain

building management system:
- monitoring presence of persons, daylight
- control of temperature, air, daylight, access

grey water recycling

close to public transport

solar warming in winter

views to the city

'chimney' effect expels used air and drives fresh air intake through floors

solar control glazing reduces heat gains & maximises daylight

blinds on auto control

air tightness of 1 & U-values of 0.12 walls & roofs

concrete ensures high thermal mass for stabilising temperatures

storage of recyclable waste including composting from kitchen area

combined heat & power uses gas to generate heat & electricity

ground source aquifer cooling

eco-information for tourists

covered bike storage

Plate 11.4 Sustainable design features (Source: Reproduced with kind permission from Buro Happold)

reduces its energy consumption and CO_2 emissions.[8] This was achieved by drilling two 150 metre deep bore holes to check the suitability of the area for the aquifer system. This confirmed the water flow rate was suitable. Using a production well, groundwater is abstracted from the aquifer and sent to the air conditioning unit to cool the building. Following heat rejection via a heat exchanger in the building, the groundwater is then discharged back to the aquifer via an injection well. Constant groundwater temperature can thereby be maintained throughout the year with this open loop system.

- *Combined heat and power plant (CHP):* CHP is the use of a heat engine or power station to simultaneously generate electricity and heat. The shaft from a gas turbine produces electricity and the excess heat from the engine is used to heat water. The excess heat is also converted through a heat exchanger to produce chilled water. The CHP system on site provides the building with a low carbon source of electricity and hot water.
- *Grey water harvesting:* The water generated from the sinks and showers is harvested and treated on site before being recycled for use in flushing toilets. This reduces the on-site demand for potable water, allowing maximum BREEAM scores to be achieved in this category.
- *Mixed mode ventilation:* This refers to the inclusion of a hybrid approach to space conditioning that uses a combination of natural ventilation from controlled vents in the façade at levels 1, 2 and 3 and a mechanical system that includes distribution and refrigeration equipment for air conditioning. Natural ventilation is used wherever possible to reduce energy use from heating and cooling systems.
- *Planting of terrace gardens and roof:* To enhance the ecological value of the development, the client was very keen to reintroduce flora and fauna into the site to encourage ecological diversity and wildlife back into the area. Landscape architects worked closely with the ecologist to ensure that all species were native to the area and would enhance the biodiversity. The building has incorporated bird and bat boxes and several species of plants. The rooftop incorporates a brown roof (a non-seeded green roof system that allows local plant species to populate the roof over time). Raise plant beds have been incorporated into the two outdoor terraces. The intent is for both these terraces to be publicly accessible but to have contrasting atmospheres and qualities.[9]
- *Alternative renewable energy:* The use of solar panels (thermal and photovoltaic) and wind turbines in addition to other technologies was considered but rejected due to the location of the building or because they did not have a realistic payback period.

Notes

1 U.N. Framework Convention on Climate Change. United Nations. 18 December 2009. Retrieved from http://unfccc.int/meetings/copenhagen_dec_2009/items/5262.php [accessed 7/6/13] and http://www.nrdc.org/international/copenhagenaccords/ [accessed 6/6/13].

2 The USA agreed to reduce CO_2 emissions by 17% below 2005 levels by 2020, by 42% below 2005 levels by 2030, and by 83% below 2005 levels by 2050.

China has committed to reduce its carbon dioxide emissions per unit of GDP by 40–45% from 2005 levels and use non-fossil fuels for about 15% of its energy. It has also committed to increasing forest cover.

India has committed to reduce its emissions per unit of GDP by 20–25% below 2005 levels by 2020. One of the measures to achieve this will be to increase the amount of electricity generated from wind, solar, and small hydro from the current 8% to 20% by 2020.

The EU announced a target to reduce emissions to 20% below 1990 levels by 2020 and would increase their commitment to 30% if other countries would commit to more ambitious efforts.

3 CSR policy from Bovis Homes Group. Retrieved from http://www.bovishomesgroup.co.uk/pdfs/CSR_2011.pdf [accessed 10/6/13].

4 Birmingham.gov.uk 2009. 'Where will the library of Birmingham be built?' Retrieved from http://libraryblog.birmingham.gov.uk/2009/09/where-will-the-library-of-birmingham-be-built-3/ [accessed 12/12/13].

5 Daden Ltd 2010. 'Library of Birmingham Case Study' Presentation. Retrieved from http://www.daden.co.uk/downloads/Library%20of%20Birmingham%20-%20Case%20Study%20Presentation.pdf [accessed 14/12/13].

6 Wills J. (no date) 'Carillion: building Birmingham's flagship library'. Retrieved from http://www.theguardian.com/sustainable-business/best-practice-exchange/carillion-building-birmingham-flagship-library [accessed 14/12/13].

7 Library of Birmingham 2011. 'Our striking new façade is installed'. Retrieved from http://libraryblog.birmingham.gov.uk/2011/08/ [accessed 12/12/13].

8 Birmingham.gov.uk (no date). 'Library of Birmingham archive'. Retrieved from http://www.birmingham.gov.uk/cs/Satellite?c=Page&childpagename=Lib-Library-of-Birmingham%2FPageLayout&cid=1223409909623&pagename=BCC%2FCommon%2FWrapper%2FWrapper [accessed 12/12/13].

9 Terry Perkins 2011. Head of Project Delivery for the Library of Birmingham. Retrieved from http://libraryblog.birmingham.gov.uk/author/terry-perkins/ [accessed 12/12/13].

Chapter 12

Emerging markets

12.1 Introduction

This chapter discusses property development in the broader context of internationalisation, created by the increasing interaction between national economies and the globalisation of businesses as a result of access to fast, reliable information due to advances in communications technology. The concept of globalisation is described and then explained within the broader context of property. The opportunities for international property development are examined, along with the drivers for international property development.

We also provide an overview of the opportunities, barriers and risks associated with international property development and outline a possible international strategy before moving on to specific examples of international property development.

12.1.1 International property trends

There has been an observed global trend towards higher levels of urbanisation which has often focused on the provision of living space within city centres. In some countries, previous planning policies and high land values led to the practice of converting residential property to office use and, as a result, many city centres in the 1990s became devoid of residential property causing many inner cities to become lifeless areas outside working hours.

Strategies to promote inner-city 'urban regeneration' represent a complete U-turn in land use and have been introduced as targeted initiatives in many global cities to rejuvenate and bring city centres back to life. Indirect advantages from increasing the residential component have included lower crime rates and increasing 'liveability' profiles which have in turn attracted both residents and visitors. These policies and initiatives have led to sustained growth in inner-city residential apartment buildings in global cities ranging from Frankfurt to Chicago and Melbourne to Birmingham.

Other external factors influence land use changes in cities and subsequent change in property development requirements. For example in some cities during downturns in property cycles and associated periods of high vacancy rates in commercial building stock, there have been conversions of high-rise property from office use to residential use – this is due to the overall higher rate per square metre for residential land use over commercial. This is a complex change of use and was relatively uncommon prior to the mid-1990s with inner-city locations typically containing office and retail buildings and residents living in outlying suburbs. However, some companies have chosen to locate in outlying office 'nodes' (business/science parks) at the same time as some residents sought inner-city locations to be close to their place of work, central transport facilities and CBD retailers. Interestingly this is an example of a global property trend replicated at approximately the same time in Canada, the USA and the UK. Such trends demonstrate the increasing levels of internationalisation of property, therefore affecting development practice and property markets at many levels. The use of 'agile' buildings is another example of an international property development trend which has now been globally adopted.

Property developers need to keep up to date with other new trends. For example one feature of changing global practice in the property industry is the relatively rapid growth in purchasers of investment property 'off the plan' from overseas investors who live in another country. Most of these purchasers are individuals and not large investment funds. For example, many seaside locations are heavily promoted in overseas markets to investors living in Asia. An indirect result can be the emergence of a new two-tiered market consisting of local buyers and overseas buyers – in this scenario the property developer, who is seeking to obtain the highest value for the completed units in the project, should market the development to both the local and international market with different targeted promotional campaigns. In the EU for example, many property transactions have occurred where 'cashed up' generation X baby boomers have been looking to purchase holiday homes or second homes as retirement properties that are sound investments. Another influencing factor has been the widespread global uptake of internet access by prospective purchasers, regardless of their location in the world. This trend has clearly made it easier for overseas investors to research property in other countries and proceed directly to purchasing, often site unseen. Arranging property transfer details (e.g. engaging a solicitor, transferring funds) is simplified due to the internet. At the same time an increased number of property agents have been able and willing to facilitate these international property transactions.

In this era of globalisation and inter-connected economies an economic event in one country can have an adverse and almost immediate effect on other global markets, especially when referring to exchange rates, equity markets and interest rate levels. Since the early 1980s, the volatility in property cycles has been increasingly caused not only by various conditions in a particular country's economy but also by global macro-economic conditions (Dehesh and Pugh 2000). It is commonly accepted that the diversity between products offered for sale in different countries has decreased. Many organisations conduct business by operating on different continents, where the actual location of the head office is often difficult to quickly identify. The expansion of these multi-national companies and their internal communication requirements have been substantially assisted by advances in technology including the internet (e.g. Skype, FaceTime, enabling free face-to-face discussions). Therefore, these companies, especially construction companies and property developers, have diversified their primary geographic location of operation in order to benefit from enhanced profit (partly due to lower operating costs with a local office and labour force) in accordance with conventional economic theory.

Profit-seeking companies and associated free market trade have led to the massive expansion of outsourced labour, such as the widely acknowledged increase in the number of call centres based in India. In other words, international companies such as telecoms are able to provide telephone help lines at a much lower cost due to cheaper labour and infrastructure costs when sourced off-shore (e.g. rather than in the UK). This in turn this has caused an expansion in demand for property development in specific countries which have embraced this trend and there is now a need to provide high-tech office space (i.e. with high speed internet cabling) which was not previously in demand (or built) in these countries.

Previously local real estate markets were always viewed as being primarily influenced by local economic activities, although with the increasing acceptance of internationalism this perception has changed over time. However, property developers have not benefited from entering global markets as much as other organisation types, even though investment in and development of real estate has traditionally been linked to wealth creation and achieving investment goals. In the past there have not been any property development companies among the largest 100 transnational companies ranked by foreign assets in the world, as this list is dominated by oil, electronic and automobile companies (Hailia 2000).

In the context of property development, the local company does not have the advantage that it once held and is no longer the company most likely be the successful bidder. After consideration is given to the scope of the property development, for example if the project is a large-scale football stadium or a small detached house, in today's competitive environment the developer is not automatically located in the immediate or local vicinity. There is a need to develop an understanding of property development processes which combine a sensitivity to the economic and social framing of development strategies with 'fine-grain' treatment of the local contingent responses of property actors (Guy and Henneberry 2000).

Following the integration of technology and the internet (i.e. utilising the worldwide web) throughout the western world, many geographical boundaries that previously existed have now been removed. For example it is practically impossible to determine whether the person you are talking to is located in Australia, India, the UK or the USA and this is certainly irrelevant for companies whose customer service is provided by phone or online. Therefore, the head office of a property development project may be located across town, across the country or across the world and still be able to compete directly with a local provider. Such increased levels of international practice have caused higher levels of competition, where a truly international property developer must be up to date with the local customs, government regulations and current state of the market in many different areas. The consultants who service property developers, such as services engineers, architects, agents and quantity surveyors for example, have also become internationalised. Sometimes this has taken the form of mergers and acquisitions of local practices, while at other times partnerships have been formed. This expansion and internationalisation has meant that consultants who work on projects in one particular country may find themselves working together in project teams in other countries at the same time.

It is necessary to discuss the concept of globalisation because it has had a substantial impact on international property development. This impact looks set to increase. Globalisation refers to the increasing worldwide connections, integration and interdependence in political, economic, environmental, social and technological areas of interest. Globalisation is an undeniable, established and irreversible trend as a result of the lowering or removing of political and trade barriers, the advent of fast, safe and cheap air travel, rapid development in new technologies, and a very high level of productivity form emerging economies such as India and China. The volume of world trade has been increasing at an accelerating rate (World Trade Organization 2013). Historically when large amounts of capital flowed across national borders usually it was deposited into the equity or government bond markets; now however there is a global real estate market in which to invest. The search for higher returns has also increased international property investment.

There is a debate about whether globalisation results in a homogenised society and culture with converging patterns of consumption or whether society can retain individual cultural differences in the long term. It has meant that identical products and services are found in an increasing range of countries and this includes property developments. There is certainly evidence of increased architectural similarity with property in global cities – for example office buildings in Toronto look much like office properties in Sydney, Frankfurt or Beijing. Such practices are further encouraged by global companies insisting that their branding is adopted globally. For example the food retailer McDonalds uses similar generic design and fit-outs for its properties throughout the world. Other examples are international hotel chains which use identical specifications in their buildings as a selling point for customers.

Economically the concept of globalisation is about the convergence of products, wages, prices, interest rates and profits towards the norms in developed countries and the International Monetary Fund (IMF) has commented on the increasing economic interdependence through the growing amount of cross-border transactions, international capital flows and more rapid and widespread diffusion of technology. However, globalisation has a wide range and number of impacts which are summarised as:

- industrial
- financial
- economic
- political
- informational
- cultural
- ecological
- social, and
- technical and legal.

All of the above impacts can affect property developments. For example one cultural impact of globalisation is that there is a raised consciousness and awareness among the peoples of the world who desire to consume foreign products and ideas. People nowadays want to participate in a world culture and increased overseas travel has added to the phenomenon. For example, in Australia European-style kitchens are considered very sophisticated and up-market in residential property developments and the marketing and promotion of the developments highlights these sought-after specifications. Equally many developments market aspects of 'feng shui'. This is the ancient Chinese practice of placement and arrangement of space to achieve harmony with the environment. Feng shui is a discipline with guidelines that are applied in architecture and property design. Advocates claim that feng shui affects health, wealth and personal relationships. Some property developers are promoting and marketing their projects stating that the principles of feng shui have been adopted throughout the design of the scheme.

The result of globalisation has been that many large corporations have become trans-national firms where companies focus on staying competitive by outsourcing services or production to developing countries with very poor labour, environmental and economic standards. Such business practices allow companies to economise, leading to larger returns for investors and cheaper services and products for consumers. There is an increasingly strong argument that such practices encourage governments in developing nations to retain poorly paid labour (and associated working conditions) and resist implementing environmental legislation. Some international corporations actively lobby governments to gain entry to developing countries, though the advent of CSR is beginning to curb some companies' less than exemplary overseas activities. These companies seek to avoid any negative publicity which results from associations with poor business and environmental practices in developing countries.

The globalisation environment is complex and countries actively seek to maintain their economic advantage. This means that many developed countries still have protectionist policies that prevent developing countries from exporting to developed markets. Critics of the expanding global economy argue that the reduction of trade barriers will create higher levels of competition for previously protected companies in developing countries, while its advocates point to the new possibilities in global markets for emerging market companies such as Haier (China) and the Tata Group (India). In 2006, the global economy continued to expand so that most individual, corporate and government borrowers are following through on their obligations, which in turn has kept financial markets performing well and thus property developers are working to meet market demands for new and or improved facilities.

12.2 Globalisation of property development

With reference to property development there are implications from increasing levels of globalisation. According to the Global Investment Strategy section of LaSalle Investment Management, capital flows in real estate around the world in 2005 were US$700 billion, 20% higher than in 2004. US$126 billion – 18% of the total – was in cross-border transactions with Asian countries which was the top market for foreign investment. To establish some context, US$700 billion is equivalent to 700 buildings of 1 million square feet each at a unit price of US$1000 per square foot. This is the volume of real estate that changed hands across national borders in only a single year.

On a more national scale, this level of activity has meant that many property developers now operate in countries other than just their own. Some have established second or satellite offices in other countries, whereas others operate solely from overseas locations and no longer have sole country operations. There has been a substantial growth in investing in and purchasing property in multiple countries and developers are taking advantage of the opportunities this presents. For example, UK buyers overseas are generally happier dealing with UK nationals who are able to explain the process of property acquisition and development in an overseas setting. The expansion of this area of property development business has been assisted by the growth and economic integration of the European Union which has led to much increased pan-European activity in the last decade. Similarly property developers from countries outside the UK will operate within the UK if the economic circumstances and business opportunities present themselves.

To give an example of the market for overseas property buyers and investors, consider residential investment properties or retirement/second homes for individuals. The UK Office of National Statistics (ONS) has the most reliable figures about UK ownership of properties outside the UK. However, even their figures are not perfect as they take account only of the number of households owning property overseas and do not take account of households owning more than one property overseas and that is a sizeable number. The UK Office for National Statistics (ONS) produced figures on 'Ownership of other property, 2008/10' and these showed 2.9% of respondents owned 'land or property overseas', being a slight decline from 3.0% as observed in 2006/08. Furthermore the increase in residents working overseas selling and promoting these developments encourages buyers because there is a sense that they will trust people from their own country. If residents buying and investing property abroad were dealing solely with indigenous or local people they would come across language and cultural barriers which would deter the more cautious buyers from completing the deal.

To cater for transnational businesses and cross-border developments, there are three key ways in which property developers can obtain general economic and property market information. First, an effective way is to use international property consultants since many well-known property companies now operate in multiple countries and have an international profile. They have local employees who know their market well and can advise developers on all aspects of property development in their locality. Furthermore these consultants have access to networks of other allied consultants that can assist with the process.

Second there are companies which provide independent, accurate, comprehensive, and up-to-date research on industries operating in their country. The data includes reports providing statistics, analysis and forecasts. These companies may prepare reports on the nationally best performing companies and provide reports with risk ratings on different industries. With increasing globalisation there has been continuing interest in investment in overseas markets. According to a report by PricewaterhouseCoopers (2013) there have been renewed levels of interest in international investment and higher risk strategies have returned or are being explored; for example large global investors were moving beyond safe cities and it was suggesting some of the gateway centres were offering good secondary investment opportunities. Another example in the same report was the belief that US investors were seeking investment in the UK due to yield compression in their home market, which in turn makes the UK market more appealing. These are example of the high level of connectivity in the property and real estate market, which has moved closer to the high level of connectivity in the equities market. Brazil, Mexico, Russia, China and India will produce five times as much real estate in the next ten years as will be created in the United States. Access to such data illustrates where opportunities will lie for international property developers. This type of information will enable developers to make more informed and therefore better business decisions.

Finally, a third way for developers to find out about market conditions and opportunities in countries outside their home country is to use affiliated professional bodies where available. For example the RICS (Royal Institution of Chartered Surveyors) is a global professional body representing land, property and construction. Members of the RICS operate in many countries and can offer a range of professional advice and services to property developers.

12.3 Development opportunities

Many property developers following the global market are seeking opportunities to identify markets which have potential for future growth or are currently under-developed. In many instances this will require the developer to be an early adopter in the marketplace, rather than waiting until the market reaches maturity with many property developers competing for the limited supply of prospective sites. Hence a developer with foresight to enter a growing market has the benefit of rapidly establishing goodwill and strong links with the local market, as opposed to entering a competitive mature market from a standing start.

At the same time as property developers have been expanding globally, there has been a parallel increase in global investment. Since ownership regulations and differential taxation are not large barriers and sources of segmentation in different property markets, many larger buildings in western civilisations are owned by international investors. It should be noted that the mix of international investors changes over time. The UK property sector has been the destination of substantial portfolio investment with the main investors from Japan, the USA, Sweden, Germany, the Netherlands, the Middle East and Asia, and the devaluation of sterling in 1992 was associated with a sharp increase in investment by German funds (McAllister 2000). However, although London still retains high levels of demand from foreign investors this trend is no longer unique to Britain and investors are seeking property internationally to invest in.

At times other countries will offer opportunities for higher returns than the host country of the property developer, although this will vary according to property cycles and conventional supply and demand interaction. Such opportunities for overseas companies in the construction industry in Asia were identified in the 1990s. The main trends observed were:

- larger private sector participation in infrastructure projects
- increasing vertical integration in the packaging of construction projects, and
- increased foreign participation in domestic construction (Raftery et al. 1998).

Since then such trends have continued in Asia and many overseas companies have benefited from the Asian development boom, especially in China. The gradual 'opening up' of previously 'closed' economies has been encouraging news to international property developers, although caution should be exercised before undertaking a large capital commitment without conducting adequate due diligence. For example in many countries there are still difficulties in accessing reliable and timely detailed property data, such as the volume of sales and actual transfer prices. Some countries still have their supply of property largely controlled by the government and this should be carefully monitored. Operating in an environment that does not operate freely can increase the exposure to risk.

International property development affects each country in a different way and to a different extent. For example, a developing country's fledging firms may confront formidable competition from corporations possessing far greater economies of scale, and global networks. However, the same argument is support by the belief that less developed countries have accessible hidden reserves of labour, savings and entrepreneurship. These strengths and weaknesses should be identified by the progressive property developer, thus opportunities do exist in many countries with each in a varying stage of development itself. For example, partnering has been occurring for quite some time between British and Chinese firms in some Asian countries (including China). This trend appears likely to continue into the foreseeable future, partly due to the increasing population base in this region.

In certain countries it has been noted that geographical clusters have formed. Cluster development has occurred rapidly in North America, Europe and newly industrialised countries where a group of similar companies (e.g. IT and computer related services) are located in a specific area. Cluster development has attracted the attention of different countries seeking to establish an identity, which in turn may provide an opportunity for an international developer who is able to bring specialised skills. Clearly each country, and in particular each region in a country, can promote a limited number of clusters. In this example a specialist property developer will have the competitive advantage of assisting to develop a cluster based on their previous experience in other countries.

A driver for establishing global links is the desire to be recognised as a truly international property company. Many organisational models are possible, ranging from (a) a management structure with the majority of the workforce sourced from the local economy, to (b) relocating an entire workforce from another country with associated relocation and housing costs. Initially the property developer will seek to reduce costs as well as their overall exposure to risk, therefore relocating the minimum number of workers required for the initial property development will often be the preferred option.

Discussion points

- What are some of the drivers encouraging the expansion of property development into other countries?
- Why are developing countries so appealing to property developers?

12.4 Barriers and limitations

The transition to a truly international property market has been faced with many challenges, especially when considering differences in currency, culture and varying levels of development in each country. As some real estate markets move through the transitional stage to a truly market-based structure, it is important that valuations (i.e. the business case for property development) align with the expectations and requirements of the international investment community (Mansfield and Royston 2007).

Knowledge about the inside workings of a property market can be one of the largest barriers to a successful property development in an overseas country. In contrast to the type of information that is freely available in general circulation in many developed countries, especially in terms of reliability, accessibility and cost, property-related information may simply not be as available as anticipated. Even if the information is available, there may be a premium attached for 'non-locals' and questions raised about the level of reliability. When considering the high importance placed on the final sale price of the property development, some companies venturing into overseas countries have been disadvantaged by this lack of local knowledge.

Cultural barriers may exist, for example if a property company used to conducting business in a western country seeks to develop property in a country with a transitional or emerging economy. Depending on the different cultures, it may take time to establish a strong trust between the property developer and the local property stakeholders. It has been demonstrated that strong trust-based relationships create advantages in conducting business, such as lowering cost, shortening duration and improving performance (Bromiley and Cummings 1993). These are all attributes that a property developer in an international marketplace is keen to improve upon. For example, the Chinese culture comprises certain core values such as trust and *guanxi* (relationship) that influence business operations; in such an environment it may be essential for project participants to know what risks are inherent in relationships, and what tools for fostering trust and managing risks should be adopted (Jin and Ling 2005). On the other hand, some overseas contractors can be accused of poor performance and low effectiveness in terms of quality and performance.

Language and cultural barriers can vary significantly between countries, although to varying degrees. For example in one study it was shown that there were no signs that differences between Swedish and UK company cultures have an adverse impact on knowledge transfer (Bröchner et al. 2004). The same study concluded that rich media, preferably face-to-face meetings, are perceived to be good for the transfer of knowledge; leaner media such as email was also seen to be efficient. In other words the language barriers can be partly broken down using media and technology including email and teleconferences. Hence communication with a distant geographical location can be maintained on a regular basis if accompanied by adequate forward planning and organisation.

12.4.1 International property development risks

The increasing internationalisation of property markets has increased the level of demand for property, although arguably it has at times exposed property investors to more risk. One view is that the international construction industry can be characterised as highly volatile, subjecting contractors to financial and geopolitical risks (ENR 2003).

For example after the Paris office market fell by two-thirds in value between 1990 and 1995, North American hedge funds were the first to enter the market and were skilled in investing against the business cycle (Nappi-Choulet 2006). These funds then benefited from large capital gains from 1995 to 1999, although this was closely followed by a 40–50% increase in rent for prime office space. This example highlights the risks associated with understanding demand for property in an overseas market, especially with respect to planning an office development. For example if a developer had examined the Paris office market (a) prior to 1990, (b) between 1990 and 1995, or (c) after 1995, they would have found three completely different development situations. After factoring in the external influence of the hedge funds in this example, it can be easily seen that a full appreciation of the international market is essential.

An international property developer is exposed to different types of risk, some of which may not occur if the developer only operates in their home country. Structural risk may come from within the property development industry itself, although growth risk may occur from the anticipated growth of the overall property market. External risk can result from forces external to the development but once again outside the control of the property developer. It is accepted that risk cannot be fully removed from the project, although identifying the varying types and levels of risk will assist to understand the threats to successful completion of the property development.

Establishing trust between stakeholders involved in the property development is absolutely critical when seeking to undertake a project in an emerging market. The importance of this relationship is emphasised when the actual property development site is located at a remote location, or at least not in the vicinity of the head office. It has been shown that as a relationship with an international partner progresses, more inherent risks are produced – in these instances trust-fostering tools are required (Jin et al. 2005). Furthermore as these tools are applied at different stages of the property development, trust develops, which in turn counterbalances the risks. This, in turn, reduces the risks associated with achieving a successful and timely completion of the project.

Many regions promote a 'buy local' culture which may indirectly create barriers for companies that are perceived as 'outsiders'. To overcome such barriers it is important to consider the culture and environment of the area surrounding the property development. Employing local workers and subcontractors, where possible, will partly overcome these concerns and improve local relations with the developer. For large projects a public relations expert may be employed to ensure interested stakeholders are kept up to date with important facets of the development, as well as emphasising how much the project will contribute to the local economy.

Another approach for addressing local responsiveness is to organise a strategic alliance with a local partner who is located in the region. The strategy presented in Figure 12.1 is designed to highlight alternative approaches to balancing global expansion with the local market. Pressures for global integration occur when a property developer is selling a standardised good or service with little ability to differentiate its products through features or quality. According to Griffin and Pustay (2007) the four strategies are as follows:

1   *Global strategy* occurs when pressures for global integration are high but the need for local responsiveness is low, such as the expansion of Japanese consumer goods into global markets.
2   *Transnational strategy* is when both global integration pressures and local responsiveness pressures are high, such as a producing a worldwide motor vehicle although designed to meet local market specifications.
3   *Home replication* is adopted when pressures for global integration and local responsiveness are low, for example a retailer who sells the same commodities successfully to all global markets.

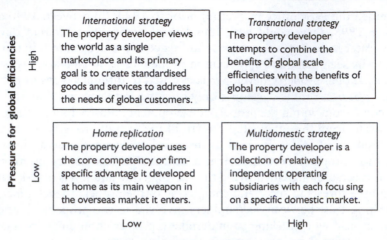

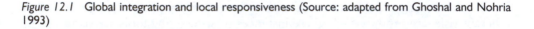

Figure 12.1   Global integration and local responsiveness (Source: adapted from Ghoshal and Nohria 1993)

4   *Multi-domestic strategy* occurs when the response to local conditions is high but pressures for global integration are low, such as where global producers sell a product known worldwide at a premium to the local market using local market resources.

Currency risk is an important consideration when undertaking a venture on an international level, such as when payments are linked to the US dollar which will fluctuate over the period of the property development. Although careful analysis can be given to the standard risks associated with a typical property development, operating in another country can dramatically increase exposure to new risk factors (currency risk, future exchange rates and more recently the threat from conflict). Other indirect risks can also be adversely affected when working in a foreign country, such as transport risks to a distant location where there is a heavy reliance on air travel. For example, travel by air is relatively expensive and may be affected by irregular services. In contrast to a local property development that is readily accessible by road, air travel is subject to flight schedules and availability with associated waiting periods of days. An urgent crisis may suffer adversely if the correct person cannot be on-site due to transport risks.

The relevant legislation and political climate are also major considerations that may hinder a prospective international property development. Although these factors are completely outside the control of the developer, careful research should be undertaken prior to entering the marketplace to ensure the developer is aware of the likely risks and can at least consider some contingency plans in the event of a particular event occurring. After completing extensive research, if a cost-benefit analysis indicates that the proposal is not viable due to the added risk involved, the project should not be pursued. Exceptions to this decision may occur, such as when the company is willing to accept a higher level of risk in order to make a strategic high profile entry into the market with a long-term perspective. The added risk should be factored into all aspects of the analysis (i.e. not only the overall level of profit and risk), including sourcing of local labour and local materials, timely completion of the project as well as demand for the finished property development, i.e. either rent or sale.

There is a view that because of the changes to weather patterns and changes in sea levels (often related to debate about 'climate change'), there will be a need for considerable international infrastructure and development in the future. Rising sea levels will create the need for improved and/or new flood defences or even the relocation of communities which are no longer viable in their current location. In some countries (e.g. Australia), planning applications have already been declined as a direct result of projected higher sea levels. Equally there will be a need for changes to the ways in which energy is produced and this may involve changing coal fired power stations to gas fired power or the construction of new power plants (for example the recently announced plan to build a nuclear power plant in the UK) or using renewables such as wind farms for example. However such developments take several years to progress from inception to completion. There remains a view that action needs to be taken immediately and therefore the extensive planning consultations previously employed may need to be shortened in the drive to reduce carbon emissions. Arguably this would partly address the effect of global warming and climate change.

Discussion points

- Identify some of the risks involved when undertaking property development in an emerging market.
- What is a strategic alliance and when is it helpful?

12.5 Developing an international strategy

A decision to develop a global profile must be accompanied by a well-planned and executed strategy. According to Griffin and Pustay (2007), the five independent steps listed below must be undertaken, as shown in Figure 12.2.

- *Step 1* – Develop a mission statement for the property developer that clarifies the organisation's purpose, value and directions. This is a means of communicating with internal and external constituents and stakeholders about the company's strategic direction.
- *Step 2* – Undertake environmental scanning and a SWOT (strengths, weaknesses, opportunities and threats) analysis. An environmental scan is a systematic collection of data about all elements of the property developer's external and internal environments, including markets, regulatory issues, competitor's actions, production costs and labour productivity.
- *Step 3* – Set strategic goals – the major objectives the developer wishes to accomplish through pursuing a particular course of action. Importantly they should be measurable, feasible and time restricted.
- *Step 4* – Develop specific tactical goals or plans which normally focus on the details of implementing the property developer's strategic goals.
- *Step 5* – A control framework is required – the set of organisational and managerial processes which keep the property developer moving towards its strategic goals.

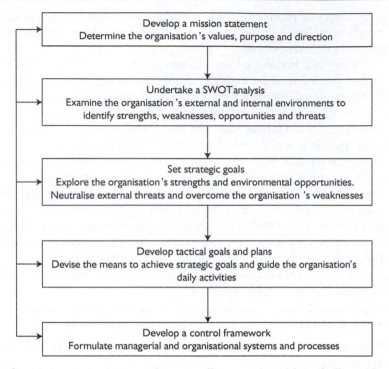

Figure 12.2   Steps in international strategy formation (Source: adapted from Griffin and Pustay 2007)

12.6 Reflective summary

A property developer can no longer work in isolation and is already exposed to global risk. Many property developers have long-term plans to expand their operations beyond their country's borders. Although this will provide new opportunities, the downside is that the increased risk (for example from a lack of knowledge about local conditions) will require the developer to be more diligent at the planning stage. Nevertheless the extent and scope of world trade has increased substantially in the 20th and 21st centuries and economic events in one country directly impact economically on other nations – the fluctuating exchange rate is a good example.

A successful property developer fully understands that a considerable amount of investment occurs in property across national borders. Globalisation has homogenised market expectations to a substantial degree in many areas, including property. For example the standards and specifications in respect of the commercial office, international hotel and retail property sectors are highly aligned in the developed countries. It is no longer a surprise that a property developer's head office may be located in a different country.

To date many property developers have embraced the concept of global business and have adopted an international perspective. This has presented many opportunities for progressive property developers who have expanded their horizons. Globalisation has provided opportunities for property developers to meet increased market requirements and demands in all property sectors in their home country as well as providing new opportunities for undertaking property development outside of their home country. Many of the aspects and theoretical approaches to property development remain unchanged (e.g. design and timing aspects); however, other aspects (e.g. labour and material supply) differ substantially. In many regions the relaxation of trade barriers has helped to encourage international property development. In addition the growth of information technology and the internet has made the operation and management of the property development process much easier to undertake, regardless of whether the developer is located in their home country or elsewhere. In some niche areas, opportunities have arisen for commercial property development supporting the expansion of firms internationally and also in the residential property development sector to service the growing numbers of buyers seeking to purchase homes outside their home country. It is now evident that buyers are becoming more used to the concept of buying and owning property outside their country of residence and as a result these buyers are becoming more confident of not only buying 'off the plan' without seeing the property but also purchasing or renting in other countries.

It is essential to undertake extensive market research prior to completing a successful international property development where access to local knowledge and expertise 'on the ground' is essential. Different emerging markets offer different opportunities within the various market sectors for property development, although the level of each opportunity varies considerably over time and should be monitored.

Along with additional opportunities there are numerous tangible and intangible risks and barriers. The key barriers are those related to language and cultural issues, local knowledge and practices, along with the benchmarking of relevant local standards of property development. This is also directly applicable to the adherence to relevant legislation which differs substantially between regions and countries. Property developers must appreciate that subtle yet significant differences in the way business is conducted can make working outside of national boundaries a challenging yet rewarding experience. Property developers who ignore or under-estimate cultural issues do so at their peril. It must be remembered that a lack of local market knowledge and market volatility can lead developers to experience much lower returns and expose them to much higher risk than anticipated. Many regions promote a 'buy local' culture, so it is important to consider the culture and environment of the area surrounding the property development. Employing local workers where possible will partly address these concerns, and for large projects a public relations expert may be employed to ensure interested stakeholders are kept up to date with important facets of the development.

Appendix
CPI by country (2004–2012)

Source: World Bank (2013), Inflation Consumer Price (%),
http://www.worldbank.org (accessed 27/11/13)

Country name	2004	2005	2006	2007	2008	2009	2010	2011	2012
World	3.59	4.14	4.42	5.03	9.01	2.92	3.55	4.98	3.69
Afghanistan		12.15	3.49	16.95	22.71	−13.23	12.88	5.69	6.80
Albania	2.28	2.37	2.37	2.93	3.36	2.28	3.55	3.45	2.03
Algeria	3.96	1.38	2.31	3.67	4.86	5.73	3.91	4.52	8.89
Angola	43.54	22.96	13.30	12.25	12.47	13.73	14.47	13.47	10.29
Antigua and Barbuda	2.03	2.10	1.79	1.42	5.33	−0.55	3.37	3.46	3.38
Armenia	6.96	0.64	2.89	4.41	8.95	3.41	8.18	7.65	2.56
Aruba	2.53	3.40	3.61	5.39	8.96	−2.14	2.08	4.38	0.56
Australia	2.34	2.67	3.54	2.33	4.35	1.82	2.85	3.39	1.76
Austria	2.06	2.30	1.45	2.17	3.22	0.51	1.81	3.27	2.49
Azerbaijan	6.71	9.68	8.37	16.60	20.79	1.40	5.86	7.85	1.06
Bahamas, The	0.98	1.59	2.39	2.49	4.49	2.06	1.34	3.17	1.99
Bahrain	2.35	2.59	2.01	3.26	3.53	2.80	1.96	−0.36	2.75
Bangladesh	7.59	7.05	6.77	9.11	8.90	5.42	8.13	10.71	8.74
Barbados	1.39	6.08	7.31	4.03	8.11	3.64	5.82	9.43	4.53
Belarus	18.11	10.34	7.03	8.42	14.84	12.95	7.74	53.23	59.22
Belgium	2.09	2.78	1.79	1.82	4.49	−0.05	2.19	3.53	2.84
Belize	3.09	3.64	4.23	2.32	6.39	−1.08	5.58	−3.65	1.31
Benin	0.87	5.36	3.78	1.30	7.95	2.16	2.31	2.71	6.75
Bhutan	9.41	5.31	5.00	5.16	8.33	4.36	7.04	8.85	10.92
Bolivia	4.44	5.39	4.29	8.71	14.00	3.35	2.50	9.81	4.59
Bosnia and Herzegovina			6.13	1.52	7.42	−0.39	2.19	3.67	2.05
Botswana	6.95	8.61	11.56	7.08	12.70	8.03	6.95	8.46	7.54

Country name	2004	2005	2006	2007	2008	2009	2010	2011	2012
Brazil	6.60	6.87	4.18	3.64	5.66	4.89	5.04	6.64	5.40
Brunei Darussalam	0.81	1.24	0.16	0.97	2.08	1.04	0.36	2.02	0.46
Bulgaria	6.35	5.04	7.26	8.40	12.35	2.75	2.44	4.22	2.95
Burkina Faso	−0.40	6.42	2.33	−0.23	10.66	2.61	−0.76	2.76	3.82
Burundi	7.85	13.52	2.81	8.34	24.11	10.98	6.40	9.74	18.01
Cambodia	3.92	6.35	6.14	7.67	25.00	−0.66	4.00	5.48	2.93
Cameroon	0.23	2.01	5.12	0.92	5.34	3.04	1.28	2.94	2.94
Canada	1.86	2.21	2.00	2.14	2.37	0.30	1.78	2.91	1.52
Cape Verde	−1.89	0.42	5.37	4.41	6.78	0.98	2.08	4.47	2.54
Central African Republic	−2.07	2.88	6.70	0.93	9.27	3.52	1.49	1.30	5.77
Chad	−5.36	7.89	8.04	−8.97	10.30	9.95	−2.08	−4.90	10.25
Chile							1.41	3.34	3.01
China	3.88	1.82	1.46	4.75	5.86	−0.70	3.31	5.41	2.65
Colombia	5.90	5.05	4.30	5.54	7.00	4.20	2.28	3.41	3.18
Comoros	4.47	3.01	3.37	4.47	1.70	4.36	3.35	1.77	1.77
Congo, Dem. Rep.	3.99	21.32	13.05	16.95	17.30	728.67	85.07		
Congo, Rep.	2.43	3.09	6.54	2.66	7.33	5.30	5.00	1.33	3.89
Costa Rica	12.32	13.80	11.47	9.36	13.42	7.84	5.66	4.88	4.50
Cote d'Ivoire	1.44	3.89	2.47	1.89	6.31	1.03	1.68	4.91	1.31
Croatia	2.05	3.32	3.21	2.86	6.09	2.40	1.04	2.25	3.42
Cyprus	2.29	2.56	2.50	2.37	4.67	0.37	2.38	3.29	2.39
Czech Republic	2.83	1.85	2.53	2.93	6.35	1.04	1.41	1.94	3.30
Denmark	1.16	1.81	1.89	1.71	3.40	1.33	2.30	2.76	2.41
Djibouti	3.12	3.10	3.48	4.97	11.96	1.68	3.95	4.39	7.88
Dominica	2.39	1.68	2.59	3.24	6.33	0.03	3.21	2.39	1.44
Dominican Republic	51.46	4.19	7.57	6.14	10.64	1.44	6.33	8.46	3.69
Ecuador	2.74	2.41	3.03	2.28	8.40	5.16	3.56	4.47	5.10
Egypt, Arab Rep.	11.27	4.87	7.64	9.32	18.32	11.76	11.27	10.05	7.12
El Salvador	4.45	4.69	4.04	4.58	6.71	1.06	1.18	5.13	1.73
Equatorial Guinea	4.22	5.63	4.42	2.80	6.55	4.69	7.79	6.95	6.15
Estonia	3.05	4.09	4.43	6.60	10.37	−0.08	2.98	4.98	3.93
Ethiopia	3.26	12.94	12.31	17.24	44.39	8.47	8.14	33.22	22.77
Fiji	2.83	2.37	2.49	4.80	7.73	3.69	5.54	8.67	4.33

Country name	2004	2005	2006	2007	2008	2009	2010	2011	2012
Finland	0.19	0.86	1.57	2.51	4.07	0.00	1.22	3.42	2.81
France	2.13	1.74	1.68	1.49	2.81	0.09	1.53	2.12	1.96
Gabon	0.41	3.71	−1.41	5.03	5.26	1.89	1.46	1.27	2.66
Gambia, The	14.21	4.84	2.06	5.37	4.46	4.55	5.05	4.80	
Georgia	5.66	8.25	9.16	9.24	10.00	1.73	7.11	8.54	−0.94
Germany	1.67	1.55	1.58	2.30	2.63	0.31	1.10	2.08	2.01
Ghana	12.62	15.12	10.92	10.73	16.52	19.25	10.71	8.73	9.16
Greece	2.90	3.55	3.20	2.90	4.15	1.21	4.71	3.33	1.50
Grenada	2.23	3.53	4.26	3.86	8.03	−0.31	3.44	3.03	2.41
Guatemala	7.58	9.11	6.56	6.82	11.36	1.86	3.86	6.22	3.78
Guinea		31.37	34.70	22.84	18.38	4.68	15.46	21.35	15.21
Guinea-Bissau	0.88	3.33	1.95	4.62	10.46	−1.65	2.52	5.05	2.13
Guyana	4.67	6.93	6.58	12.30	8.10	2.91	2.09	4.98	2.39
Haiti	22.81	15.73	13.07	8.53	15.52	−0.01	5.70	8.41	6.28
Honduras	8.11	8.81	5.58	6.94	11.41	5.49	4.70	6.76	5.20
Hong Kong SAR, China	−0.45	0.90	2.11	1.96	4.26	0.61	2.34	5.26	4.06
Hungary	6.78	3.55	3.88	7.94	6.07	4.21	4.88	3.96	5.71
Iceland	3.15	4.00	6.68	5.06	12.68	12.00	5.40	3.99	5.19
India	3.77	4.25	6.15	6.37	8.35	10.88	11.99	8.86	9.31
Indonesia	6.24	10.45	13.11	6.41	9.78	4.81	5.13	5.36	4.28
Iran, Islamic Rep.	14.76	13.43	11.94	17.21	25.55	13.50	10.14	20.63	27.34
Iraq	26.96	36.96	53.23	−10.07	12.66	6.87	2.88	5.80	
Ireland	2.19	2.43	3.94	4.88	4.05	−4.48	−0.95	2.58	1.69
Israel	−0.41	1.33	2.11	0.51	4.60	3.33	2.69	3.46	1.71
Italy	2.22	2.00	2.07	1.82	3.38	0.75	1.54	2.74	3.04
Jamaica	13.63	15.30	8.59	9.29	22.02	9.57	12.61	7.53	6.90
Japan	−0.01	−0.27	0.24	0.06	1.37	−1.35	−0.72	−0.28	−0.03
Jordan	3.36	3.49	6.25	5.39	14.93	−0.68	5.01	4.41	4.77
Kazakhstan	6.88	7.58	8.59	10.77	17.15	7.31	7.12	8.35	5.11
Kenya	11.62	10.31	14.45	9.76	26.24	9.23	3.96	14.02	9.38
Korea, Rep.	3.59	2.75	2.24	2.53	4.67	2.76	2.96	4.00	2.21
Kosovo	−1.06	−1.39	0.62	4.36	9.35	−2.41	3.48	7.34	2.48
Kuwait	1.25	4.14	3.06	5.48	10.58	3.97	4.02	4.75	2.92

Country name	2004	2005	2006	2007	2008	2009	2010	2011	2012
Kyrgyz Republic	4.11	4.35	5.56	10.18	24.52	6.90	7.97	16.50	2.69
Lao PDR	10.46	7.17	6.80	4.52	7.63	0.04	5.98	7.58	4.26
Latvia	6.19	6.74	6.53	10.11	15.40	3.53	−1.09	4.38	2.25
Lebanon						1.19	3.99		
Lesotho	5.02	3.44	6.07	8.01	10.72	7.38	3.60	5.02	6.10
Liberia	7.83	10.83	7.34	11.39	17.49	7.43	7.29	8.49	6.83
Libya	−2.20	2.65	1.46	6.25	10.36	2.46	2.80	15.52	6.07
Lithuania	1.18	2.64	3.75	5.73	10.93	4.44	1.33	4.13	3.08
Luxembourg	2.23	2.49	2.68	2.30	3.40	0.37	2.27	3.41	2.66
Macao SAR, China	0.98	4.40	5.15	5.57	8.62	1.17	2.81	5.80	6.11
Macedonia, FYR	0.93	0.16	3.22	2.24	8.27	−0.74	1.61	3.90	3.31
Madagascar	13.81	18.51	10.77	10.30	9.22	8.96	9.25	9.48	6.36
Malawi	11.43	15.41	13.97	7.95	8.71	8.42	7.41	7.62	21.27
Malaysia	1.52	2.96	3.61	2.03	5.44	0.58	1.71	3.20	1.66
Maldives				7.37	12.26	3.98	4.74	14.85	11.29
Mali	−3.10	6.40	1.54	1.41	9.17	2.46	1.11	2.86	5.43
Malta	2.79	3.01	2.77	1.25	4.26	2.09	1.52	2.72	2.42
Mauritania	10.37	12.13	6.24	7.25	7.35	2.22	6.28	5.64	4.94
Mauritius	4.71	4.94	8.93	8.80	9.73	2.55	2.89	6.53	3.85
Mexico	4.69	3.99	3.63	3.97	5.13	5.30	4.16	3.41	4.11
Moldova	12.48	11.77	12.87	12.14	12.90	−0.11	7.35	7.74	4.68
Mongolia	8.24	12.72	5.10	9.05	25.06	6.28	10.15	9.48	14.98
Montenegro			2.92	4.35	8.76	3.47	0.65	3.18	
Morocco	1.49	0.98	3.28	2.04	3.71	0.99	0.99	0.92	1.28
Mozambique	12.66	7.17	13.24	8.16	10.33	3.25	12.70	10.17	1.15
Myanmar	4.53	9.37	20.00	35.02	26.80	1.47	7.72	5.02	1.47
Namibia	4.15	2.26	5.05	6.73	10.35	8.78	4.47	5.05	6.54
Nepal	2.84	6.84	7.56	6.10	10.91	11.61	9.98	9.55	9.45
Netherlands	1.24	1.67	1.17	1.61	2.49	1.19	1.28	2.35	2.45
New Zealand	2.29	3.04	3.37	2.38	3.96	2.12	2.30	4.43	0.88
Nicaragua	8.47	9.60	9.14	11.13	19.83	3.69	5.45	8.08	7.19
Niger	0.26	7.80	0.04	0.05	11.31	0.58	0.80	2.94	0.46
Nigeria	15.00	17.86	8.24	5.38	11.58	11.54	13.72	10.84	12.22
Norway	0.47	1.52	2.33	0.73	3.77	2.17	2.40	1.30	0.71

Country name	2004	2005	2006	2007	2008	2009	2010	2011	2012
Oman	0.76	1.86	3.20	5.96	12.09	3.94	3.20	4.07	2.91
Pakistan	7.44	9.06	7.92	7.60	20.29	13.65	13.88	11.92	9.69
Panama	0.18	3.18	2.10	4.17	8.76	2.41	3.49	5.88	5.70
Papua New Guinea	2.10	1.84	2.37	0.91	10.76	6.92	6.02	8.44	2.24
Paraguay	4.32	6.81	9.59	8.13	10.15	2.59	4.65	8.25	3.68
Peru	3.66	1.62	2.00	1.78	5.79	2.94	1.53	3.37	3.65
Philippines	4.83	6.52	5.49	2.90	8.26	4.13	3.88	4.65	3.17
Poland	3.58	2.11	1.11	2.39	4.35	3.83	2.71	4.22	3.75
Portugal	2.36	2.29	2.74	2.81	2.59	−0.84	1.40	3.65	2.77
Qatar	6.80	8.81	11.84	13.76	15.05	−4.86	−2.43	1.92	1.87
Romania	11.88	8.99	6.58	4.84	7.85	5.59	6.09	5.79	3.33
Russian Federation	10.86	12.68	9.68	9.01	14.11	11.65	6.86	8.44	5.07
Rwanda	12.25	9.01	8.88	9.08	15.44	10.36	2.31	5.67	6.27
Samoa	16.31	1.86	3.70	5.58	11.57	6.32	0.78	5.20	2.05
San Marino	1.43	1.69	2.10	2.50	4.29	2.20	2.59	2.01	2.83
Sao Tome and Principe	15.23	17.21	24.56	27.56	24.83	16.09	12.89	11.94	10.41
Saudi Arabia	0.33	0.70	2.21	4.17	9.87	5.07	5.34	5.82	2.89
Senegal	0.51	1.70	2.11	5.85	5.77	−1.05	1.25	3.38	1.42
Serbia	11.03	16.12	11.72	6.39	12.41	8.12	6.14	11.14	7.33
Seychelles	3.86	0.91	−0.35	5.32	36.97	31.76	−2.41	2.56	7.11
Sierra Leone				11.65	14.83	9.25	16.64	16.19	12.87
Singapore	1.66	0.43	1.02	2.10	6.52	0.60	2.80	5.25	4.53
Slovak Republic	7.55	2.71	4.48	2.76	4.60	1.62	0.96	3.92	3.61
Slovenia	3.59	2.48	2.46	3.61	5.65	0.86	1.84	1.81	2.60
Solomon Islands	6.99	7.33	11.22	7.67	17.32	7.09	1.05	7.34	2.56
South Africa	1.39	3.40	4.64	7.10	11.54	7.13	4.26	5.28	5.41
South Sudan						5.01	1.17	47.28	
Spain	3.04	3.37	3.52	2.79	4.08	−0.29	1.80	3.20	2.45
Sri Lanka	7.58	11.64	10.02	15.84	22.56	3.46	6.22	6.72	6.83
St. Kitts and Nevis	2.31	3.38	8.47	4.51	5.34	2.03	0.51	7.07	1.37
St. Lucia	1.46	3.91	2.34	3.07	7.18	−1.67	3.25	2.77	4.18
St. Vincent and the Grenadines	2.96	3.73	3.05	6.92	10.07	0.42	1.48	3.19	2.60
Sudan	8.42	8.52	7.20	7.98	14.31	11.25	13.25	22.11	37.39

Country name	2004	2005	2006	2007	2008	2009	2010	2011	2012
Suriname	9.99	9.90	11.28	6.43	14.67	−0.11	6.94	17.71	5.01
Swaziland	3.45	4.77	5.30	8.08	12.66	7.45	4.51	6.11	9.40
Sweden	0.37	0.45	1.36	2.21	3.44	−0.49	1.16	2.96	0.89
Switzerland	0.80	1.17	1.06	0.73	2.43	−0.48	0.70	0.23	−0.67
Syrian Arab Republic	4.43	7.24	10.02	3.91	15.75	2.92	4.40	4.75	36.70
Tajikistan	7.14	7.09	10.01	13.15	20.47	6.45	6.42	12.43	5.83
Tanzania	4.74	5.03	7.25	7.03	10.28	12.14	6.20	12.69	16.00
Thailand	2.76	4.54	4.64	2.24	5.47	−0.85	3.27	3.81	3.01
Timor-Leste	3.24	1.11	3.94	10.30	9.06	0.67	6.77	13.50	11.80
Togo	0.39	6.80	2.23	0.96	8.68	3.31	1.83	3.57	2.63
Tonga	10.98	8.32	6.44	5.89	10.44	1.42	3.55	6.26	1.21
Trinidad and Tobago	3.72	6.89	8.32	7.89	12.05	6.97	10.55	5.10	9.26
Tunisia	3.63	2.02	4.49	3.42	4.92	3.52	4.42	3.61	5.50
Turkey	10.58	10.14	10.51	8.76	10.44	6.25	8.57	6.47	8.89
Uganda	3.72	8.45	7.31	6.14	12.05	13.02	3.98	18.69	14.02
Ukraine	9.05	13.57	9.06	12.84	25.23	15.89	9.38	7.96	0.56
United Arab Emirates					12.25	1.56	0.88	0.88	
United Kingdom	1.34	2.05	2.33	2.32	3.61	2.17	3.29	4.48	2.82
United States	2.68	3.39	3.23	2.85	3.84	−0.36	1.64	3.16	2.07
Uruguay	9.16	4.70	6.40	8.11	7.86	7.10	6.68	8.09	8.10
Vanuatu	1.42	1.20	2.04	3.96	4.83	4.25	2.81	0.86	1.36
Venezuela, RB						27.08	28.19	26.09	21.07
Vietnam	7.76	8.28	7.39	8.30	23.12	7.05	8.86	18.68	9.09
West Bank and Gaza	3.00	3.47	3.88	1.83	9.89	2.75			
Yemen, Rep.	12.52	11.81	10.84	7.91	18.98	5.41	11.17	16.39	17.29
Zambia	17.97	18.32	9.02	10.66	12.45	13.40	8.50	6.43	6.59
Zimbabwe	282.38	302.12	1,096.68						

References

Chapter 1 – References and useful websites

Alonso, W. (1964), *Location and Land Use: Toward a General Theory of Land Rent*, Harvard University Press.

Carbon Trust (2009), 'Building the future, today', http://www.carbontrust.com/resources/reports/technology/building-the-future (accessed 11/12/13).

Commercial Real Estate Finance Council (2014), 'Educational initiative: property lending and development finance explained'. CREFC Europe Spring Conference, Wednesday 2/4/14.

English Partnerships http://www.englishpartnerships.co.uk/

Forster-Kraus, S., Reed, R. and Wilkinson, S. (2009), 'Affordable housing in the context of social sustainability' in *Proceedings of the ISA International Housing Conference*, 1–4 September 2009, Glasgow.

Percy Williams (2008), 'Building Homes in Cornwall'. Percy Williams and Sons Ltd, Redruth, Cornwall, UK. Retrieved from http://www.new-homes-cornwall.com/ (accessed 18/12/13).

Royal Institution of Chartered Surveyors (RICS) www.rics.org

Scottish Enterprise http://www.scottish-enterprise.com/sedotcom_home/about_se.htm

Smarter Homes (2013), 'Passive Heating'. Retrieved from http://www.smarterhomes.org.nz/design/passive-heating/ (accessed 17/12/13).

Von Thunen, J.H. (1826), 'The Isolated State'.

Chapter 2 – References and useful websites

British Urban Regeneration Association http://www.bura.org.uk/

CABE Commission for Architecture Built Environment http://www.cabe.org.uk/

Communities and Local Government http://www.communities.gov.uk

Communities and Local Government http://www.communities.gov.uk/pub/769/CommercialandIndustrialFloorspaceandRateableValueStatistics2005PDF6100Kb_id1163769.pdf UK government statistics on commercial and industrial floorspace and rateable value statistics.

Department for Transport http://www.dft.gov.uk

Department for Transport – Highways Agency. http://www.highways.gov.uk

English Partnerships http://www.englishpartnerships.co.uk/

Heritage Victoria (2013), http://vhd.heritage.vic.gov.au/places/result_detail/172?print=true (accessed 11/12/2013).

Renewal.net http://www.renewal.net/

Chapter 3 – References

Baum and Mackmin, D. (2011), *The Income Approach to Property Valuation*, 6th edition, Elsevier.

International Valuation Standards Committee (2013), *International Valuation Glossary*, London http://www.ivsc.org/glossary

Chapter 4 – References

IPD (2013), *IPD UK Annual Returns 2012*, Investment Property Databank, London.

KPMG (2007), *Real Estate Investment Trusts (REITs)*, London.

London Stock Exchange (2008) 'How to Qualify as a Reit'. Retrieved from http://www.londonstockexchange.com/specialist-issuers/reits/howtoqualifyasauk-reit.pdf (accessed 5/3/13).

RICS (2007), *REITs on the rise – Has a revolution begun?* Royal Institution of Chartered Surveyors, 17 January, London.

Stockton, K. (2012), 'Housebuilders preferred to REITs', Investors' Soapbox 29/8/2012 Barrons Online Retrieved from http://online.barrons.com/article/SB50001424053111190390490457 7619424065955022.html) (accessed 29/8/2012).

U.S. Securities and Exchange Commission (2013), http://www.sec.gov/answers/reits.htm (accessed 5/3/13).

Chapter 5 – References and useful websites

Barras, R. (1994), 'Property and the economic cycle: building cycles revisited', *Journal of Property Research*, (11), pp. 183–187.

Burns, A. and Mitchell, W (1946), *Measuring Business Cycles*, National Bureau of Economic Research, New York.

Keynes, J.L. (1936), *The General Theory of Unemployment*, Macmillan Cambridge University Press.

Modelski, G. (1987), 'The study of long-term cycles' in *Exploring Long Term Cycles*, edited by G. Modelski, pp. 1–16, Frances Pinter, London.

Niehans, J. (1992), 'Juglar's credit cycles', *History of Political Economy*, 24(3), pp. 545–569.

Scott, P. and Judge, G. (2000), 'Cycles and steps in British commercial property values', *Applied Economics*, 32, pp. 1287–1297.

Sims, S., Dent, P. and Ennis-Reynolds (2009), 'Calculating the cost of overheads: the real impact of HVOTLs on house prices', *Property Management*, 27(5), pp. 319–347.

Solomou, S. (1987), *Phases of Economic Growth 1850–1973*, Cambridge University Press, Cambridge.

Stanca (1998), 'Are business cycles all alike? Evidence from long-run international data', *Applied Economic Letters*, 6, pp. 765–769.

Tvede, L. (1997), *Business Cycles*, Harwood Academic Publishers, Amsterdam.

Tylecote, A. (1994), 'Long waves, long cycles and long swings', *Journal of Economic Issues*, 28(2), pp. 477–479.

Chapter 6 – References and useful websites

Planning Portal www.planningportal.gov.uk

Planning Advisory Service (PAS) www.pas.gov.uk

Planning Aid www.planningaid.rtpi.org.uk

Planning Inspectorate www.planning-inspectorate.gov.uk/pins/index.htm

Planning Officer's Society www.planningofficers.org.uk

Planning Resource www.planning.haynet.com

Royal Institution of Chartered Surveyors www.rics.org.

Royal Town Planning Institute www.rtpi.org.uk

Town and Country Planning Act 1990, http://www.opsi.gov.uk/ACTS/acts1990/Ukpga_19900008_en_1.htm

Town and Country Planning Act (Scotland) 1997, http://www.opsi.gov.uk/acts/acts1997/1997008.htm

Town and Country Planning Association www.tcpa.org.uk

WCED (1987), World Commission on Environment and Development. *Our Common Future.* Oxford, UK: Oxford University Press.

Chapter 7 – References and useful websites

Kelly, J., Morledge, R. and Wilkinson, S. (eds) 2002, *Best Value in Construction*, Blackwells ISBN 0 632 05611 8.

UDAYA Pty Ltd, 2012. UDAYA, available at: www.UDAYA.com.au (accessed 20/9/13).

Chapter 8 – References

Reed, R.G. and Krajinovic-Bilos (2013), 'An examination of international sustainability rating tools: an update', *Proceedings of the 19th Annual Pacific Rim Real Estate Society Conference,* 13–16/01/2013, Melbourne.

Chapter 9 – References

Abanda F.H., Tah, J.H.M. and Keivani, R. (2013), 'Trends in built environment semantic Web applications: Where are we today?' *Expert Systems with Applications*, 40(2013), 5563–5577.

Argus (2013), 'Argus Software'. Retrieved from http://www.argussoftware.com (accessed 27/07/13).

Caldes (2013), 'Caldes Software'. Retrieved from http://www.caldes.co.uk/home.asp (viewed 27/7/13).

Estate Master (2013), 'Development Feasibility'. Retrieved from http://estate-master-df.software.informer.com/ (accessed 12/12/13]).

Havard, T. (2013), *Financial Feasibility Studies For Property Development Theory and Practice*. Routledge, London, p91.

Koudounas, V. and Iqbal, O. (no date), 'Mobile computing: past present and future'. Retrieved from http://www.doc.ic.ac.uk/~nd/surprise_96/journal/vol4/vk5/report.html#step6 (accessed 12/12/13).

Mahdjoubi, L., Moobela, C. and Laing, R. (2013), 'Providing real-estate services through the integration of 3D laser scanning and building information modelling', *Computers in Industry*, 64(2013) 1272–1281.

Microsoft (2013), 'Microsoft Corporation'. Retrieved from http://office.microsoft.com/en-au/novice/what-is-excel-HA010265948.aspx (viewed 7/7/13).

National Institute of Building Sciences (2013), 'National BIM Standard – United States Version 2'. Retrieved from http://www.nationalbimstandard.org/faq.php#faq1 (accessed 20/11/13).

NBS (2011), 'BIM Roundtable Discussion'. Retrieved from http://www.thenbs.com/roundtable/ (accessed 11/1/14).

NBS (2012), 'NBS National BIM Survey'. Retrieved from http://www.thenbs.com/topics/bim/articles/nbsNationalBimSurvey_2012.asp (accessed 12/12/13).

Revit Services (no date), 'BIM Model Clash Detection'. Retrieved from http://www.revitservices.com/revit-clash-detection.htm (accessed 12/12/13).

Wikipedia Project Management Software. Retrieved from http://en.wikipedia.org/wiki/Project_management_software#Scheduling (accessed 26/11/13).

Zeiss, G. (2013a), 'The future of national mapping agencies over the next 5–10 years'. Retrieved from http://geospatial.blogs.com/geospatial/interoperability/ (accessed 12/12/13).

Zeiss, G. (2013b), 'Towards spatial standards for the Internet of Things (IoT)'. Retrieved from http://geospatial.blogs.com/geospatial/interoperability/ (accessed 12/12/13).

Relevant websites

Argus Software	http://www.argussoftware.com,
Caldes Software	http://www.caldes.co.uk/home.asp
Estatemaster Software	http://www.estatemaster.net/index.html
Feastudy software	http://www.devfeas.com.au/
Microsoft Excel	http://office.microsoft.com
SketchUp	http://www.sketchup.com/

Chapter 10 – Reference

PricewaterhouseCoopers (2003), *5th Annual Global CEO Survey*, London: Pricewaterhouse-Coopers.

Chapter 11 – References and useful websites

AufbauinternationalerBewertungsmethodeninAbhängigkeitvoneinander,Zertifizierungssysteme für Gebäude, p. 25, Thilo Ebert, Nathalie Eßig, Gerd Hauser, DETAIL Green Books 2010.

Awbi, H.B. (2013), *Ventilation of Buildings*, Spon Press, London.

BRE. www.bre.co.uk/envprofiles (accessed 24/5/07).

Bowen, H.R. (1953), *Social responsibilities of the businessman*. New York: Harper and Row. page 6, cited in Schreck, P. (ed.) (2009), *The Business Case for Corporate Social Responsibility*, Physica Verlag, p.10.

Brammer, S. and Millington, A. (2004), 'The development of corporate charitable contributions in the UK: a stakeholder analysis', *Journal of Management Studies*, vol.41 no.8, pp. 1412–1434.

Cone Communications (2013), *Cone Communications/Echo Global CSR Study*, http://www.conecomm.com/2013-global-csr-study-release (accessed 7/3/14).

DEFRA, Department of Environment Food and Rural Affairs http://www.defra.gov.uk/environment/sustainable/index.htm

Department for Communities and Local Government (2013), https://www.gov.uk/government/organisations/department-for-communities-and-local-government (accessed 31/10/2013).

DECC, Department of Energy and Climate Change (2013), *Climate Change* Retrieved from https://www.gov.uk/government/topics/climate-change (accessed 6/6/13).

Environmental Agency (2006), *Sustainable Construction Position Statement 2006*.

EPA (2013), Climate Change: Basic Information. Retrieved from http://www.epa.gov/climatechange/ (accessed 6/6/2013).

Frankental, P. (2001) Corporate social responsibility – a PR invention? *Corporate Communications*, 6(1), pp. 18–23.

International Sustainable Development http://www.sustainable-development.gov.uk//international/index.htm

IPCC (2007), *Inter-Governmental Panel on Climate Change Working Group II Fourth Assessment Report*, 4 April 2007, http://www.ipcc.ch/SPM6avr07.pdf (accessed 27/4/07).

IPCC, Intergovernmental Panel on Climate Change (2013) Preparations for AR5 enter final stage. Retrieved from http://www.ipcc.ch/ (accessed 6/6/13).

Lantos, G.P. (2001), 'The ethicality of altruistic corporate social responsibility', *Journal of Consumer Marketing*, 19(3), pp. 205–230.

Lindgreen, A. and Swaen, V. (2010), 'Corporate social responsibility', *International Journal of Management Review*, Special Edition Blackwell Publishing and British Academy of Management.

Lindgreen, A., Swaen, V. and Johnston, W.J. (2009), 'Corporate social responsibility: An empirical investigation of US organizations', *Journal of Business Ethics*, 85 (Suppl.2), pp. 303–323. Cited in Lindgreen, A. and Swaen, V. (2010).

Moir, L. (2001), 'What do we mean by corporate social responsibility?' *Corporate Governance*, vol.1 iss.2, pp. 16–22.

Myers, G., Reed, R.G. and Robinson, J. (2008), 'Sustainable property – the future of the New Zealand Market', *Pacific Rim Property Research Journal*, vol.14 no.3, pp. 298–321.

Nidumolu, R., Prahalad, C. and Rangaswami, M. (2009), 'Why sustainability is now the key driver of innovation' *Harvard Business Review,* 87(9), pp. 56–64.

Rangan, K., Chase, L.A and Karim, S. (2012), Why Every Company Needs a CSR Strategy and How to Build It. Working Paper. HBS working paper number 12-088 5 April 2012. Retrieved from http://www.hbs.edu/faculty/Publication%20Files/12-088.pdf (accessed 28/5/13).

Reed, R.G. and Krajinovic-Bilos (2013), 'An examination of international sustainability rating tools: an update' *Proceedings of the 19th Annual Pacific Rim Real Estate Society Conference*, 13–16/01/13, Melbourne.

Reverte C. (2009), 'Determinants of corporate social responsibility disclosure rating by Spanish listed firms', *Journal of Business Ethics* 88: 351–366.

RICS (2013), 'Sustainability'. Retrieved from http://www.rics.org/uk/about-rics/what-we-do/corporate-responsibility/sustainability/

Rodriguez, L.C., and LeMaster, J. (2007) 'Voluntary corporate social responsibility disclosure: SEC 'CSR seal of approval'', *Business and Society*, Sep 2007, 46(3), pp. 370–385 (accessed 11/6/13).

UNFCCC (2007), *United Nations Framework Convention on Climate Change,* 20 http://unfccc.int/resource/docs/convkp/kpeng.html (accessed 27/4/07).

U.S. Green Building Council (2013), http://www.usgbc.org/ (accessed 01/12/2013).

Van Aaken, D., Splitter, V. and Seidl (2013), 'Why do corporate actors engage in pro-social behaviour? A Bourdieusian perspective on corporate social responsibility', *Organization*, 20(3), pp. 349–371.

Vurro, C. and Perrini, F. (2011), 'Making the most of corporate social responsibility reporting: disclosure structure and its impact on performance', *Corporate Governance*, 11(4), pp. 459–474.

Warren-Myers, G. (2013), 'Is the valuer the barrier to identifying the value of sustainability', *Journal of Property Investment and Finance,* 31(4), pp. 345–359.

Water UK (2011), http://www.water.org.uk/ (accessed 01/11/11).

WCED (1987), World Commission on Environment and Development. *Our Common Future*. Oxford: Oxford University Press.

Weber, M. (2008) 'The business case for corporate social responsibility: A company-level measurement approach for CSR', *European Management Journal*, 26(4), pp. 247–261.

Whetton, D.A., Rands, D.P. and Godfrey, P. (2002), 'What are the responsibilities of business to society?' in *Handbook of Strategy and Management,* A.M. Pettigrew, H. Thomas and R. Whittington (eds), London, Thousand Oaks and New Delhi, Sage: pp. 373–408.

Williams, S.M. and Pei, C.A. (1999), 'Corporate social disclosures by listed companies on their web sites: An international comparison', *International Journal of Accounting*, 34 (3), pp. 389–419.

World Business Council for Sustainable Development. (2000), 'Making Good Business Sense'. www.wbcsd.org. January 2000 (accessed 5/5/07).

Chapter 12 – References

Bröchner, J., Rosander, S. and Waara, F. (2004), 'Cross-border post-acquisition knowledge transfer among construction consultants', *Construction Management and Economics*, 22, pp.421–427.

Bromiley, P. and Cummings, L. (1993), 'Transition costs in organisations with trust', *Research and Negotiation in Organisations*, 5, pp. 219–247.

Dehesh, A. and Pugh, C. (2000), 'Property Cycles in a Global Economy', *Urban Studies*, 37(13), pp. 2581–2602.

ENR (Engineering News-Record) (2003), *The Top International Contractors*, 251(8), pp. 36–41.

Ghoshal, S. and Nohria, N. (1993), 'Horses for courses: organizational forms for multinational corporations', *Sloan Management Review*, Winter.

Griffin, R.W. and Pustay, M.W. (2007), *International Business*, Pearson Prentice Hall, Sydney.

Guy, S. and Henneberry, J. (2000), 'Understanding urban development processes: integrating the economic and the social in property research' *Urban Studies*, 37, pp. 2399–2416.

Hailia, A. (2000), 'Why is Shanghai building a giant speculative property bubble?' *International Journal of Urban and Regional Research*, 37(12), pp. 2241–2256.

Jin, X. and Ling, Y.Y.L. (2005), 'Constructing a framework for building relationships and trust in project organizations: two case studies of building projects in China', *Construction Management and Economics*, 23, pp. 685–696.

Mansfield, J.R. and Royston, P.J. (2007), 'Aspects of valuation practice in Central and Eastern European economies', *Property Management*, 25(2).

McAllister, P. (2000), 'Is direct investment in international property markets justifiable?' *Property Management*, 18(1), pp. 25–33.

Nappi-Choulet, I. (2006), 'The role and behaviour of commercial property investors and developers in French urban regeneration: the experience of the Paris region', *Urban Studies*, 43(9), pp. 1511–1535.

Office for National Statistics (2011), *Wealth in Great Britain 2008/10*. Newport: Office of National Statistics.

PricewaterhouseCoopers (2013), 'Emerging trends in real estate', Urban Land Institute – PricewaterhouseCoopers, http://www.pwc.com (accessed 11/12/2013).

Raftery, J., Pasadilla, B., Chiang, Y.H., Hui, E.C.M. and Tang, B.S. (1998), 'Globalization and construction industry development: implications of recent developments in the construction sector in Asia', *Construction Management and Economics*, vol.16, pp. 729–737.

World Trade Organization (2013), http://wto.org (accessed 1/12/13).

Index

Page references in *italic* indicate illustrations, which are also listed in full after the Contents.

access to sites 59–60
accountants 29, 98 *see also* quantity surveyors (QS)
acquisition: decision pathway, in preparation for 9; as part of development process 8–10; of sites 8–10, 67–9, 90
advertising 52; marketing through 259–61 *see also* marketing; regulations 174
affordable land purchase price approach 78–9
AIDA marketing strategy 257
amenity societies 30, 170, 173
anchor tenants 15, 46, 105
appeals, planning 179–80
architects 27, 28, 29, 41, 52, 62, 87, 185, 186, 195–6, 200, 259; certificates for building costs 133; ecologist partnership 308; and handover of completed development 204–5; issuing of project changes 197; resident 202; Royal Institution of British Architects 185; snagging list preparation 203–4
architectural drawings 11, 204, 259
Argus software 248
Asian Financial Crisis (AFC) 158
Assisted Areas 70
auctions 55–6

Bank of England ix
banks ix, 21, 23, 25, 26, 112, 113, 114, 121–4, 137–40; bailing out of ix, 113; clearing 112, 113, 122–3, 137, 138; corporate loans 138; environment-related policies 286–7; foreign 114, 123; globalisation of the banking sector 137; interest rate options 141–2; investment loans 139, 140; merchant 112, 113, 115, 122–3, 124, 137, 141; project loans 138–40, 141;

syndicated project loans 141; syndicates of 122
bar charts 205–6, *207*
Barras, R. *156*, 157
base rent 133, 134, *134*
Bills of Quantities 186, 188–9, 201, 202, 210
BIM (building information model) 245–6, 247
Birmingham library, case study in sustainability 302–8, *303–4*, *306–7*; background 302–3; competition for choosing design 304; design 305–6; location 304; planning permission 304–5; site conditions 305; sustainable features 306–8
bonds 143–4
bore holes 62
Bowness, William 163–4
Bradley–Craven brick press 71, 72
BREEAM 291, 295, 301
British Property Federation 185
brochures, marketing 256–9
brownfield sites 35, 48, 63
Bruntland Commission/Report 166, 276, 277
building contractors *see* contractors
building contracts *see* contracts
building costs 12, 79, 80–3, 87, 93–4, 97, 103, 104–5, 106, 107–8, 116, 121; architect's certificates for 133; financial reports 206–10, *209 see also* development appraisal, financial perspective; development finance
building information model (BIM) 245–6, 247
building regulation fees 88
buildings: complicated, design and build erection 194; construction *see* construction; demolition work on 290; environmental impact 284; green 284, 292; handover of completed development 203–5; leased *see* leaseholds/leases;

listed 13, 169, 173–4; low carbon 284;
name and identity of 254; occupiers *see*
occupiers; opening ceremonies 263; repair
obligations 272; sustainable 31, 285, 288–
91, 292 *see also* sustainable development
building societies 21, 114, 123, 142
business cycles 156–8, *156*
'buy local' 319

CAD (computer aided design) 259, 263
Caldes Development Valuer 242
carbon: embedded 41; emissions 39, 283,
284–5, 287, 289; low carbon building 284
Carnegie, Andrew 278
CASBEE 295
cashflow method 92–9, *93–7*, *100*;
discounted 99–102, *101*, 120, 124
cashflow tables and graphs 206, *208*
checklists for development activities 210, *211*
Clash Detection 245–6
clearing banks 112, 113, 122–3, 137, 138
climate change ix, 180, 283, 284–5, 321 *see
also* global warming
cluster development 317
collateral warranties 29, 272–3
combined heat and power plant (CHP) 308
commitment phase 12
competition process 54–5
compulsory purchase powers 58–9
computer technology 240–8, 251; 3D
laser scanning 246; 3D modelling/
design software 244, 245, 247; 4D
modelling 244; Argus software 248; BIM
(building information model) 245–6;
CAD 259, 263; email *see* email; future
trends 246–7; information tools 243–4;
installation options 247; and marketing
250, 261, 262–3; mobile computing
245; and program users 247–8; project
management software 242–4; property
development appraisal software 242;
real estate agent, realtor or letting agent
software 244; Revit Building Design
Suite 248; scheduling tools 243; software
package options 247–8; spreadsheets
241–2 *see also* internet
conditional contracts 104, 169
construction 183–221; contractors *see*
contractors; contract terms 191–2;
cost calculation 190–1; design and
build approach 193–6, *196*; design-
bid-build traditional approach 185–93,
193; economists 185; hybrid modular
215, *216*, 220; low carbon 284;
management contracting 196–9, *199*;
modular (case study) 215–21, *215*,

216; partnering 213–14; procurement
183–5, *184*; professional team member
roles 186–9; project management *see*
project management/managers; public-
private partnerships *see* public-private
partnerships (PPPs); and sustainability
288–91
Consumer Price Index, global inflation 115,
324–9
container modules 215–21, *215*, *216*
contamination 62–4, 172, 286; case study
of due diligence and contaminated
sites 236–9, *237–9*; pollution from
construction process 289
contingency allowance 91
contractors 24; calculating the cost 190–1;
choosing 188–9; design and build
approach 194–5; duration of contract
191–2; management contractors 196–9,
199; paying 190; project manager's
appointment of 201–2
contracts: conditional 104, 169; duration of
191–2; exchange of 55, 61; packaging of
198; PFI 212; risks of 183; signing of 14;
Subject to Council Approval status of 14
Copenhagen Accord 284
corporate loans 138
corporate responsibility (CR) 277
corporate social responsibility (CSR): and
large corporation behaviour 314; and
planning 166; and sustainability 277–82,
283–4, 288, 293; types of 281–2
costing 10–12
CPI, global inflation 115, 324–9
CSR *see* corporate social responsibility
currency risk 320

debentures 144
debt finance 22, 114, 126, 137, 143–4;
bonds 143–4; debentures 144; unsecured
loan stock 144
defects: latent defects insurance 187–8;
liability period 183, 204; snagging lists
203–4
demand: site-specific analysis 230–1; supply
and 32–3
demise 269
design 10–12; Birmingham library 305–8;
commitment phase for design changes
12; computer aided (CAD) 259, 263;
drawings 11, 12, *36*, 169; Fairglen Low
Energy Housing 37–9, *38*; plans 11; and
sustainability 288–91, 305–8
design and build approach 193–6, *196*
design-bid-build approach 185–93, *193*
desk research 231

developers: appraisal by *see* development appraisal, financial perspective; development appraisal, risk perspective; base rent arrangement 134, *134*; communication with agents 268; construction *see* construction; and contractors *see* contractors; corporate social responsibility *see* corporate social responsibility (CSR); and the design and build approach 193–6, *196*; and the environment *see* environment; finance *see* development finance; guarantees and performance obligations 135–6; joint venture partners 129–30; lettings by *see* leaseholds/leases; and management contractors 196–9, *199*; marketing *see* marketing; market research *see* market research; on-going responsibilities 273; planning *see* planning; priority yield arrangement 135, *135*; profit/risk allowance 92; profit with forward-funding 134–5, *134–5*; public-private partnerships with *see* public-private partnerships (PPPs); site acquisition *see* development sites: acquisition; site initiation by 48–51; as stakeholders 18–19; visionary skills of 34, 44, 152; Wilbow Corporation case study 162–4, *164*

development appraisal, financial perspective 7–8, 77–102; affordable land purchase price approach 78–9; building costs *see* building costs; building regulation fees 88; cashflow method 92–9, *93–7, 100*; contingency allowance 91; conventional technique of financial evaluation 77–92, *80–3*; developer's profit/risk allowance 92; discounted cashflow method 99–102, *101*, 120, 124; evaluation of profit *80–2*; funding fees 88; interest costs 88–90; investment risk/return model 78; land costs 86; letting agent's fees 90; net development value *see* net development value; net terminal approach 99, *100*; planning fees 88; professional fees 87; promotion costs 91; purchaser's costs 86; residual valuation 77–9, *80, 93*; sale costs 91; site investigation fees 87; stamp duty *80, 82*, 86

development appraisal, risk perspective 102–9; building costs 104–5; investment yield 106–7; land costs 104; rental value 105; sensitivity analysis 107–9, *108, 109*, 226; short-term interest rates 106

Development Corporations (UDCs) 70

development finance 10, 111–47; bank loans 121–4, 137 *see also* banks; base rent arrangement 134, *134*; bonds 143–4; building society loans 21, 114, 123, 142; collateral warranties 29, 272–3; corporate finance 138, 142–4; debentures 144; debt finance 22, 114, 126, 137, 143–4; design and costing 10–12; and development site investigation 66; eco finance 286; economic consultants 27; and the environment 286–7; equity finance *see* equity finance; equity partners 68–9; evaluation *see* development appraisal, financial perspective; financial institutions *see* financial institutions; forward-funding *see* forward-funding; future trends 145–6; and the global market 33–4; government assistance *see* government assistance; ground rent 19, 57, 60, 68, 69, 129; historical perspective 112–14; interest costs 88–90; interest rate options 141–2; joint venture partners 129–30; Loan to Cost (LTC) 33–4; loan-to-value ratio (LVR/LTV) 33, 89, 123, 124; long-term funding 10, 21, 22, 106, 111, 112, 113, 124–6, 147; methods 130–46; mezzanine finance 141; mortgages 142; overseas investors 127–8; priority yield arrangement 135, *135*; private investors 128–9; profit-seeking 7; project loans 138–40; property companies 124–6; public/private project distinctions 7; securitisation 144–5; short-term 10, 21, 22, 79, 106, 111, 112, 113, 121, 133; sources 111–30; and supply and demand 32–3; and sustainability 286–7; syndicated project loans 141; unitisation 144–5; unsecured loan stock 144

development process 2–16; acquisition 8–10 *see also* development sites: acquisition; case study of low energy housing *see* Fairglen Low Energy Housing scheme; commitment 13–14; 'commitment phase' for design changes 12; consent and permission 12–13; contract signing 14; decision pathway, in preparation for acquisition 9; design and costing 10–12; financing *see* development finance; implementation 14–15; initiation 3–6, *4, 5*; investigation and analysis of viability 7–8 *see also* planning; leasing/ management/disposal 15–16 *see also* leaseholds/leases; legal investigation 8–9; main stakeholders in 16–31 *see also* stakeholders; physical inspection and examination 9–10

development sites 42–76; access 59–60; acquisition 8–10, 67–9, 90; advertising

52, 174, 259–61 *see also* marketing;
auctions 55–6; boards and hoardings on
site for marketing 254–6; brownfield
35, 48, 63; case study of due diligence
and contaminated sites 236–9, *237–9*;
case study of redevelopment of a pottery
kiln into residential accommodation
see Hoffman Brickworks development;
competition process 54–5; contamination
62–4, 172, 236–9, *237–9*, 286; developer
initiation 48–51; ecological impact and
issues 288 *see also* environment; finance
and site investigation 66; formal tenders
53; government assistance *see* government
assistance; greenfield 48, 168; ground
investigation 62; handover inspection
204; identification 43–8; industrial 15,
32, 47–8, 86, 265; informal tenders and
invitations to offer 53; infrastructure 59–
60; initiation avenues 48–61; investigation
fees 87; landowner initiatives 52–6;
legal title 65–6; local authority initiatives
see local authority initiatives; office
development 45–6; open 'for sale' listings
53–4; and public-private partnerships *see*
public-private partnerships (PPPs); real
estate agent approach 51–2 *see also* real
estate agents; residential 44–5; retail 46,
230; services 64–5; site-specific analysis
230–2; supervision by project managers
202; surveys 61–2; and sustainability
285–6 *see also* sustainable development;
waste management 290; and zoning 47
development timetable 88–90, *89*
direct marketing 262
discounted cashflow (DCF) method 99–102,
101, 120, 124
disposal 15–16 *see also* leaseholds/leases
drop lock loans 142

Earth Summit, 1992, Rio de Janeiro 166
eco finance 286
economic consultants 27
economy: global *see* global economy; local
see local economy; national 32–3
eco-system destruction 288
electricity production 283
email 49, 231; addresses 254, 255, 260;
alerts 25; databases 250; marketing 250,
251, 257, 258, 262–3
embedded carbon 41
embodied energy 289, 291
emerging markets and internationalism *see*
international property development
energy: embodied 289, 291; Low Energy
Housing scheme case study *see* Fairglen

Low Energy Housing scheme; low energy
lighting 219, 306; renewable 283–4, 289,
308
enforcement notices 178
engineers 28, 196, 205
environment: brownfield sites 35, 48, 63;
climate change ix, 180, 283, 284–5,
321; consultations with pressure groups
173; contamination *see* contamination;
and development finance 286–7; Earth
Summit, 1992, Rio de Janeiro 166; eco-
system destruction 288; environmental
impact assessments/studies 172–4;
environmental rating tools 291, 295–301,
295, 296–300; environmental statements
172–4; green agenda ix; greenfield sites
48, 168; greenhouse gas emissions 39,
283, 284–5, 287, 289; liabilities 286;
and planning 166, 172–4; pollution 47,
170, 172, 289 *see also* contamination;
risks 286–7; and sustainable development
see sustainable development; World
Commission on Environment and
Development 166, 275
Environmental Health Officers 170
Environmental Management System (EMS)
283
equity finance 125–6, 143–4; new shares
143; retained earnings 143; rights issues
143
equity partners 68–9
Estate Master DF 242
Estates Gazette 52
Estimated Rental Value (ERV) *80, 82, 93*
evaluation process *see* development appraisal,
financial perspective; development
appraisal, risk perspective

Face Time 311
Factories Acts 170
Fairglen Low Energy Housing scheme
35–41, *35, 36*; background 35–7; design
37–9, *38*; location 35; site issues 37;
sustainability measures 39–41
fair market rent 79
fees: building regulation 88; funding 88;
letting agent's 88; planning application
169; professional 87; site investigation 87
feng shui 314
finance *see* development finance
Finance Act 2006 (UK) 126
financial institutions 21–3, 25, 26, 114–21;
and the hedge against inflation 116,
120; and illiquidity of property 117;
and indivisibility of property 117;
institutional leases 116–17; insurance

companies *see* insurance companies; and the lack of a centralised marketplace 117–18; life assurance companies 115, 142; management 118; pension funds *see* pension funds; real estate investment trusts 126–7; research and performance measurement 118–21, *119*; unit trusts 115, 122, 145
financial reports 206–10, *209*
flyers 256–9
Ford, Henry 278
forecasting 232–4
formal tenders 53
forward-funding 22, 112, 115, 131–7; costs 133–4; developer's guarantees and performance obligations 135–6; developer's profit 134–5, *134–5*; and lettings 136; rent 133; sale and leaseback arrangement 136–7; yield 132–3
fossil fuels 283
four dimensional modelling 244
Frankental, P. 280
freehold titles 67, 68
funding fees 88

Gantt charts 205–6, *207*
gardens 308
gearing 124–5
Geographical Information Systems (GIS) 7, 228, 244
global economy 33, 127, 311, 314; CPI 115, 324–9; GFC *see* Global Financial Crisis
Global Financial Crisis (GFC) 1, 15, 27, 114, 125, 234, 251; and property cycles 158–61, *159–61*
global inflation, CPI 115, 324–9
globalisation 1, 180, 310, 311, 313–14, 322–3; of the banking sector 137; global markets 32, 33–4, 144, 225, 228, 311–12, 314, 316, 319 *see also* international property development; of property development 314–16 *see also* international property development
Global Reporting Initiative (GRI) 282–3
global warming 283, 284, 321
government agencies: as stakeholders 19–20; for urban regeneration 69–70
government assistance 69–70, 130; agencies 69–70; grants 40, 70
grants 40, 70
Greece 33
'green agenda' ix *see also* environment
greenfield sites 48, 168
greenhouse gases 39, 283, 284–5, 287, 289
Greenstar 293, 295
grey water harvesting 308

Griffin, R.W. and Pustay, M.W 319–20, 321, *322*
ground investigation 62
ground leases 69
ground rent 19, 57, 60, 68, 69, 129
ground source aquifier cooling system 306–8
ground source heat pumps (GSHPs) 40, 41

Haier 314
Harvard Business Review 292
health 292
Health and Safety Executive 170
heating: under-floor 39; ground source heat pumps 40, 41; mechanical heat recovery 39; passive 39; solar hot water panels 40, 308; solar photovoltaic panels 40, 308; wood burning stoves 39
hedge against inflation 116, 120
hedge funds 318–19
hedging, interest rate 142
Highway Authorities 170
hoardings 254–6
Hoffman Brickworks development 71–6, *73–5*; challenges for redevelopment of the kilns 73–6; challenges from an overall development perspective 72–3; historical site background 71–2
hot-desking 45
in-house land buyers 49
housing bubbles ix
HQE 301
hybrid modular construction 215, *216*, 220

ICT *see* computer technology
illiquidity, property 117
IMF (International Monetary Fund) 313
indivisibility, property 117
industrial development sites 15, 32, 47–8, 86, 265
informal tenders 53
information and communications technology *see* computer technology
infrastructure 13, 23, 49, 105, 172, 177; and economic development 57; and government agencies 19, 20, 69, 113; international 321; local authority initiatives 59; off-shore costs 312; and PPPs 212; railway 129; and site access 59–60
initial purchase offering (IPO) 126
insulation 37, 39, 216–17, 219
insurance: and handover of completed development 204, 205; latent defects 187–8; public liability 205
insurance companies 18, 21, 112–13, 115, 131, 141, 142, 187, 189

interest costs 88–90; short-term interest rates 106
interest rate options 141–2
inter-generational equity 276
internal rate of return (IRR) 102, 120
International Monetary Fund (IMF) 313
international property development 310–23; approaches to balancing global expansion with local market 319–20, 320; barriers and limitations 317–21; cluster development 317; developing an international strategy 321, 322; development opportunities 316–17; globalisation of property development 314–16 see also globalisation; global strategy 319; home replication 319; multi-domestic strategy 320; risks 318–21; transnational strategy 319; trends 310–14 see also globalisation
International Valuation Standards Committee 79
internet: and internationalism/globalisation 311, 312; and multi-national companies 311; online/email marketing 250, 251, 257, 258, 260, 261, 262–3; social media 244, 273 see also email
intra-generational equity 276
investment loans 139, 140
investment risk/return model 78
investment yield: and risk 106–7; variable in net development value 85–6
IPD Index 228
ISO 14001 283

Joint Contractors Tribunal (JCT) 185, 203, 204
joint venture (JV) companies 129–30
Juglar, Clement 154
'just in time' (JIT) approaches 47

Kondratieff cycle 154
Kuznet cycle 154
Kyoto Protocol 284

land assembly 57–9
land banking 14
land contamination 62–4, 172, 236–9, 237–9, 286
land costs 86, 104
land for development see development sites
landlords 19, 116–17, 271, 272, 273; landlord's fixtures and fittings 270
landowners 17–18, 49; initiatives by 52–6; and planning obligations 177
Lantos, G.P. 279, 281
LaSalle Investment Management 314

latent defects insurance 187–8
leaseholds/leases 6, 15–16, 19, 50, 65, 68, 69, 79, 118, 268–73; and agents 265–9, 270; demise 269; and forward-funding 136; ground leases 69; institutional leases 116–17; lease terms 271; and market information 227–8; monitored by agents 265; rent 270–1; and risk 121; sale and leaseback 136–7; and tenants 271–2 see also rent
LEED 295, 301
legal investigation 8–9
legal title 65–6
lenders 21–3, 66, 146; banks see banks; building societies 21, 114, 123, 142; environment-related policies 286–7; financial institutions see financial institutions; funding fees 88
letting agent's fees 90
lettings see leaseholds/leases
life assurance companies 115, 142
lighting, low energy 219, 306
Lindgreen, A. and Swaen, V. 279, 280
Linked Data 246, 249
listed buildings 13, 169, 173–4
Loan to Cost (LTC) 33–4
loan-to-value ratio (LVR/LTV) 33, 89, 123, 124, 138, 142
local authority initiatives 56–9; economic development 57–9; as equity partners 68–9; infrastructure 59; land assembly 57–9; planning application 56–7
Local Development Framework (LDF) 181
local economy ix, 32, 161, 225, 227, 228, 230, 235, 312, 319
Localism Act (2011) 180–1
Local Land Charges register 66
local markets 319
local planning authorities (LPAs) see planning: authorities
LTC (Loan to Cost) 33–4
LTV see loan-to-value ratio
LVR see loan-to-value ratio

McDonalds 313
magazine advertising 260
management 15–16; contracting 196–9, 199; financial institutions 118; project management/managers see project management/managers
marketing 250–68; advertisements 259–61; AIDA strategy 257; approaches 252–65; brochures 256–9; with computer simulations 251; direct mail 262; email 250, 251, 257, 258, 262–3; internet 250, 260, 261, 262–3; and legislation

257–8; and the naming of buildings/developments 254; opening ceremonies 263; particulars 256–9; promotion costs 91, 111; and public relations 264–5; radio 260–1; show suites and offices 263–4; with site boards and hoardings 254–6; a sustainable development 292, 293; television 252, 260; of unique selling points 253, 288
market rent 79
market research 7, 26, 44, 222–39; agent's role in 265–8; analysts 27; best/worst case scenarios 226; broad to specific approaches 223, 224; case study of due diligence and contaminated sites 236–9, 237–9; cross-tabulation 225; desk research 231; forecasting 232–4; impact 234–5; meanings of 223; portfolio analysis 234; and the prevailing real estate market conditions 232; qualitative surveys 231; quantitative analysis 231; real estate market information 227–8; relationship between areas of 224, 224; site-specific analysis 230–2; sourcing real estate market information 226–7; strategic analysis 229; supporting information 228–9; and sustainability 292–3; SWOT analysis 232
mechanical heat recovery 39
merchant banks 112, 113, 115, 122–3, 124, 137, 141
Methods of Measurement (PCA publication) 84
mezzanine debt 141
Microsoft: Excel 241, 242, 247; Word 261
mixed mode ventilation 308
mobile computing 245
Modelski, G. 155
modularisation 215–21, 215, 216
mortgages 142

national economy 32–3
National Planning Policy Framework (NPPF) 180–1
net asset value (NAV) 125, 126, 143
net development value (NDV) 79–86, 91, 92; and cashflow method 93–4; and investment yield variable 85–6; and rent variable 79–85, 80–3
net lettable area (NLA) 84
net terminal approach 99, 100
networking 52
Network Rail 129
newspaper advertising 260
Niehans, J. 154

non-government organisations (NGOs), sustainability reporting 282

objectors 29–30
occupiers 16, 30–1, 116–17, 204; adaptive intervention 40; anchor tenants 15, 46, 105; leasing *see* leaseholds/leases; repair obligations 272; tenant covenant 121; tenant risk 8, 269, 271; tenant's guide 258; tenants's financial position 271–2
office development sites 45–6
Office for National Statistics (ONS, UK) 315
opening ceremonies 263
open market selling 53–4
Our Common Future 275, 276
overseas investors 127–8

partnering 213–14 *see also* public-private partnerships (PPPs)
party wall agreements 91, 200
passive heating 39
Peabody, George 278
pension funds 18, 21, 113, 115, 117, 125, 131, 145, 146
peppercorn rent 68
PFI (Private Finance Initiative) 212
PINCs (Property Income Certificates) 145
planners, as stakeholders 20–1
planning 165–82; appeals 179–80; applications 11–13, 56–7, 167–72, 174–5; authorities 6, 12–13, 21, 26, 88, 167, 168, 169, 172–3, 177, 178, 181; consent and permission 12–13, 67, 175–6; consultants 26, 50; control breaches 178–9; design plans 11; enforcement notices 178; and the environment 166, 172–4; evaluation *see* development appraisal, financial perspective; development appraisal, risk perspective; fees 88; fees for applications 169; future perspectives 180–1; and the Localism Act (2011) 180–1; market research *see* market research; National Planning Policy Framework (NPPF) 180–1; obligations/agreements 176–8; process in England 171; of promotional campaign 253; recent changes to a planning system in the UK 180–1; satisfactory planning consent 67; and sustainability 21, 165, 166, 287–8 *see also* development process
pollution 47, 170, 172, 289 *see also* contamination
portfolio analysis 234
poster advertising 261
PPPs *see* public-private partnerships
pressure groups 173

PricewaterhouseCoopers 279, 315
priority yield arrangement 135, *135*
Private Finance Initiative (PFI) 212
private investors 128–9
private-public partnerships *see* public-private partnerships (PPPs)
procurement 183–5, *184*
professional fees 87
profit-seeking 7
program users 247–8
project changes 197
project loans 138–40; syndicated 141
project management/managers 28–9, 199–211; appointing the contractor 201–2; checklists 210, *211*; financial reports 206–10, *209*; handover of completed development 203–5; monitoring of construction progress 205–11; pre-contract preparations 200; preparation of contract documents 201; site supervision 202; software for 242–4
promotion costs 91, 111
property companies 124–6
property cycles 148–64; and business cycles 156–8, *156*; case study of Wilbow property developer 162–4, *164*; and the changing nature of the real estate market 152; characteristics of a typical cycle phase *150*; data limitations 149–50; definition 149; and equilibrium in a real estate market *159*; existence of 149–53; and external shift in demand for real estate *160*; and the fixed nature of land parcels 152; global extremely long-term 155; and investment by infrequent traders 152–3; Kondratieff cycle 154; Kuznet cycle 154; long-term 154–5; and the 'lumpiness' of property and real estate assets 151–2; medium-term 154; multiple cycles in a single market *155*; seven-year cycle notion 153–4; short-term 154; and structural change 158; surviving market downturns/GFCs 158–61, *159–61*; time lag between transaction and release of information 150–1; types of 153–6; and uniqueness of land parcels and buildings 151
property developers *see* developers
property development: appraisal *see* development appraisal, financial perspective; development appraisal, risk perspective; cluster development 317; construction *see* construction; definition 2, 166–7; economic context 31–4; and the environment *see* environment; finance *see* development finance;

globalisation of 314–16 *see also* international property development; handover of completed development 203–5; infrastructure *see* infrastructure; international *see* international property development; marketing *see* marketing; naming of buildings/developments 254; objectors 29–30; opening ceremonies 263; planning *see* planning; process *see* development process; project management *see* project management/managers; sales and lettings 268–73 *see also* leaseholds/leases; sites *see* development sites; software and computer technology *see* computer technology; stakeholders *see* stakeholders; sustainability *see* sustainable development; timetable 88–90, *89*
Property Income Certificates (PINCs) 145
property market research and performance measurement 118–21, *119*
property unit trusts (PUTs) 145
Property Weekly 52
Public Health Acts 170
public liability insurance 205
public-private partnerships (PPPs) 19, 23, 54, 60–1, 212; and PFI 212
public relations 264–5
public sector stakeholders 19–20
public transport *see* transport
PUTs (property unit trusts) 145

qualitative surveys 231
quantitative analysis 231
quantity surveyors (QS) 11, 12, 28, 52, 62, 87, 97, 103, 185, 189, 201, 203, 204

radio marketing 260–1
rainwater harvesting 39
rating tools, sustainability 291, 295–301, *295*, *296–300*
real estate agents 16, 50, 51–2; and leasing 265–9, 270; marketing role 265–8; software for 244; as stakeholders 24–6
real estate investment trusts (REITs) 126–7, 146
Real Estate Magazine 52
real estate market 6–7, 15; and the barriers of illiquidity and indivisibility 117; changes in 6, 31, 32, 49, 127, 145, 152, 159; cyclical nature of 144, 153, *156*, 158, 159 *see also* property cycles; emerging markets and internationalism *see* international property development; equilibrium *159*;

external shift in demand *160*; indexes 228; information 227–8; 'knowledge is power' adage 7, 160; lack of a centralised marketplace 150; lumpiness of 151–2, 160; and marketing approaches 252; and the national economy 32; planning *see* planning; prevailing conditions 232; property market research and performance measurement 118–21, *119*; research *see* market research; surviving market downturns/GFCs 158–61, *159–61*

Real Estate Times 52

realtors 24, 50, 244; software for 16 *see also* real estate agents

recycling 290–1

REDD Programme, UN 284

REITs (real estate investment trusts) 126–7, 146

renewable energy 283–4, 289, 308

rent 270–1; base rent arrangement 133, 134, *134*; fair market 79; ground 19, 57, 60, 68, 69, 129; peppercorn 68; reviews 31, 44, 65, 116, 121, 136–7, 270; variable, net development value 79–85; zoning for retail rent 84–5

rental value 51, 56, 90, 103, 105, 121, 159, 270; Estimated Rental Value (ERV) *80, 82, 93*

repair obligations 272

residential development sites 44–5

residual valuation 77–9, *80, 93*

restrictive covenants 65

retail development 15, 46, 84–5, 86, 228, 230, 235

retained earnings 143

re-use 290

Revit Building Design Suite 248

RICS (Royal Institution of Chartered Surveyors) 84, 185, 316

risk: appraisal *see* development appraisal, risk perspective; of building contracts 183; currency 320; decreasing exposure to 146; developer's profit/risk allowance 92; environmental 286–7; international property development 318–21; investment risk/return model 78; with leases 121; with project loans 138–40; systemic 146; tenant 8, 269, 271; transport risks 320; unsystemic 146

roads 37, 44, 45, 47, 59, 168, 170, 172, 200; private 62; and the Private Finance Initiative 212

roof gardens 308

Rowntree, Joseph 278

Royal Institution of British Architects 185

Royal Institution of Chartered Surveyors (RICS) 84, 185, 316

sale and leaseback 136–7

sales 268–73; costs 91; sale and leaseback 136–7; selling a sustainable development 293–4; unique selling points 253, 288

Salt, Titus 278

satisfactory planning consent 67

scenario modelling software 7

scheduling tools 243

securitisation of property investment 144–5

Semantic Web 246, 249n1

sensitivity analysis 107–9, *108, 109*, 226

services 64–5; access to 44 *see also* transport

shareholders 7, 98, 115, 116; of property companies 125; rights issues 143

shares 98, 112, 114, 115, 116, 125–6, 143

the Shop Acts 170

show suites and offices 263–4

Single Property Ownership Trusts (SPOTs) 145

sites *see* development sites

site-specific analysis 230–2

Skype 311

snagging lists 203–4

social media 244, 273

software *see* computer technology

solar hot water panels 40, 308

solar photovoltaic panels 40, 308

solicitors 29, 65, 66

SPOTs (Single Property Ownership Trusts) 145

spreadsheets 241–2

stakeholders 6; building contractors 24; commitment phase acknowledgement 12; developers 18–19; landowners 17–18; lenders *see* lenders; main stakeholders in development process 16–31; objectors 29–30; occupiers 30–1; planners 20–1; professional team 26–9; public sector and government agencies 19–20; real estate agents 24–6; trust between 319

stamp duty *80, 82*, 86

stock market 125–6, 143; crash 125 *see also* Global Financial Crisis (GFC) *see also* shares

strategic analysis 229

structural change 158

structural engineers 28

Subject to Council Approval (STCA) 14

supply: and demand 32–3; site-specific analysis 231–2
surveys: qualitative 231; site 61–2
sustainable development 1, 45, 59–60, 275–308; case for 284–5; case study of Birmingham library *see* Birmingham library, case study in sustainability; and construction 288–91; and corporate social responsibility 277–82, 283–4, 288, 293 *see also* corporate social responsibility (CSR); definitions and models 275–7, 277, 278; and demolition work on existing buildings 290; and design 288–91, 305–8; and development finance 286–7; development land and sustainability 285–6; and health and well-being 292; issues in development stages *294*; marketing and selling a development 292, 293–4; and market research 292–3; and planning 21, 165, 166, 287–8; and recycling 290–1; and re-use 290; sustainability measures 39–41; sustainability rating tools 291, 295–301, *295*, *296–300*; sustainability reporting 282–3; sustainable buildings 31, 285, 288–91, 292; sustainable features 306–8; and water 291–2, 306–8
SWOT analysis 232
syndicated project loans 141
syndicates: of banks 122; private 21, 111, 117; syndicated project loans 141

Tata Group 314
television marketing 252, 260
tenant covenant 121
tenant occupation *see* leaseholds/leases; occupiers
tenant's guide 258
terraced gardens 308
three dimensional laser scanning 246
three dimensional modelling/design software 244, 245, 247
transport 59–60; access 35, 44, 45, 47, 78, 118, 258, 260, 287–8, 311; carbon emissions 287; low carbon forms of 60; networks 4; noise 47; public transport funding 60; public transport links 44, 45, 59; public transport use 48, 60, 287; railway line infrastructure 129; risks 320
Tree Preservation Orders (TPOs) 174
triple bottom line (TBL) accounting 277

UDAYA 220–1
under-floor heating 39
unique selling points (USPs) 253, 288
United Nations: Bruntland Commission/Report 166, 276, 277; Kyoto Protocol 284; *Our Common Future* report 275, 276; REDD Programme 284; World Commission on Environment and Development 166, 275, 276
unitisation of property investment 144–5
unit trusts 115, 122, 145; property (PUTs) 145
unsecured loan stock 144
Urban Development Corporations (UDCs) 70
Urban Priority Areas (UPAs) 70
urban regeneration 48, 310; government agencies for 69–70
valuation surveyors 27

Van Aaken, D. et al. 278
viability analysis 7–8
volatile organic compounds (VOCs) 291
Von Thunen, J.H. 4

waste management 290
water 291–2, 306–8
WBCSD (World Business Council for Sustainable Development) 277, 278
WCED (World Commission on Environment and Development) 166, 275, 276
well-being 292
Whetton, D.A. et al. 279
Wilbow Corporation 162–4, *164*
wood burning stoves 39
World Business Council for Sustainable Development (WBCSD) 277, 278
World Commission on Environment and Development (WCED) 166, 275, 276

year's purchase (YP) 85

Zeiss, Geoff 246–7
zoning: land 47, 288; for retail rent 84–5

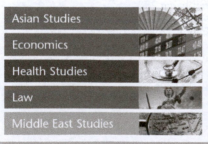